METHODS IN MOLECULAR BIOLOGY™

AF769106

Series Editor
John M. Walker
School of Life Sciences
University of Hertfordshire
Hatfield, Hertfordshire, AL10 9AB, UK

For further volumes:
http://www.springer.com/series/7651

Plant Organogenesis

Methods and Protocols

Edited by

Ive De Smet

School of Biosciences, University of Nottingham, Loughborough, UK

Humana Press

Editor
Ive De Smet
School of Biosciences
University of Nottingham
Loughborough, UK

ISSN 1064-3745 ISSN 1940-6029 (electronic)
ISBN 978-1-4939-5931-0 ISBN 978-1-62703-221-6 (eBook)
DOI 10.1007/978-1-62703-221-6
Springer New York Heidelberg Dordrecht London

© Springer Science+Business Media New York 2013
Softcover re-print of the Hardcover 1st edition 2013
This work is subject to copyright. All rights are reserved by the Publisher, whether the whole or part of the material is concerned, specifically the rights of translation, reprinting, reuse of illustrations, recitation, broadcasting, reproduction on microfilms or in any other physical way, and transmission or information storage and retrieval, electronic adaptation, computer software, or by similar or dissimilar methodology now known or hereafter developed. Exempted from this legal reservation are brief excerpts in connection with reviews or scholarly analysis or material supplied specifically for the purpose of being entered and executed on a computer system, for exclusive use by the purchaser of the work. Duplication of this publication or parts thereof is permitted only under the provisions of the Copyright Law of the Publisher's location, in its current version, and permission for use must always be obtained from Springer. Permissions for use may be obtained through RightsLink at the Copyright Clearance Center. Violations are liable to prosecution under the respective Copyright Law.
The use of general descriptive names, registered names, trademarks, service marks, etc. in this publication does not imply, even in the absence of a specific statement, that such names are exempt from the relevant protective laws and regulations and therefore free for general use.
While the advice and information in this book are believed to be true and accurate at the date of publication, neither the authors nor the editors nor the publisher can accept any legal responsibility for any errors or omissions that may be made. The publisher makes no warranty, express or implied, with respect to the material contained herein.

Printed on acid-free paper

Humana Press is a brand of Springer
Springer is part of Springer Science+Business Media (www.springer.com)

Dedication

I dedicate this book to my mother and father. Without their support I would not have reached this stage.

Preface

1. Building Plant Organs

1.1. Plant Organs: What Are They? Where Are They From? How Are They Connected?

Next to the clearly visible, above-ground parts—the leaves, flowers, fruits and stems—plants also comprise a less-visible half, hidden below ground: root systems. The green or colorful above-ground parts are essential for photosynthesis and reproduction (1, 2), while roots are important for nutrient and water uptake, anchoring, mechanical support, and storage (3–5). In contrast to mammals, plants generate new organs and tissues throughout their whole life. This often results in enormous organisms, such as giant sequoias or the Trembling Giant, a clonal colony of a single quaking aspen with a massive underground root system. Surprisingly, organs, such as leaves and lateral roots, are positioned at fairly regular spatial and temporal intervals and this requires tight coordination of the underlying molecular processes (3, 6). Especially since these organs develop from a subset of cells, often deeply embedded between various plant tissues. In addition, to control overall plant growth and reproduction, the various above- and below-ground organs need to communicate (7).

1.2. Model Systems

A good model is simple in structure, easy to study, to grow and to multiply, amenable to genetic analyses, and can increase our understanding of plant organogenesis fast. In the past decades, a lot of progress has been made by studying the model plant *Arabidopsis thaliana* or Thale Cress, and a recent special Plant Journal Issue was dedicated to this (8). However, for example, Arabidopsis has a small shoot apical meristem that is deeply buried between rosette leaves, is virtually impossible to access, and cannot be grown in culture. Thus, most studies on Arabidopsis organ initiation concern the induction of floral meristems from the inflorescence apex, which is more easily accessed (9–11). An alternative system is tomato because its vegetative shoot apical meristem is relatively large and therefore can be dissected without problems, grows vigorously under defined culture conditions, and is well suited for a wide variety of micromanipulations (6, 12, 13).

The first chapters of this book will give an up-to-date overview of the above- (Chapters 1 and 2) and below-ground parts (Chapters 3 and 4) in monocot and dicot plants. We especially highlight aspects of why monocot and dicot roots are ideal model systems for organogenesis (Chapters 2 and 3). However, on the one hand we need to translate this information to economically interesting crops, or further investigate this directly in these crops (4, 14, 15). On the other hand, we need other simpler systems to understand the evolution of organs and to provide insight in the underlying molecular networks (e.g. Chapters 5 and 6). The chapters in this book provide exactly that, focusing on tools to study organogenesis in *Arabidopsis*, but also taking it further to cereal crops and highlighting emerging model systems.

While it is not always straightforward to translate particular techniques and approaches that work well in, for example, Arabidopsis to crop plants, several examples are discussed in this book on the level of shoot kinematics (Chapter 17), immunolocalization (Chapters 14 and 15), 3D root systems (Chapter 11) and a lateral root-inducible system (Chapter 9). In addition, to address specific questions, for example on the level of evolutionary biology, we need to start using other model systems. A few of these emerging model systems are introduced here, such as brown algae (Chapter 6), *Physcomitrella* (Chapter 2), and Podostemaceae (Chapter 5). More details on how to grow and study the brown alga *Ectocarpus* are provided in Chapter 22. There are obviously a number of other models for plant organogenesis that are not addressed here, but that have recently been reviewed, such as the *Arabidopsis* petal (16).

1.3. Plant Organogenesis

Organogenesis entails the regulation of cell division, cell expansion, cell- and tissue-type differentiation, and patterning of the organ as a whole. De novo organogenesis is especially important in plants, as most of plant development takes place post-embryonically. Therefore it is essential to gain insight into how organs are initiated and how they develop. However, this very often is subject to technical difficulties as these processes take place embedded deep in tissues or are difficult to access or visualize. Furthermore, plant cells are enclosed in a rigid wall making a tight control of the direction of polar cell growth and of the positioning of cell division planes very important for plant organogenesis. To study this, we need specialized techniques that are described in this book.

One of the very first steps in the development of a plant is the formation of ovules and embryos. The ovule and embryo of *Arabidopsis thaliana* have been established as an excellent model system with which to study organogenesis at the molecular and genetic level (17–19). How to study and image these structures is addressed in Chapters 9 and 18.

A new plant organ develops from a subset of cells that has been specified, primed, etc. and which will undergo a series of cell divisions to give rise to a new plant part, such as a leaf, a flower, and a lateral root. To visualize the contribution of each cell and cell division to building the mature organ, it is necessary to establish cell lineages. An elegant tool to achieve this is described in Chapter 13.

The totipotency of several plant cells is reflected in their ability to regenerate tissues and organs. An approach to study this is described in Chapter 21.

2. Novel Techniques

Due to the difficulties associated with studying particular processes, the development of novel, more sensitive techniques is essential. For example, the use of fluorescence-activated cell sorting (FACS) brought about a revolution in cell-specific analyses of transcriptomes and hormone levels in Arabidopsis (20, 21). Here, the use of this approach in the shoot apical meristem is described (Chapter 16). However, it is also important to get closer to the proteins, and as cell-specific proteome analyses are still difficult, other techniques have been developed. For example, ribosome pull down provides insight into the translatome (Chapter 19), and localizing RNAs and proteins in plants is useful (Chapters 20 and 21). In addition, classical genetics has its limitations, as exemplified through redundancy and embryo lethal mutations. To circumvent this, chemical genetics was put forward as an ideal tool, as described in Chapter 12.

3. Mathematical Modelling

Finally, as our knowledge increases, we need computer-based approaches to bring everything together. In several areas of plant organogenesis, this has been used successfully. Auxin has been a major focus of mathematical modelling, and this is reflected in a wide range of models describing the distribution and role of auxin (22–25). However, these in silico approaches are not always easy to use by wet-lab scientists. We therefore also need simpler, user-friendly systems, such as the one in Chapter 23.

References

1. Fletcher JC (2002) Shoot and floral meristem maintenance in arabidopsis. Annu Rev Plant Biol 53:45–66
2. Niinemets U (2007) Photosynthesis and resource distribution through plant canopies. Plant Cell Environ 30:1052–1071
3. De Smet I (2012) Lateral root initiation: one step at a time. New Phytol 193:867–873
4. De Smet I, White PJ, Bengough AG, Dupuy L, Parizot B, Casimiro I, Heidstra R, Laskowski M, Lepetit M, Hochholdinger F, Draye X, Zhang H, Broadley MR, Peret B,

Hammond JP, Fukaki H, Mooney S, Lynch JP, Nacry P, Schurr U, Laplaze L, Benfey P, Beeckman T, Bennett M (2012) Analyzing lateral root development: how to move forward. Plant Cell 24:15–20

5. Smith S, De Smet I (2012) Root system architecture—insights from Arabidopsis and cereal crops. Phil Trans R Soc B 367:1441–1452
6. Reinhardt D, Mandel T, Kuhlemeier C (2000) Auxin regulates the initiation and radial position of plant lateral organs. Plant Cell 12:507–518
7. Suarez-Lopez P (2005) Long-range signalling in plant reproductive development. Int J Dev Biol 49:761–771
8. Issue S (2010) Arabidopsis: a rich harvest 10 years after completion of the genome sequence. Plant J 61
9. Hamant O, Heisler MG, Jonsson H, Krupinski P, Uyttewaal M, Bokov P, Corson F, Sahlin P, Boudaoud A, Meyerowitz EM, Couder Y, Traas J (2008) Developmental patterning by mechanical signals in Arabidopsis. Science 322:1650–1655
10. Heisler MG, Ohno C, Das P, Sieber P, Reddy GV, Long JA, Meyerowitz EM (2005) Patterns of auxin transport and gene expression during primordium development revealed by live imaging of the Arabidopsis inflorescence meristem. Curr Biol 15:1899–1911
11. Reddy GV, Heisler MG, Ehrhardt DW, Meyerowitz EM (2004) Real-time lineage analysis reveals oriented cell divisions associated with morphogenesis at the shoot apex of Arabidopsis thaliana. Development 131:4225–4237
12. Fleming AJ, Mandel T, Roth I, Kuhlemeier C (1993) The patterns of gene expression in the tomato shoot apical meristem. Plant Cell 5:297–309
13. Reinhardt D, Frenz M, Mandel T, Kuhlemeier C (2003) Microsurgical and laser ablation analysis of interactions between the zones and layers of the tomato shoot apical meristem. Development 130:4073–4083
14. Den Herder G, Van Isterdael G, Beeckman T, De Smet I (2010) The roots of a new green revolution. Trends Plant Sci 15:600
15. Lynch JP (2007) Roots of the Second Green Revolution. Aust J Bot 55:493–512
16. Irish VF (2008) The Arabidopsis petal: a model for plant organogenesis. Trends Plant Sci 13:430–436
17. Lau S, Slane D, Herud O, Kong J, Jurgens G (2012) Early embryogenesis in flowering plants: setting up the basic body pattern. Annu Rev Plant Biol 63:483–506
18. De Smet I, Lau S, Mayer U, Jurgens G (2010) Embryogenesis—the humble beginnings of plant life. Plant J 61:959–970
19. Schneitz K (1999) The molecular and genetic control of ovule development. Curr Opin Plant Biol 2:13–17
20. Petersson SV, Johansson, AI, Kowalczyk M, Makoveychuk A, Wang JY, Moritz T, Grebe M, Benfey PN, Sandberg G, Ljung K (2009) An auxin gradient and maximum in the Arabidopsis root apex shown by high-resolution cell-specific analysis of IAA distribution and synthesis. Plant Cell 21:1659–1668
21. Birnbaum K, Shasha DE, Wang JY, Jung JW, Lambert GM, Galbraith DW, Benfey PN (2003) A gene expression map of the Arabidopsis root. Science 302:1956–1960
22. Grieneisen VA, Xu J, Maree AF, Hogeweg P, Scheres B (2007) Auxin transport is sufficient to generate a maximum and gradient guiding root growth. Nature 449:1008–1013
23. Merks RM, Van de Peer Y, Inze D, Beemster GT (2007) Canalization without flux sensors: a traveling-wave hypothesis. Trends Plant Sci 12:384–390
24. Bayer EM, Smith RS, Mandel T, Nakayama N, Sauer M, Prusinkiewicz P, Kuhlemeier C (2009) Integration of transport-based models for phyllotaxis and midvein formation. Genes Dev 23:373–384
25. Smith RS, Guyomarc'h S, Mandel T, Reinhardt D, Kuhlemeier C, Prusinkiewicz P (2006) A plausible model of phyllotaxis. Proc Natl Acad Sci USA 103:1301–1306

Contents

Contributors

Alok Arun • *Centre National de la Recherche Scientifique, Unité Mixte de Recherche 7139, Laboratoire International Associé Dispersal and Adaptation in Marine Species, Station Biologique de Roscoff, Roscoff Cedex, France; Laboratoire International Associé Dispersal and Adaptation in Marine Species, CNRS, UMR 7139, Roscoff, France*

Dominique Audenaert • *Department of Plant Systems Biology, VIB, Ghent, Belgium; Department of Plant Biotechnology and Genetics, Ghent University, Ghent, Belgium*

Julia Bailey-Serres • *Center for Plant Cell Biology, University of California, Riverside, CA, USA; Department of Botany and Plant Sciences, University of California, Riverside, CA, USA*

Tom Beeckman • *Department of Plant Systems Biology, VIB, Ghent, Belgium; Department of Plant Biotechnology and Bioinformatics, Ghent University, Ghent, Belgium*

Gerrit T.S. Beemster • *Department of Biology, University of Antwerpen, Antwerpen, Belgium*

Philip N. Benfey • *Department of Biology, Duke University, Durham, NC, USA; Center for Systems Biology, Duke University, Durham, NC, USA*

Kenny A. Bogaert • *Phycology Research Group, Department of Biology, Center for Molecular Phylogenetics and Evolution, Ghent University, Ghent, Belgium*

Sandrine Bonhomme • *Institut Jean-Pierre Bourgin, UMR1318 INRA-AgroParisTech, INRA Centre de Versailles-Grignon, Versailles, France*

Yogev Burko • *The Robert H. Smith Institute of Plant Sciences and Genetics in Agriculture, Hebrew University, Rehovot, Israel*

The Otto Warburg Minerva Center for Agricultural Biotechnology, Hebrew University, Rehovot, Israel

Pilar Bustos-Sanmamed • *Institut des Sciences du Végétal, Centre National de la Recherche Scientifique, Gif-sur-Yvette, France*

Nicola Carraro • *Department of Horticulture and Landscape Architecture, Purdue University, West Lafayette, IN, USA*

Bénédicte Charrier • *Morphogenesis of Macroalgae, UMR7139 Marine Plants and Biomolecules, CNRS-UPMC, Station Biologique, Roscoff, France*

Susana M. Coelho • *Centre National de la Recherche Scientifique, Unité Mixte de Recherche 7139, Laboratoire International Associé Dispersal and Adaptation in Marine Species, Station Biologique de Roscoff, Roscoff Cedex, France; Laboratoire International Associé Dispersal and Adaptation in Marine Species, UMR 7139, Station Biologique de Roscoff, Roscoff, France*

Frederik Coppens • *Department of Plant Systems Biology/Department of Plant Biotechnology and Bioinformatics, VIB/University of Ghent, Ghent, Belgium*

MARTIN CRESPI • *Institut des Sciences du Végétal, Centre National de la Recherche Scientifique, Gif-sur-Yvette Cedex, France*

OLIVIER DE CLERCK • *Phycology Research Group, Department of Biology, Center for Molecular Phylogenetics and Evolution, Ghent University, Ghent, Belgium*

BERT DE RYBEL • *Laboratory of Biochemistry, Wageningen University, Wageningen, The Netherlands*

STIJN DHONDT • *Department of Plant Systems Biology/Department of Plant Biotechnology and Bioinformatics, VIB/University of Ghent, Ghent, Belgium*

BALAJI ENUGUTTI • *Entwicklungsbiologie der Pflanzen, Wissenschaftszentrum Weihenstephan, Technische Universität München, Freising, Germany*

FABIO FIORANI • *Forschungszentrum Jülich GmbH, IBG-2: Plant Sciences, Jülich, Germany*

CRISTIAN FORESTAN • *Department of Agronomy, Food, Natural Resources, Animals and Environment, University of Padova, Padova, Italy*

FLORIAN FRUGIER • *Institut des Sciences du Végétal, Centre National de la Recherche Scientifique, Gif-sur-Yvette Cedex, France*

DANNY GEELEN • *Department of Plant Production, Faculty of bioscience engineering, Ghent University, Ghent, Belgium*

THOMAS GIRKE • *Department of Botany and Plant Sciences, Center for Plant Cell Biology (CEPCEB), Institute of Integrative Genome Biology (IIGB), University of California, Riverside, CA, USA*

MICHAEL A. GURAVAGE • *Centrum Wiskunde & Informatica (CWI), XG Amsterdam, The Netherlands*

Netherlands Consortium for Systems Biology/Netherlands Institute for Systems Biology (NCSB-NISB), XG Amsterdam, The Netherlands

SOAZIG GUYOMARC'H • *Université Montpellier 2, UMR DIADE, IRD, Montpellier, France*

FRANK HOCHHOLDINGER • *Institute of Crop Science and Resource Conservation (INRES), Crop Functional Genomics, University of Bonn, Bonn, Germany*

ANJALI S. IYER-PASCUZZI • *Department of Biology, Duke University, Durham, NC, USA; Center for Systems Biology, Duke University, Durham, NC, USA*

LEEN JANSEN • *Department of Plant Systems Biology, VIB, Ghent, Belgium; Department of Plant Biotechnology and Bioinformatics, Ghent University, Ghent, Belgium*

NATSU KATAYAMA • *Division of Life Sciences, Graduate School of Natural Science and Technology, Kanazawa University, Kakuma, Kanazawa, Japan*

SATOSHI KOI • *Graduate School of Biological Sciences, Nara Institute of Science and Technology, Ikoma, Nara, Japan*

SMITA KURUP • *Plant Biology and Crop Sciences Department, Rothamsted Research, Harpenden, Hertfordshire, UK*

CAROLE LAFFONT • *Institut des Sciences du Végétal, Centre National de la Recherche Scientifique, Gif-sur-Yvette Cedex, France*

LAURENT LAPLAZE • *Institut de Recherche pour le Développement, UMR DIADE, IRD, Montpellier, France; Institut de Recherche pour le Développement, Laboratoire Commun de Microbiologie IRD/ISRA/UCAD, Centre de Recherche de Bel Air, Dakar, Senegal*

Julien Lavenus • *Université Montpellier 2, UMR DIADE, IRD, Montpellier, France; School of Biosciences and Centre for Plant Integrative Biology, University of Nottingham, Nottingham, UK; Institut de Recherche pour le Développement, UMR DIADE, IRD, Montpellier, France*

Aude Le Bail • *Cell Biology Division, Department of Biology, University of Erlangen-Nuremberg, Erlangen, Germany*

Christine Lelandais-Brière • *Institut des Sciences du Végétal, Centre National de la Recherche Scientifique, Gif-sur-Yvette, France; Université Paris Diderot-Paris 7, Paris, France*

Cristina Llavata-Peris • *Laboratory of Biochemistry, Wageningen University, HA Wageningen, The Netherlands*

Annemarie Lokerse • *Laboratory of Biochemistry, Wageningen University, HA Wageningen, The Netherlands*

Mikaël Lucas • *Institut de Recherche pour le Développement, UMR DIADE, IRD, Montpellier, France*

Caroline Marcon • *Institute of Crop Science and Resource Conservation (INRES), Crop Functional Genomics, University of Bonn, Bonn, Germany*

Roeland M.H. Merks • *Centrum Wiskunde & Informatica (CWI), XG Amsterdam, The Netherlands; Netherlands Consortium for Systems Biology/Netherlands Institute for Systems Biology (NCSB-NISB), XG Amsterdam, The Netherlands*

Elliot M. Meyerowitz • *Division of Biology 156-29, California Institute of Technology, Pasadena, CA, USA*

Barbara Möller • *Laboratory of Biochemistry, Wageningen University, HA Wageningen, The Netherlands*

Angelika Mustroph • *Department of Plant Physiology, University of Bayreuth, Bayreuth, Germany*

Hilde Nelissen • *Department of Plant Systems Biology/Department of Plant Biotechnology and Bioinformatics, VIB/University of Ghent, Ghent, Belgium*

Long Nguyen • *VIB Compound Screening Facility (VIB-CSF), Ghent, Belgium*

Fabien Nogué • *Institut Jean-Pierre Bourgin, UMR1318 INRA-AgroParisTech, INRA Centre de Versailles-Grignon, Versailles, France*

Maxi Oelschner • *Entwicklungsbiologie der Pflanzen, Wissenschaftszentrum Weihenstephan, Technische Universität München, Freising, Germany*

Naomi Ori • *The Robert H. Smith Institute of Plant Sciences and Genetics in Agriculture and The Otto Warburg Minerva Center for Agricultural Biotechnology, Hebrew University, Rehovot, Israel*

Boris Parizot • *Department of Plant Systems Biology, VIB, Ghent, Belgium; Department of Plant Biotechnology and Bioinformatics, Ghent University, Ghent, Belgium*

Anja Paschold • *Institute of Crop Science and Resource Conservation (INRES), Crop Functional Genomics, University of Bonn, Bonn, Germany*

Catherine Rameau • *Institut Jean-Pierre Bourgin, UMR1318 INRA-AgroParisTech, INRA Centre de Versailles-Grignon, Versailles, France*

G. Venugopala Reddy • *Department of Botany and Plant Sciences, Center for Plant Cell Biology (CEPCEB), Institute of Integrative Genome Biology (IIGB), University of California, Riverside, CA, USA*

John Runions • *Department of Biological and Medical Sciences, Oxford Brookes University, Gypsy Lane, Oxford, UK*

Bart Rymen • *Plant Science Center, RIKEN Yokohama Institute, Tsurumi, Yokohama, Kanagawa, Japan*

Didier G. Schaefer • *Laboratory of molecular and cell biology, Institute of Biology, University of Neuchatel, Neuchatel, Switzerland*

Kay Schneitz • *Entwicklungsbiologie der Pflanzen, Wissenschaftszentrum Weihenstephan, Technische Universität München, Freising, Germany*

Stephen Snipes • *Department of Botany and Plant Sciences, Center for Plant Cell Biology (CEPCEB), Institute of Integrative Genome Biology (IIGB), University of California, Riverside, CA, USA*

Kaoru Sugimoto • *Division of Biology 156-29, California Institute of Technology, Pasadena, CA, USA*

Serena Varotto • *Department of Agronomy, Food, Natural Resources, Animals and Environment, University of Padova, Padova, Italy*

Inge Verstraeten • *Department of Plant Production, Faculty of bioscience engineering, Ghent University, Ghent, Belgium; Department of Plant Systems Biology, VIB (Flanders Institute for Biotechnology), Ghent, Belgium*

Dolf Weijers • *Laboratory of Biochemistry, Wageningen University, HA Wageningen, The Netherlands*

Ram Kishor Yadav • *Department of Botany and Plant Sciences, Center for Plant Cell Biology (CEPCEB), Institute of Integrative Genome Biology (IIGB), University of California, Riverside, CA, USA*

M. Eugenia Zanetti • *Department of Botany and Plant Sciences, University of California, Riverside, CA, USA; Center for Plant Cell Biology, University of California, Riverside, CA, USA; Facultad de Ciencias Exactas, Instituto de Biotecnología y Biología Molecular, Universidad Nacional de La Plata, La Plata, Argentina*

Paul R. Zurek • *Department of Biology, Duke University, Durham, NC, USA; Center for Systems Biology, Duke University, Durham, NC, USA*

Chapter 1

The Tomato Leaf as a Model System for Organogenesis

Yogev Burko and Naomi Ori

Abstract

Compound tomato leaves are composed of multiple leaflets that are generated gradually during leaf development, and each resembles a simple leaf. The elaboration of a compound leaf form requires the maintenance of transient organogenic activity at the leaf margin. The developmental window of organogenic activity is defined by the antagonistic activities of factors that promote maturation, such as TCP transcription factors, SFT and gibberellin, and factors that delay maturation, such as KNOX transcription factors and cytokinin. Leaflet initiation sites are specified spatially and temporally by spaced and specific activities of CUCs, auxin and ENTIRE, as well as additional factors. The partially indeterminate growth of the compound tomato leaf makes it a useful model to understand the balance between determinate and indeterminate growth, and the mechanisms of organogenesis, some of which are common to many developmental processes in plants.

Key words: Tomato, Indeterminate growth, Compound leaf, KNOX, TCP, Auxin, Cytokinin, Gibberellin, CUC, Marginal blastozone

1. Introduction

Leaves evolved from lateral branches by acquisition of determinate growth and a flat structure (1–3). Simple and compound leaves can be viewed as representing different levels of determination and flattening (4, 5). Simple leaves differentiate and flatten relatively fast on a developmental timescale, while compound leaves are in some ways intermediate forms between lateral branches and simple leaves. As a result, simple leaves have an entire, continuous lamina, while compound leaves are "branched" with leaflets, each resembling a simple leaf. The development of compound leaves can therefore serve as a model to study general developmental principles in plants. In addition, studying compound-leaf development can shed light on the mechanisms that regulate and tune the gradient from a simple leaf, through a compound leaf, to a shoot.

Ive De Smet (ed.), *Plant Organogenesis: Methods and Protocols*, Methods in Molecular Biology, vol. 959,
DOI 10.1007/978-1-62703-221-6_1, © Springer Science+Business Media New York 2013

Fig. 1. Tomato leaf development and definitions of terms used to describe the leaf parts. The fifth leaf produced by the plant is shown at successive stages of leaf development. P3–P9 designate the developmental stage, for example at the P3 stage there are two younger leaf primordia. m-P3, the shoot apical meristem and three youngest leaf primordia, the fifth leaf is at the P3 stages. Note that the leaf maintains leaflet and lobe organogenesis until very late stages of its development, long after it has expanded. The mature leaf contains a lobed terminal leaflet (TL), three or more pairs of primary lateral leaflets (pr leaflet), secondary leaflets (sec leaflet), and intercalary leaflets that were initiated after the primary lateral leaflets. Scale bars: 0.5 mm (*left panels*), 1 cm (*right panels*).

Tomato leaves show considerable shape diversity and their shape is sensitive to environmental and developmental circumstances. They consist of a terminal leaflet and usually three pairs of primary lateral leaflets, each consisting of a petiolule and a lamina. The primary leaflets are separated by a rachis, and some of them develop secondary leaflets. Intercalary leaflets are additional lateral leaflets that develop later from the rachis between the primary leaflets (Fig. 1). Tomato mutants, natural species and cultivars display many variations on this basic form, including more simple leaves and much more compound leaves, with many orders of leaflets (6–12).

The variability of leaf sizes and forms in nature reflects a corresponding variability in leaf ontogeny, which results from flexible tuning of common principles shared by all leaves. Leaf development follows three continuous and overlapping phases: initiation (I), primary morphogenesis (PM), and secondary morphogenesis (SM) (13, 14). At I, the leaf emerges from the flanks of the shoot apical meristem (SAM). During PM, the lamina is initiated and leaf marginal structures such as leaflets, lobes, and serrations are formed. In SM, the leaf area grows substantially, the various leaf tissues mature and differentiate, and the final leaf form is determined by the relative growth of the different zones. Leaf development is followed by plastochrons, such that P1 is the most recently initiated leaf primordium, and it becomes P2, P3, etc. as new primordia arise. Tomato leaves initiate as entire primordia, and leaflets initiate in a sequential basipetal order only from the P3 stage on (Fig. 1).

In this chapter we review the current knowledge of tomato leaf development. We discuss the mechanisms that enable organogenesis at the tomato leaf margins and the specification and separation of leaflets.

2. Balancing Indeterminate and Determinate Growth

While leaf growth is determinate, the elaboration of a compound leaf forms, such as that of the tomato leaf, requires the maintenance of transient partially indeterminate growth. During early stages of leaf development, a specific region at the leaf margin, termed marginal blastozone, possesses the organogenic potential and is thus responsible for this transient indeterminate growth (13, 15). The marginal blastozone is responsible for lamina initiation and elaboration of marginal structures such as serrations in both simple and compound leaves. In tomato as well as other species with compound leaves, its activity is maintained for a longer time and within a longer portion of the leaf margin, enabling the organogenesis of leaflets and lobes (13, 15). Recent studies in tomato revealed that the developmental window of organogenic activity during leaf development, which is characterized by the maintenance of the marginal blastozone, underlies much of the variability in leaf shape. The extent of this organogenesis window is defined by the antagonistic activities of maturation promoting and maturation delaying factors. Several factors that affect this balance between organogenesis and differentiation in tomato have been recently identified.

2.1. Promoting Maturation

The CIN-TCP transcription factor LANCEOLATE (LA) plays a central role in promoting leaf maturation and differentiation.

The semi-dominant mutant *La* has small and simple leaves in comparison to the wild-type compound leaves (16, 17). Homozygote *La* mutants show a range of phenotypes, in the most severe of which the SAM aborts. Detailed phenotypic analysis led to the hypothesis that *La* SAM and leaves lose their meristematic activity precociously (17, 18). This was supported by a detailed analysis of early leaf development, which showed that lamina expansion and growth, cell expansion, development of trichomes, termination of marginal blastozone activity, and accelerated leaf growth occur earlier in *La/+* leaves than in the wild type (19–21). Dosage analysis of *La* in tetraploid and diploid states revealed a quantitative effect for the *La* allele on a range of phenotypes including the rate of leaf differentiation and the resulting leaf shape (18).

Identification of the *LA* gene as a *miR319*-regulated *CIN-TCP* gene provided a molecular mechanism for this quantitative effect. *LA* is expressed at relatively low levels in the SAM and young leaf primordia, and a steep increase in its expression occurs during the transition from the P4 and P5 stages of leaf development, just before the leaf primordium starts its accelerated growth phase. *miR319*, a negative regulator of *LA* and three closely related *CIN-TCP* genes (*LA-like*), shows an opposite expression gradient: it is expressed at relatively high levels in the SAM and young leaf primordia, and its expression declines sharply concurrently with the increase in *LA* expression. Dominant *La* alleles are partially resistant to *miR319* regulation, which leads to their precocious elevated expression and consequently to precocious leaf maturation and differentiation, resulting in a simpler and smaller leaf in comparison to the wild type. Conversely, the loss-of-function allele *la-6* shows increased activity of the leaf margins. Furthermore, downregulation of the expression of *LA* and *LA-like* genes by leaf-specific overexpression of *miR319* results in a substantial delay in leaf maturation and in the maintenance of extremely prolonged indeterminate growth in the leaf margin (20, 21). These observations indicate that LA facilitates determinate leaf growth by promoting leaf maturation and differentiation. The maintenance of the organogenic window during which the marginal blastozone is active depends on low LA activity during early stages of leaf development. The quantitative nature of LA activity provides an attractive potential mechanism for the control of leaf shape diversity by tuning the extent of the morphogenetic activity of the leaf margin.

The hormone gibberellin (GA) negatively regulates leaf complexity in tomato. Upon exogenous GA application only primary leaflets are formed, and the primary leaflets have smooth margins without lobes or serrations (22–26). In agreement, leaves of the *procera* (*pro*) mutant, in which there is a constitutive GA response due to a mutation in the single tomato DELLA-type GA-response inhibitor, have only primary leaflets with smooth margins (26–29).

solanifolia (*sf*) mutants similarly produce primary and intercalary leaflets only, with smooth margins (30). *sf* leaves resemble GA-treated wild-type leaves, and inhibition of GA biosynthesis suppressed the *sf* simple leaf phenotype, suggesting that elevated GA levels mediate the *sf* leaf shape phenotype (22). Examination of early leaf development in *pro* mutants revealed that the effect of *pro* on leaf shape results from a combination of faster growth during early stages, and a delay in leaflet initiation relative to the wild type (28). Leaflet initiation is similarly delayed in *sf* mutants (30). Young *pro* leaf primordia also acquire an upright position earlier than the wild type. Some of these phenotypes, including the precocious fast growth and upright position, are reminiscent of the *La/+* phenotypes. This similarity suggests that *pro* leaves mature faster than wild-type leaves and that GA promotes leaf maturation. It should be noted that in some species GA has an opposite effect of inducing more compound leaves (8, 31, 32).

The ratio between SINGLE FLOWER TRUSS (SFT), the tomato FT homolog, and SELF-PRUNING (SP), which affect the induction of flowering in tomato, has recently been shown to be involved in general regulation of growth and determination, including the control of leaf shape. In the leaf, a high SFT/FT ratio promotes leaf maturation, leading to a simplified leaf form. This effect is further enhanced in the background of the *trifoliate* (*tf*) mutant (33).

Several additional tomato mutants have been described with precocious leaf maturation that leads to lack of higher order leaflets and smooth margins. These include for example *expelled shoot* (*exp*) (6). Further characterization of these mutants will enhance the understanding of how the timing and rate of leaf maturation is controlled.

Successive leaves on the plant differ from each other in size and shape (8, 34). In tomato, early leaves are simpler and smaller than later leaves. Recent work has shown that differences in the rate of maturation, which is correlated with the timing of expression of *LA-like* genes, may underlie some of these differences (21).

2.2. Delaying Maturation

ClassI KNOTTED1-LIKE HOMEBOX (KNOXI) is a family of transcription factors that play important roles in many developmental processes in plants, including SAM maintenance and leaf development. Loss-of-function mutations in maize and *Arabidopsis* *KNOXI* genes result in aborted SAM (35, 36). In many species with simple leaves, *KNOXI* expression is restricted to the SAM and is downregulated in initiating leaf primordia (35, 37–39). *KNOXI* overexpression in these species leads to variable phenotypes, depending on the species, including knot-like structures on the leaves, curled or lobed leaves, and formation of ectopic meristems on leaves (40–47).

In many plants with compound leaves, including tomato, *KNOXI* expression is restored in developing leaf primordia (48). The expression pattern of *KNOXI* genes and proteins in tomato leaves was investigated by several groups by in situ hybridization and immuno-localization, but is still not completely clear, as the different reports are not completely consistent, possibly due to complex and dynamic expression patterns, relatively low expression, and the use of diverse methodology (48–58). For example, it is not clear whether the tomato *KNOXI* proteins are expressed at the site of future primordium initiation at the flanks of the SAM (P0). In most reports tomato *KNOXI* genes or proteins were detected in the SAM and vasculature, in initiating leaflets, and in the margins of young leaf and leaflet primordia. The expression in initiating leaflets is in contrast to the case of *Cardamine hirsuta*, where the promoter of the *KNOXI* gene *STM* is downregulated in initiating leaflet primordia (51, 53, 59–61).

Two dominant tomato mutants, *Mouse ears* (*Me*) and *Curl* (*Cu*), show dramatic alterations in leaf shape that includes substantially increased leaf complexity (62, 63). Identification of the genetic lesion in these mutants revealed that in both mutants the tomato *KNOXI* gene *Tomato Knotted2* (*TKN2*)/*Let6* is misexpressed (52, 54). Furthermore, overexpression of *KNOXI* genes in tomato resulted in variable phenotypes, including highly compound leaves, due to delayed leaf maturation (50, 52, 54, 64). Overexpression of *TKN2* at different stages of leaf development indicated that the leaf responds to TKN2 in a developmental-stage dependent manner: overexpression at I results in narrow simple leaves, expression during PM leads to highly compound leaves, but the leaf is insensitive to *TKN2* overexpression at later developmental stages. Conversely, expression of a putative dominant repressor for TKN2 targets, TKN2-SRDX, resulted in accelerated leaf maturation and in the development of a reduced and simple leaf (64). The cumulative observations indicate that KNOXI proteins inhibit the transition from the PM to the SM stages of the tomato leaf development (Fig. 2), and that their activity is restricted to the developmental window of the PM (64). Thus, precise spatial and temporal regulation of the levels and patterns of *KNOXI* expression may contribute to defining leaf structure.

Several factors that regulate *KNOXI* expression have been described (58, 65–67). The recessive *clausa* (*clau*) and *tripinnate* (*tp*) mutants (68) show increased *KNOXI* expression as well as increased leaf complexity, and these are enhanced in the double mutant *cla tp*, indicating that these genes restrict the level of leaf complexity by negative regulation of *KNOXI* expression (58, 69). The ARP (ASYMETRIC LEAVES1, ROUGH SHEAT2, and PHANTASTICA) proteins were proposed to be conserved negative regulators of KNOXI expression in leaves (65, 70). In tomato, *ARP* and *KNOXI* genes were shown to be co-expressed, suggesting

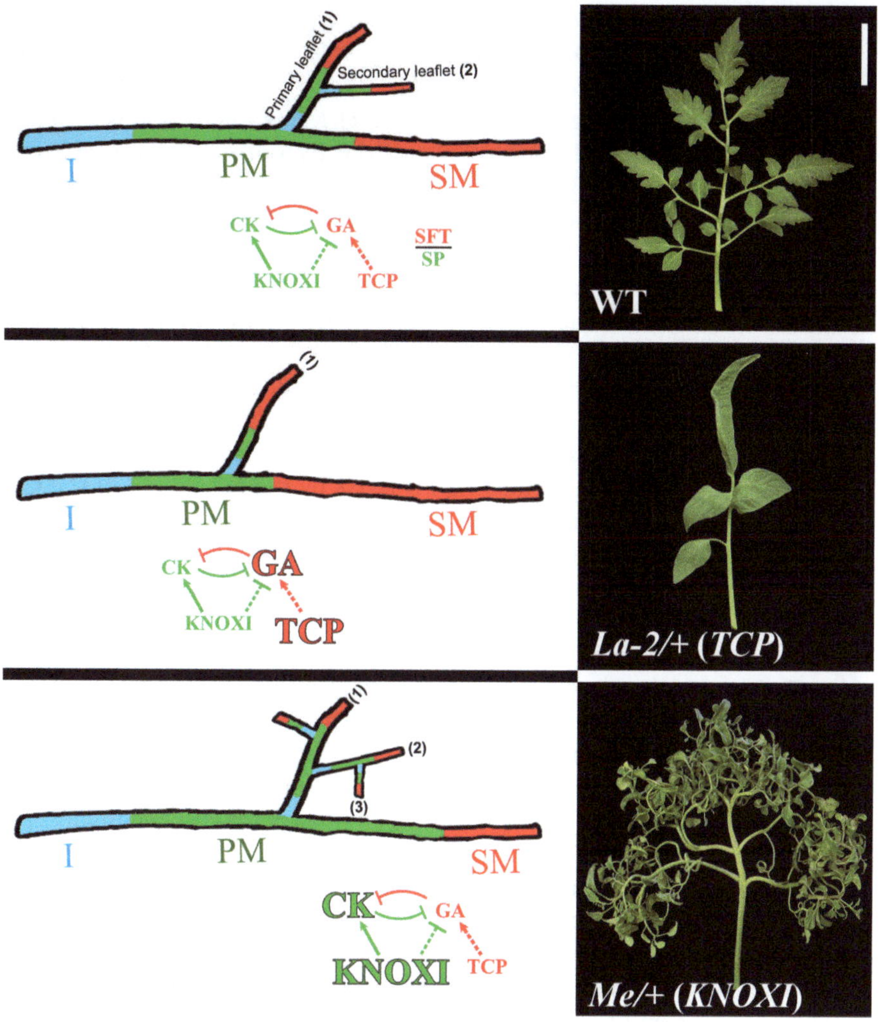

Fig. 2. Balancing indeterminate and determinate growth. The *left panels* show a scheme of the stages of leaf development in tomato and the factors that balance determinate and indeterminate growth, color coded for the process they promote. The *right panels* show mature leaves. *I* initiation, *PM* primary morphogenesis, *SM* secondary morphogenesis, *CK* cytokinin, *GA* gibberellin. *Solid lines* represent known genetic interactions, and dashed lines putative interactions. In the wild type (*top panels*), KNOXI proteins and cytokinin promote PM, and TCPs, GA and high SFT/SP promote SM. *Lanceolate-2* (*La-2*, *middle panels*) is a gain of function mutant in the *TCP* gene *LA*. Its precocious activation promotes early maturation. *Mouse ears* (*Me*, *bottom panels*) is a dominant mutant in which The *KNOXI* gene *Tkn2* is upregulated, leading to prolonged PM and a highly compound leaf. Scale bar: 5 cm.

a more complex interaction between these factors (55, 67, 71). Another regulation on KNOXI activity is achieved by their interaction with the *BEL-LIKE HOMEODOMAIN (BELL)* proteins. In *Arabidopsis*, the interaction with different BELLs was proposed to determine whether the complex KNOXI-BELL will serve as an activator or a repressor (reviewed in 65). In tomato, the BELL protein BIPINNATE (BIP) was shown to interact with TKN2/LeT6, and the classic mutant *bip* (72), is characterized by a more compound leaf than the wild type. These results suggest that the BIP-KNOXI interaction repress KNOXI activity (66). The dominant mutant *Petroselinum* (*Pts*) shows increased leaf complexity (73). The *PTS* gene encodes a novel *KNOX* gene that lacks a homeodomain, and the presence of the PTS protein can interfere with the KNOXI-BELL interaction. Kimura et al. proposed that the increased expression of *PTS* in the *Pts* mutant leads to the release of Class I KNOX proteins from the interaction with BELL, and to the consequent increase in leaf complexity (66).

The plant hormone cytokinin (CK) has recently been shown to be involved in the maintenance of prolonged organogenic activity in the tomato leaf margin (74) (Fig. 2). Overexpression of the CK biosynthesis gene *ISOPENTENYLTRANSFERASE 7* (*IPT7*) in tomato leaves led to the formation of super-compound leaves, and conversely, reducing CK levels by the expression of the CK degradation gene *cytokinin oxidase* (*CKX*) resulted in reduced leaf complexity. Genetic and molecular analysis indicated that CK acts downstream of KNOXI activity in delaying maturation.

2.3. Interactions Among Factors That Balance Organogenesis and Maturation

How do the factors that promote and the factors that inhibit maturation interact to balance determinate and indeterminate growth and define the extent of the window of organogenic activity in the tomato leaf margin? This question is mostly still open. However, some directions emerge from research in tomato and other species.

The *La* gain-of-function phenotype is substantially epistatic to phenotypes of transgenic *KNOXI* overexpression or *KNOXI* gain-of-function mutations (9, 20, 50). This suggests that properly controlled LA activity is required for the KNOXI-mediated extended organogenic activity. However, phenotypes of *Tkn2/LeT6* and *miR319* overexpression are additive (20), and the expression of *Tkn2/LeT6* mRNA or protein is not reduced in *La* mutants (53), suggesting that KNOXI and LA-like affect maturation via at least partially parallel pathways. The mechanism underlying the epistasis of *La* to *KNOXI* overexpression is thus likely developmental: downregulation of LA and KNOXI activity are both required for the extended organogenic activity of the tomato leaf margin. This is in agreement with the finding that the *La* phenotype is epistatic to the phenotypes of nearly all leaf shape mutants, and to the general tendency of simpler leaf phenotypes to be epistatic to phenotypes of increased complexity. It is also in accordance with the fact that no single genetic manipulation can convert a simple leaf into a compound one. However, it is still possible that KNOXI and LA-like affect each other's expression or activity.

While both TCP and the SFT/SP balance promote maturation and the loss of TF activity enhances the effect of SFT/SP, the relationship between these factors is still not known. The complexity of the mechanism underlying the timing and pace of maturation suggest that it is a highly regulated process, which facilitates its flexibility and sensitivity to changing circumstances.

KNOXI proteins have been shown to oppositely affect the homeostasis of GA and CK. KNOXI proteins negatively regulate the expression of the GA biosynthesis gene *GA-20-oxidase* (*GA20ox*) and positively regulate the GA deactivation gene *GA-2-oxidase* (*GA2ox*) in several species (25, 75–77). The tobacco KNOXI protein NTH15 directly binds an intron sequence in the *GA20ox* gene *Ntc12* (76), and in maize, the KNOXI protein kn1 binds in vivo regulatory sequences upstream of the *GA2ox1* gene (77). Conversely, KNOXI proteins activate CK biosynthesis genes and promote CK accumulation in *Arabidopsis* and rice (75, 78, 79). Furthermore, GA was shown to repress CK signaling, and the *GA2ox* gene was activated by CK (75). Importantly, CK was shown to affect the organogenic activity of the tomato leaf margin downstream of KNOXI proteins.

As mentioned above, some aspects of the early leaf development of *La/+* mutants resemble that of *pro* mutants, and a subset of phenotypes are common to *La/+* mutants and GA-treated plants. It is therefore possible that LA-like proteins affect GA and CK responses in an opposite manner with KNOXI proteins.

Many developmental processes are affected by the balance between hormones rather than their individual absolute amounts. GA and CK antagonistically affect many developmental processes, in which the outcome is determined by their ratio (23, 80). It is therefore tempting to speculate that one of the ways that the window of organogenic activity in the tomato leaf margin is determined by the dynamic distribution of these two hormones, which is in turn controlled by the dynamic spatial and temporal activities of LA-like, KNOXI and additional factors yet to be identified.

2.4. Late Organogenic Activity

In addition to the organogenic activity during early stages of leaf and leaflet development, the tomato leaf displays a unique feature of sustained organogenesis until very late stages of its development, long after the leaf has started to expand (21, 28, 53, 74). This late organogenic activity is manifested, for example, by the generation of intercalary and basal lateral leaflets after leaf expansion (Fig. 1). CK and LA have been recently shown to play central roles in balancing this late organogenic activity (21, 74). Downregulation of *LA*-like genes at late stages of leaf development revealed that the tomato leaf margin has a potential for indeterminate growth, which is inhibited by the prolonged expression of LA-like proteins. GA response also appears to inhibit the late organogenic activity, as *pro* mutant leaves were reported to cease leaflet formation earlier than wild-type leaves (28). While CK, GA, and LA affect both the early and the late organogenic activity, other factors affect only one of these organogenic phases. For example, manipulating KNOXI activity only affects early stages of leaf development (64).

Interestingly, some mutants with smooth margins and no secondary leaflets, such as *sf*, feature late initiation of basal primary leaflets and/or intercalary leaflets. This exemplifies the extreme developmental flexibility of the tomato leaf. It remains to be seen whether the late organogenic activity is continuous with the early activity or whether it represents reactivation of the organogenic potential.

3. The Involvement of Adaxial–Abaxial Polarity in Compound-Leaf Development

Tomato *wiry* mutants show a gradient of leaf-shape aberrations, with milder phenotypes in early leaves, and very severe phenotype in late leaves. Typical *wiry* leaf phenotypes include almost normal leaves, narrow leaves that are broader in their distal part, leaves with trumpet-shape distal part, and very narrow, wiry-like leaves (81, 82). Several allelic and nonallelic mutants with the *wiry* syndrome have been described (10, 67). *wiry* leaves show reduced lamina and reduced levels of leaf complexity, but often have increased number of intercalary leaflets. The tomato *PHANTASTICA* (*PHAN*) gene (which belongs to the ARP group) is expressed in the adaxial side of the developing tomato leaf, and its downregulation led to a *wiry*-like phenotype (55). Furthermore, analysis of many species revealed a correlation between *PHAN* expression, the integrity of the adaxial domain, and leaf shape. These observations suggest that proper adaxial–abaxial polarity is important for compound-leaf development. Further research is required to understand the role of leaf polarity in compound-leaf development. As the leaf margin is defined by the juxtaposition between the adaxial and the abaxial domains of the leaf, it is possible that the maintenance of the marginal blastozone depends on the preservation of the abaxial–adaxial polarity. In agreement, *Arabidopsis YABBY* genes were implicated in the interpretation of abaxial–adaxial polarity, the specification of the leaf margin, the negative regulation of *KNOXI* genes and the promotion of determinate growth in the leaves (3, 83).

4. Specification and Separation of Leaflets

As discussed above, maintaining transient partially indeterminate growth in the tomato leaf margin enables the organogenesis of marginal structures such as leaflets and lobes from the marginal blastozone. Leaflet primordia that initiate from the tomato leaf margin substantially resemble leaf primordia that initiate at the flanks of the SAM, despite the very different developmental context (1, 15, 84). In agreement, some of the mechanisms that mediate leaf initiation and separation at the SAM also play a role in leaflet initiation and separation.

4.1. Leaflet Organogenesis

How are leaflets localized, initiated and separated within this spatial and temporal window of organogenic activity? Recent studies have shown that the hormone auxin, the putative auxin response inhibitor ENTIRE and the transcription factors GOBLET and LYRATA are involved in these processes.

Auxin: Classical and recent studies have revealed a role for auxin in nearly all developmental processes in plants. In the SAM, auxin was shown to be a central coordinator of phylotaxis. Auxin maxima are formed at sites of future leaf initiation at the flanks of the SAM. One of the mechanisms that mediate the formation of these auxin maxima is auxin transport by PIN1. Auxin is transported to the point of future leaf initiation, and the resulting auxin maximum is thought to facilitate leaf initiation. This process depletes auxin from flanking regions, such that the next primordium will initiate as far as possible from existing primordia. Application of auxin transport inhibitors inhibits leaf initiation, and auxin application to SAMs that were pretreated with auxin transport inhibitors leads to leaf initiation at the sites of auxin application. These observations point to a central role for auxin in the specification of the sites of leaf initiation. However, only specific regions of the SAM respond to auxin application with leaf initiation, suggesting that additional factors are involved in conferring auxin sensitivity (85–89).

Auxin was similarly implicated in the initiation of leaflets and lobes from the margin of compound-leaf primordia. Application of auxin transport inhibitors to tomato seedlings resulted in the development of simpler leaves relative to non-treated plants (90, 91), and mutants that affect polar auxin transport also influence leaf shape (92). Inhibition of auxin transport or activity also caused leaf simplification in pea (93), and mutations in the auxin transporter PIN1 simplified leaf shape in *C. hirsute* (59). In tomato and *C. hirsuta*, *PIN1* expression, thought to be positively regulated by auxin, is upregulated in initiating leaflet primordia, and PIN1 subcellular localization was found to form convergence points at the sites of future leaflet initiation (59, 91). In agreement, peaks of expression of the auxin response sensor *DR5* were shown to mark initiating leaflets as well as vascular strands in *C. hirsuta* and pea (59, 94). External auxin application led to ectopic lamina growth in both tomato and *C. hirsuta*, and spotted microapplication of auxin to specific regions of the tomato leaf margin led to the initiation of ectopic leaflets, depending on the developmental context and the size of the application domain (59, 91). These observations indicate that discrete auxin response maxima promote leaflet initiation.

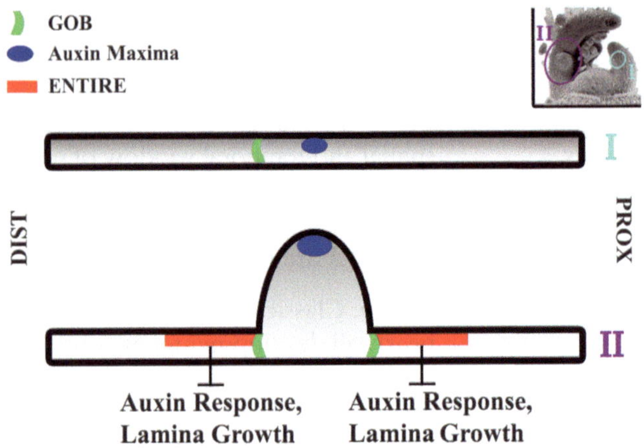

Fig. 3. Model for leaflet initiation and separation. *Top right*: an SEM of a tomato SAM with young leaf primordia. I and II designate developmental stages prior (I) and following (II) leaf initiation. *DIST* distal, *PROX* proximal. Juxtaposition of *GOB* and an auxin response maxima leads to leaflet initiation. The inhibition of auxin response by ENTIRE (E), a sufficient distance from the next combination of auxin maxima and *GOB* expression, and a sharp *GOB* expression domain inhibit lamina growth between leaflets thus promoting leaflet separation.

Entire: Leaves of the tomato mutant *entire* (*e*) are simpler than wild-type leaves (19, 62), with a single, lobed lamina, and no secondary leaflets. *e* leaf primordia initiate leaflets similar to although slightly later than the wild type, but these fuse to form the final *e* leaf form (19, 92). *E* (*SlIAA9*) encodes a protein from the Aux/IAA family of auxin response repressors (84, 95, 96). In *e* leaf primordia, the expression of the transgenic *Arabidopsis PIN1:PIN1-GFP* reporter is upregulated, mainly in the intercalary regions between initiating leaflet primordia. Application of auxin transport inhibitors to wild-type tomato plants led to ectopic lamina growth between leaflets, similar to *e* mutants, and to leaves that were subjected to auxin microapplication throughout their margins (91). These observations suggest that E/SlIAA9 restricts lamina growth between developing leaflets by inhibiting auxin response (Fig. 3).

Lyrate: Leaves of the recessive mutant *lyrate* (*lyr*) have more secondary and higher order leaflets in comparison to the wild type. The leaflets are smaller and the rachis, petiole, and petiolules are longer (68). The expression of the *PIN1:PIN1-GFP* reporter is downregulated in *lyr* mutants. The *LYR* gene is an ortholog of the *Arabidopsis* gene *JAGGED*, which is a transcription factor that promotes lateral organ outgrowth (97–99). The *LYR* mRNA is expressed in initiating leaflets, and its overexpression leads to leaflet fusion. These observations suggest that LYR promotes lamina growth, possibly by affecting auxin response or distribution (97).

Goblet: Mutations in the *GOBLET* (*GOB*) gene, encoding a *NAM* transcription factor that is most similar to the *Arabidopsis CUC2*, lead to aborted SAM and fused cotyledons, similar to the *Arabidopsis cuc1 cuc2* double mutants. Recovered *gob* mutants produce leaves that are much simpler than the wild-type leaves and are similar to *e* leaves (6, 84, 100). *GOB* mRNA is expressed in narrow stripes within the margin of the leaf primordia, immediately adjacent to initiating leaflet primordia. These stripes of *GOB* expression mark the boundary between the leaf margin and the future leaflet, and precede leaflet initiation (84, 100). *GOB/CUC2* and *CUC3* orthologs are expressed in a similar manner in an array of species with compound leaves, and silencing *CUC2* and/or *CUC3* orthologs led to leaf simplification in these species (100). A subset of *CUC* genes, including *GOB*, is negatively regulated by *miR164*. The gain-of-function tomato mutant *Gob-4d*, in which *GOB* is insensitive to *miR164*-mediated inhibition, produces leaves with reduced number of deeply lobed leaflets, possibly due to leaflet fusion. Transgenic expression of a *miR164*-insensitive *GOB* form leads to numerous initiation events in the margins of the primary leaflets, which later fuse to produce a final leaf form that is simpler and more deeply lobed in comparison to the wild-type leaf. Thus, both reduced and expanded expression domains of *GOB* lead to leaflet fusion (84). These observations suggest that sharp, distinct and sufficiently distant stripes of *GOB* expression are essential for leaflet separation, but that fluctuations in *GOB* expression are sufficient to induce leaflet initiation in specific developmental contexts.

4.2. Coordination of Leaflet Initiation and Separation

Auxin, LYR, E and GOB thus collaborate to specify leaflet initiation and promote leaflet separation. A possible model of their collaborative activities in leaflet development is that a leaflet is specified by a distinct auxin response maximum, which is co-localized with, and possibly regulated by *LYR* expression, and is flanked by a sharp and distinct stripe of GOB activity (Fig. 3). For proper leaflet initiation and separation the subsequent occurrence of such a combination has to be sufficiently distant in time and space. The distinct auxin response maxima are generated by a combination of auxin accumulation, auxin transport, and inhibition of auxin response between leaflets by E/SlIAA9 (Fig. 3). When either of these components is absent, secondary leaflets, lobes, and serrations fail to form. When their boundaries are less sharp, or when they occur "too close" to each other, leaflets fail to separate. In the context of primary leaflet development, GOB and E appear to function partially redundantly with respect to leaflet initiation, but they are both required for proper leaflet separation. It remains to be seen how the loss of both activities will affect primary leaflet initiation.

Interestingly, similar mechanisms function in patterning of serrations in the margins of the simple *Arabidopsis* leaf (91, 101, 102). It thus appears that the coordinated activities of CUCs and auxin are utilized within the context of the simple or compound leaf rather than distinguish between them. One possible exception is the *CUC* expression domain. *CUC2* expression domains in the *Arabidopsis* leaf margin are much broader and relatively closer to each other than *GOB* expression in tomato (84, 100, 101). In addition, while *CUC2* was shown to be negatively regulated by auxin in the context of leaf and serration development in *Arabidopsis*, in tomato leaves the *e* mutation did not affect *GOB* expression.

While in this chapter the regulation of the transient indeterminate growth and leaflet specification are discussed separately, this separation is somewhat artificial. For example, in addition to their role in leaflet initiation and separation, GOB and possibly also E also affect the rate of leaf maturation (83). Conversely, overexpression of *KNOXI* genes has been shown to lead to reiteration of leaflet formation (50, 71).

4.3. Similarities and Differences Between Leaves, Primary Leaflets, Higher Order Leaflets, Lobes, and Serrations

Despite the striking similarities in the development of leaves, primary leaflets, higher order leaflets, lobes, and serrations, important differences exist.

The initiation of primary leaflets is more robust than that of secondary leaflets and serrations. This is manifested by the fact that loss of proper distribution of either GOB, E or auxin activities is sufficient to eliminate the formation of secondary and higher order leaflets and serrations but not primary leaflets (6, 84, 91, 100). Thus, in *gob* and *e*, primary leaflets are initiated but are often fused. In addition, many mutations with simplified leaves affect secondary leaflets more severely than primary leaflets (6). Furthermore, several mutations affect secondary leaflets and serrations in an opposite manner than primary and intercalary leaflets. For example, *sl* and *wiry* mutants lack secondary leaflets and serrations but have additional primary and intercalary leaflets, respectively (10, 53). Conversely, *Me* mutants have numerous leaflet orders but lack intercalary leaflets. *Me* mutants also have smooth margins, indicating that the development of higher order marginal structures can also be differentially regulated. Lobes can arise either from leaflet fusion, as in the *e* mutant, or from initiation events that occur after the lamina has expanded, in which case they can be viewed as deep serrations.

Kaplan (2001) followed Hofmeister in viewing branches, leaves, and hairs as qualitatively similar outgrowths that differ in hierarchy rather than having essentially different developmental properties (2, 103). Thus, these outgrowths are gradually more determinate and flat. This approach is also useful in the discussion of the different outgrowths from the leaf margin: leaves, primary leaflets, higher order leaflets, and serrations are gradually more determinate and arise from gradually more flattened structures. This view can also explain part of the difference between simple and compound leaves: compound leaves are "higher" in this hierarchy in that they are less determinate and flatten more gradually, enabling the existence of additional degrees of elaboration. Conversely, simple leaves become determinate and flatten faster, allowing only for the development of serrations and lobes.

5. Concluding Remarks

Organogenesis of marginal structures during compound-leaf development has a lot in common with the organogenesis of lateral organs from the SAM (1, 4, 6, 15, 84, 103, 104). This principle can be further generalized to many developmental processes in plants: throughout development, plants utilize common mechanisms in a context-specific manner to produce numerous variations of plant form. Thus, similar building blocks are used in different contexts to balance indeterminate and determinate growth, and to specify the location of lateral organs. Important differences also exist between compound-leaf- and shoot-development. Many of the differences result from the leaf being more determinate and flat. As the leaf is more flexible and more resistant to developmental changes, compound-leaf development, and in particular the tomato leaf, has served as an attractive model to study these principles.

As the tomato leaf is in some ways an intermediate between a branch and a simple leaf with respect to the maturation paste and level of indeterminacy, it is also a useful tool to study the mechanisms that regulate the balance between indeterminate and determinate growth.

Acknowledgments

We would like to thank Eilon Shani and members of the Ori group for critical reading of the manuscript. The work on related topics in the Ori group is supported by grants from ISF (60/10), BARD (IS04140-08C), and the Israeli ministry of Agriculture (837-0055-09).

References

1. Floyd SK, Bowman JL (2010) Gene expression patterns in seed plant shoot meristems and leaves: homoplasy or homology? J Plant Res 123:43–55
2. Kaplan DR (2001) Fundamental concepts of leaf morphology and morphogenesis: a contribution to the interpretation of molecular genetic mutants. Int J Plant Sci 162: 465–474
3. Sarojam R, Sappl PG, Goldshmidt A, Efroni I, Floyd SK, Eshed Y, Bowman JL (2010) Differentiating arabidopsis shoots from leaves by combined YABBY activities. Plant Cell Online 22:2113–2130

4. Bourque L, Lacroix C (2011) Lobe-generating centres in the simple leaves of Myriophyllum aquaticum: evidence for KN1-like activity. Ann Bot 107:639–651
5. Lacroix CR (1995) Changes in leaflet and leaf lobe form in developing compound and finely divided leaves. Bot Jahrb Syst 117:317–331
6. Brand A, Shirding N, Shleizer S, Ori N (2007) Meristem maintenance and compound-leaf patterning utilize common genetic mechanisms in tomato. Planta 226:941–951
7. Frary A, Fritz LA, Tanksley SD (2004) A comparative study of the genetic bases of natural variation in tomato leaf, sepal, and petal morphology. TAG Theor Appl Genet 109:523–533
8. Goliber T, Kessler S, Chen JJ, Bharathan G, Sinha N (1999) Genetic, molecular, and morphological analysis of compound leaf development. Curr Top Dev Biol 43:259–290
9. Kessler S, Kim M, Pham T, Weber N, Sinha N (2001) Mutations altering leaf morphology in tomato. Int J Plant Sci 162:475–492
10. Menda N, Semel Y, Peled D, Eshed Y, Zamir D (2004) In silico screening of a saturated mutation library of tomato. Plant J 38:861–872
11. Rick CM, Harrison AL (1959) Inheritance of five new tomato seedling characters. J Hered 50:91–98
12. Stevens AM, Rick CM (1986) Genetic and breeding. In: Atherton JG, Rudich J (eds) The tomato crop, a scientific basis for improvement. New York: Chapman and Hall London. pp 35–59
13. Dengler NG, Tsukaya H (2001) Leaf morphogenesis in dicotyledons: current issues. Int J Plant Sci 162:459–464
14. Holtan HE, Hake S (2003) Quantitative trait locus analysis of leaf dissection in tomato using *Lycopersicon pennellii* segmental introgression lines. Genetics 165:1541–1550
15. Hagemann W, Gleissberg S (1996) Organogenetic capacity of leaves: the significance of marginal blastozones in angiosperms. Plant Syst Evolut 199:121–152
16. Mathan DS, Jenkins JA (1960) Chemically induced phenocopy of a tomato mutant. Science 131:36–87
17. Mathan DS, Jenkins JA (1962) A morphogenetic study of Lanceolate, a leaf shape mutant in the tomato. Am J Bot 49:504–514
18. Stettler RF (1964) Dosage effects of the Lanceolate gene in tomato. Am J Bot 51:253–264
19. Dengler NG (1984) Comparison of leaf development in Normal (+/+), Entire (E/E), and Lanceolate (La/+) plants of tomato, Lycopersicon esculentum Ailsa Craig. Bot Gaz 145:66–77
20. Ori N, Cohen AR, Etzioni A, Brand A, Yanai O, Shleizer S, Menda N, Amsellem Z, Efroni I, Pekker I, Alvarez JP, Blum E, Zamir D, Eshed Y (2007) Regulation of LANCEOLATE by miR319 is required for compound-leaf development in tomato. Nat Genet 39:787–791
21. Shleizer-Burko S, Burko Y, Ben-Herzel O, Ori N (2011) Dynamic growth program regulated by LANCEOLATE enables flexible leaf patterning. Development 138:695–704
22. Chandra-Shekhar KN, Sawhney VK (1991) Regulation of leaf shape in the solanifolia mutant of tomato (Lycopersicon esculentum) by plant growth substances. Ann Bot 67:3–6
23. Fleishon S, Shani E, Ori N, Weiss D (2011) Negative reciprocal interactions between gibberellin and cytokinin in tomato. New Phytol 190:609–617
24. Gray RA (1957) Alteration of leaf size and shape and other changes caused by gibberellins in plants. Am J Bot 44:674–682
25. Hay A, Kaur H, Phillips A, Hedden P, Hake S, Tsiantis M (2002) The gibberellin pathway mediates KNOTTED1-type homeobox function in plants with different body plans. Curr Biol 12:1557–1565
26. Jones GM (1987) Gibberellins and the procera mutant of tomato. Planta 172:280–284
27. Bassel GW, Mullen RT, Bewley JD (2008) Procera is a putative DELLA mutant in tomato (Solanum lycopersicum): effects on the seed and vegetative plant. J Exp Bot 59:585–593
28. Jasinski S, Tattersall A, Piazza P, Hay A, Martinez-Garcia JF, Schmitz G, Theres K, McCormick S, Tsiantis M (2008) PROCERA encodes a DELLA protein that mediates control of dissected leaf form in tomato. Plant J 56:603–612
29. Van Tuinen A, Peters AHLJ, Kendrick RE, Zeevaart JAD, Koornneef M (1999) Characterisation of the procera mutant of tomato and the interaction of gibberellins with end-of-day far-red light treatments. Physiol Plant 106:121–128
30. Chandra-Shekhar KN, Sawhney VK (1990) Leaf development in the normal and solanifolia muant of tomato (Lycopersicon esculentum). Am J Bot 77: 46–53
31. Robbins WJ (1957) Gibberellic acid and the reversal of adult hedera to a juvenile state. Am J Bot 44:743–746
32. Rogler CE, Hackett WP (1975) Phase change in hedera helix: induction of the mature to juvenile phase change by gibberellin A3. Physiol Plant 34:141–147

33. Shalit A, Rozman A, Goldshmidt A, Alvarez JP, Bowman JL, Eshed Y, Lifschitz E (2009) The flowering hormone florigen functions as a general systemic regulator of growth and termination. Proc Natl Acad Sci USA 106: 8392–8397
34. Poethig RS (1997) Leaf morphogenesis in flowering plants. Plant Cell 9:1077–1087
35. Long JA, Moan EI, Medford JI, Barton MK (1996) A member of the KNOTTED class of homeodomain proteins encoded by the STM gene of Arabidopsis. Nature 379:66–69
36. Vollbrecht E, Reiser L, Hake S (2000) Shoot meristem size is dependent on inbred background and presence of the maize homeobox gene, knotted1. Development 127:3161–3172
37. Jackson D, Veit B, Hake S (1994) Expression of maize KNOTTED1 related homeobox genes in the shoot apical meristem predicts patterns of morphogenesis in the vegetative shoot. Development 120:405–413
38. Smith LG, Hake S (1992) The initiation and determination of leaves. Plant Cell 4: 1017–1027
39. Smith LG, Jackson D, Hake S (1995) Expression of knotted1 marks shoot meristem formation during maize embryogenesis. Dev Genet 16:344–348
40. Chuck G, Lincoln C, Hake S (1996) KNAT1 induces lobed leaves with ectopic meristems when overexpressed in Arabidopsis. Plant Cell Online 8:1277–1289
41. Hake S, Meyerowitz EM (1998) Growing up green. Curr Opin Plant Biol 1:9–11
42. Schneeberger RG, Becraft PW, Hake S, Freeling M (1995) Ectopic expression of the knox homeo box gene rough sheath1 alters cell fate in the maize leaf. Genes Dev 9:2292–2304
43. Sinha NR, Williams RE, Hake S (1993) Overexpression of the maize homeo box gene, KNOTTED-1, causes a switch from determinate to indeterminate cell fates. Genes Dev 7:787–795
44. Tamaoki M, Kusaba S, Kano-Murakami Y, Matsuoka M (1997) Ectopic expression of a tobacco homeobox gene, NTH15, dramatically alters leaf morphology and hormone levels in transgenic tobacco. Plant Cell Physiol 38:917–927
45. Vollbrecht E, Veit B, Sinha N, Hake S (1991) The developmental gene Knotted-1 is a member of a maize homeobox gene family. Nature 350:241–243
46. Ori N, Eshed Y, Chuck G, Bowman JL, Hake S (2000) Mechanisms that control knox gene expression in the Arabidopsis shoot. Development 127:5523–5532
47. Lincoln C, Long J, Yamaguchi J, Serikawa K, Hake S (1994) A knotted1-like homeobox gene in Arabidopsis is expressed in the vegetative meristem and dramatically alters leaf morphology when overexpressed in transgenic plants. Plant Cell 6:1859–1876
48. Bharathan G, Goliber TE, Moore C, Kessler S, Pham T, Sinha NR (2002) Homologies in leaf form inferred from KNOXI gene expression during development. Science 296:1858–1860
49. Koltai H, Bird DM (2000) Epistatic repression of PHANTASTICA and class 1 KNOTTED genes is uncoupled in tomato. Plant J 22:455–459
50. Hareven D, Gutfinger T, Parnis A, Eshed Y, Lifschitz E (1996) The making of a compound leaf: genetic manipulation of leaf architecture in tomato. Cell 84:735–744
51. Janssen BJ, Lund L, Sinha N (1998) Overexpression of a homeobox gene, LeT6, reveals indeterminate features in the tomato compound leaf. Plant Physiol 117:771–786
52. Parnis A, Cohen O, Gutfinger T, Hareven D, Zamir D, Lifschitz E (1997) The dominant developmental mutants of tomato, Mouse-ear and Curl, are associated with distinct modes of abnormal transcriptional regulation of a Knotted gene. Plant Cell 9:2143–2158
53. Kang J, Sinha NR (2010) Leaflet initiation is temporally and spatially separated in simple and complex tomato (Solanum lycopersicum) leaf mutants: a developmental analysis. Botany 88:1916–2790
54. Chen JJ, Janssen BJ, Williams A, Sinha N (1997) A gene fusion at a homeobox locus: alterations in leaf shape and implications for morphological evolution. Plant Cell 9: 1289–1304
55. Kim M, McCormick S, Timmermans M, Sinha N (2003) The expression domain of PHANTASTICA determines leaflet placement in compound leaves. Nature 424:438–443
56. Reinhardt D, Frenz M, Mandel T, Kuhlemeier C (2003) Microsurgical and laser ablation analysis of interactions between the zones and layers of the tomato shoot apical meristem. Development 130:4073–4083
57. Stieger PA, Meyer AD, Kathmann P, Frundt C, Niederhauser I, Barone M, Kuhlemeier C (2004) The orf13 T-DNA gene of Agrobacterium rhizogenes confers meristematic competence to differentiated cells. Plant Physiol 135:1798–1808
58. Jasinski S, Kaur H, Tattersall A, Tsiantis M (2007) Negative regulation of KNOX expression in tomato leaves. Planta 226:1255–1263

59. Barkoulas M, Hay A, Kougioumoutzi E, Tsiantis M (2008) A developmental framework for dissected leaf formation in the Arabidopsis relative Cardamine hirsuta. Nat Genet 40:1136–1141
60. Hay A, Tsiantis M (2006) The genetic basis for differences in leaf form between Arabidopsis thaliana and its wild relative Cardamine hirsuta. Nat Genet 38:942–947
61. Koenig D, Sinha N (2010) Evolution of leaf shape: a pattern emerges. Curr Top Dev Biol 91:169–183
62. Rick CM, Butler L (1956) Cytogenetics of tomato. Adv Genet 7:267–382
63. Young PA (1955) Curl, a mutant teratism of the tomato. J Hered 46:243–244
64. Shani E, Burko Y, Ben-Yaakov L, Berger Y, Amsellem Z, Goldshmidt A, Sharon E, Ori N (2009) Stage-specific regulation of Solanum lycopersicum leaf maturation by class 1 KNOTTED1-LIKE HOMEOBOX proteins. Plant Cell 21:3078–3092
65. Hay A, Tsiantis M (2010) KNOX genes: versatile regulators of plant development and diversity. Development 137:3153–3165
66. Kimura S, Koenig D, Kang J, Yoong FY, Sinha N (2008) Natural variation in leaf morphology results from mutation of a novel KNOX gene. Curr Biol 18(9):672–677
67. Kim M, Pham T, Hamidi A, McCormick S, Kuzoff RK, Sinha N (2003) Reduced leaf complexity in tomato wiry mutants suggests a role for PHAN and KNOX genes in generating compound leaves. Development 130:4405–4415
68. Clayberg CD, Butler L, Kerr EA, Rick CM, Robinson RW (1966) Third list of known genes in the tomato. J Hered 57:189–196
69. Avivi Y, Lev-Yadun S, Morozova N, Libs L, Williams L, Zhao J, Varghese G, Grafi G (2000) Clausa, a tomato mutant with a wide range of phenotypic perturbations, displays a cell type-dependent expression of the homeobox gene LeT6/TKn2. Plant Physiol 124: 541–552
70. Harrison CJ, Corley SB, Moylan EC, Alexander DL, Scotland RW, Langdale JA (2005) Independent recruitment of a conserved developmental mechanism during leaf evolution. Nature 434:509–514
71. Efroni I, Eshed Y, Lifschitz E (2010) Morphogenesis of simple and compound leaves: a critical review. Plant Cell 22:1019–1032
72. Stubbe H (1959) Mutanten der Kulturtomate Lycopersicon esculentum Miller III. Genet Resour Crop Evolut 7:82–112
73. Rick CM (1980) Petroselinum (Pts), a new marker for chromosome 6. Tomato Genet Coop 20:32
74. Shani E, Ben-Gera H, Shleizer-Burko S, Burko Y, Weiss D, Ori N (2010) Cytokinin regulates compound leaf development in tomato. Plant Cell 22:3206–3217
75. Jasinski S, Piazza P, Craft J, Hay A, Woolley L, Rieu I, Phillips A, Hedden P, Tsiantis M (2005) KNOX action in Arabidopsis is mediated by coordinate regulation of cytokinin and gibberellin activities. Curr Biol 15:1560–1565
76. Sakamoto T, Kamiya N, Ueguchi-Tanaka M, Iwahori S, Matsuoka M (2001) KNOX homeodomain protein directly suppresses the expression of a gibberellin biosynthetic gene in the tobacco shoot apical meristem. Genes Dev 15:581–590
77. Bolduc N, Hake S (2009) The maize transcription factor KNOTTED1 directly regulates the gibberellin catabolism gene ga2ox1. Plant Cell 21:1647–1658
78. Yanai O, Shani E, Dolezal K, Tarkowski P, Sablowski R, Sandberg G, Samach A, Ori N (2005) Arabidopsis KNOXI proteins activate cytokinin biosynthesis. Curr Biol 15: 1566–1571
79. Sakamoto T, Sakakibara H, Kojima M, Yamamoto Y, Nagasaki H, Inukai Y, Sato Y, Matsuoka M (2006) Ectopic expression of KNOTTED1-like homeobox protein induces expression of cytokinin biosynthesis genes in rice. Plant Physiol 142:54–62.
80. Weiss D, Ori N (2007) Mechanisms of cross talk between gibberellin and other hormones. Plant Physiol 144:1240–1246
81. Edwardson JR, Corbett MK (1962) A virus-like syndrome in tomato caused by a mutation. Am J Bot 49:571–575
82. Lesley JW, Lesley MM (1928) The "WIRY" tomato. J Hered 19:337–344
83. Kumaran MK, Bowman JL, Sundaresan V (2002) YABBY polarity genes mediate the repression of KNOX homeobox genes in Arabidopsis. Plant Cell 14:2761–2770
84. Berger Y, Harpaz-Saad S, Brand A, Melnik H, Sirding N, Alvarez JP, Zinder M, Samach A, Eshed Y, Ori N (2009) The NAC-domain transcription factor GOBLET specifies leaflet boundaries in compound tomato leaves. Development 136:823–832
85. Bayer EM, Smith RS, Mandel T, Nakayama N, Sauer M, Prusinkiewicz P, Kuhlemeier C (2009) Integration of transport-based models for phyllotaxis and midvein formation. Genes Dev 23:373–384

86. Braybrook SA, Kuhlemeier C (2010) How a plant builds leaves. Plant Cell 22:1006–1018
87. Reinhardt D, Mandel T, Kuhlemeier C (2000) Auxin regulates the initiation and radial position of plant lateral organs. Plant Cell 12:507–518
88. Heisler MG, Ohno C, Das P, Sieber P, Reddy GV, Long JA, Meyerowitz EM (2005) Patterns of auxin transport and gene expression during primordium development revealed by live imaging of the Arabidopsis inflorescence meristem. Curr Biol 15:1899–1911
89. de Reuille PB, Bohn-Courseau I, Ljung K, Morin H, Carraro N, Godin C, Traas J (2006) Computer simulations reveal properties of the cell-cell signaling network at the shoot apex in Arabidopsis. Proc Natl Acad Sci USA 103:1627–1632
90. Avasarala S, Yang J, Caruso JL (1996) Production of phenocopies of the lanceolate mutant in tomato using polar auxin transport inhibitors. J Exp Bot 47:709–712
91. Koenig D, Bayer E, Kang J, Kuhlemeier C, Sinha N (2009) Auxin patterns Solanum lycopersicum leaf morphogenesis. Development 136:2997–3006
92. Al-Hammadi AS, Sreelakshmi Y, Negi S, Siddiqi I, Sharma R (2003) The polycotyledon mutant of tomato shows enhanced polar auxin transport. Plant Physiol 133:113–125
93. DeMason D, Chawla R (2004) Roles for auxin during morphogenesis of the compound leaves of pea (*Pisum sativum*). Planta 218:894–894
94. DeMason DA, Polowick PL (2009) Patterns of DR5::GUS expression in organs of pea (Pisum sativum). Int J Plant Sci 170:1–11
95. Wang H, Jones B, Li Z, Frasse P, Delalande C, Regad F, Chaabouni S, Latche A, Pech JC, Bouzayen M (2005) The tomato Aux/IAA transcription factor IAA9 is involved in fruit development and leaf morphogenesis. Plant Cell 17:2676–2692
96. Zhang J, Chen R, Xiao J, Qian C, Wang T, Li H, Ouyang B, Ye Z (2007) A single-base deletion mutation in SlIAA9 gene causes tomato (Solanum lycopersicum) entire mutant. J Plant Res 120:671–678
97. David-Schwartz R, Koenig D, Sinha NR (2009) LYRATE is a key regulator of leaflet initiation and lamina outgrowth in tomato. Plant Cell 21:3093–3104
98. Dinneny JR, Yadegari R, Fischer RL, Yanofsky MF, Weigel D (2004) The role of JAGGED in shaping lateral organs. Development 131:1101–1110
99. Ohno CK, Reddy GV, Heisler MGB, Meyerowitz EM (2004) The Arabidopsis JAGGED gene encodes a zinc finger protein that promotes leaf tissue development. Development 131:1111–1122
100. Blein T, Pulido A, Vialette-Guiraud A, Nikovics K, Morin H, Hay A, Johansen IE, Tsiantis M, Laufs P (2008) A conserved molecular framework for compound leaf development. Science 322:1835–1839
101. Bilsborough GD, Runions A, Barkoulas M, Jenkins HW, Hasson A, Galinha C, Laufs P, Hay A, Prusinkiewicz P, Tsiantis M (2011) Model for the regulation of Arabidopsis thaliana leaf margin development. Proc Natl Acad Sci USA 108:3424–3429
102. Hasson A, Plessis A, Blein T, Adroher B, Grigg S, Tsiantis M, Boudaoud A, Damerval C, Laufs P (2011) Evolution and diverse roles of the CUP-SHAPED COTYLEDON genes in Arabidopsis leaf development. Plant Cell 23:54–68
103. Arber A (1950) The partial-shoot theory of the leaf. In: The natural philosophy of plant form. Cambridge University Press, London, pp 70–92
104. Stattler R, Rutishauser R (1992) Partial homology of pinnate leaves and shoots: orientation of leaflet inception. Bot Jahrb Syst 114:61–79

Chapter 2

Usefulness of *Physcomitrella patens* for Studying Plant Organogenesis

Sandrine Bonhomme, Fabien Nogué, Catherine Rameau, and Didier G. Schaefer

Abstract

In this chapter, we review the main organogenesis features and associated regulation processes of the moss *Physcomitrella patens (P. patens)*, the model plant for the Bryophytes. We highlight how the study of this descendant of the earliest plant species that colonized earth, brings useful keys to understand the mechanisms that determine and control both vascular and non vascular plants organogenesis. Despite its simple morphogenesis pattern, *P. patens* still requires the fine tuning of organogenesis regulators, including hormone signalling, common to the whole plant kingdom, and which study is facilitated by a high number of molecular tools, among which the powerful possibility of gene targeting/replacement. The recent discovery of moss cells reprogramming capacity completes the picture of an excellent model for studying plant organogenesis.

Key words: *Physcomitrella patens*, Bryophytes, Organogenesis, Morphogenesis, Gene Targeting

1. Introduction

In contrast to metazoans, organogenesis in metaphytae occurs postembryonically and throughout the organism's life. Plant organogenesis is an iterative process: it is accomplished through the activity of multicellular meristems, which continuously divide and differentiate to form the basic unit of plant architecture, the phytomer. In shoots, each phytomer corresponds to a leaf and a stem internode, and these units are piled up along the growth axis, frequently displaying phyllotaxis, to give rise to the plant body. As plants cannot move to escape to unfavorable developmental conditions, coordinated divisions and differentiation steps ensuring both plasticity and adaptability are critical for plant life (1).

Bryophytes represent the extant-living descendants of the earliest plant species that colonized earth ca. 470 million years ago (2).

Ive De Smet (ed.), *Plant Organogenesis: Methods and Protocols*, Methods in Molecular Biology, vol. 959,
DOI 10.1007/978-1-62703-221-6_2, © Springer Science+Business Media New York 2013

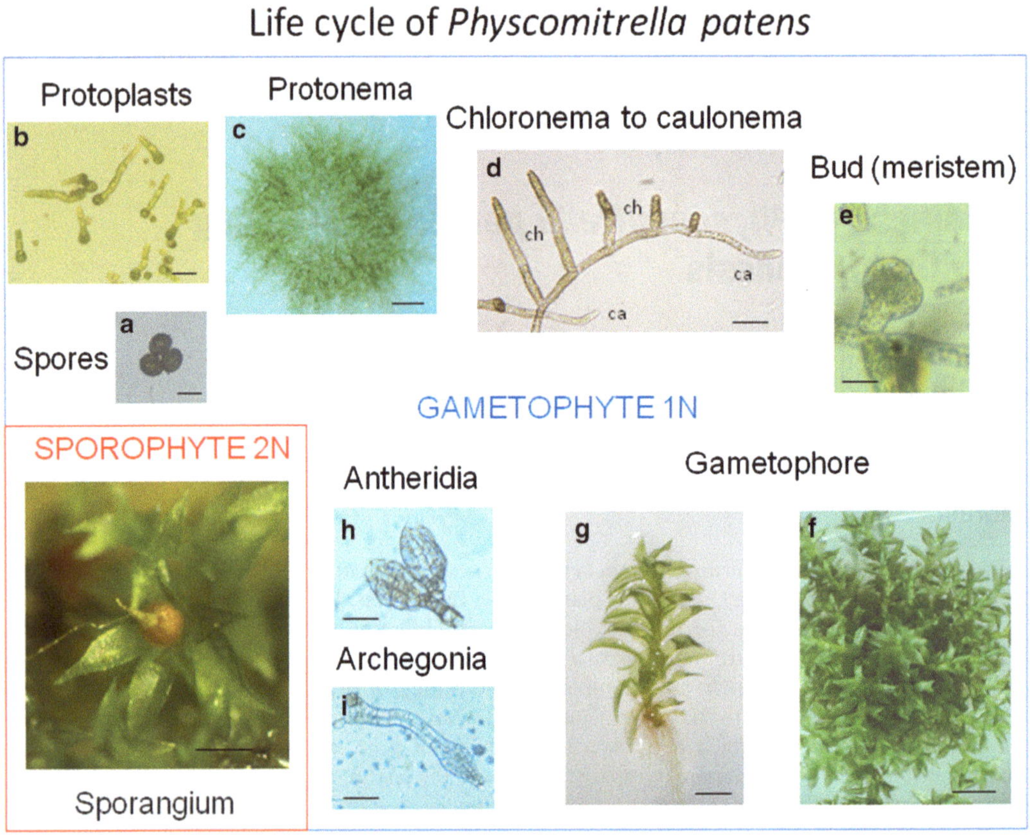

Fig. 1. Spores (**a**) or regenerating protoplasts (**b**) will give rise to a filamentous protonema (**c**) formed by chloronemata and caulonemata (ch and ca in **d**). The chloronema to caulonema transition is visible at the apex of the filaments. Transition to bushy growth is initiated with the differentiation of meristematic buds (**e**) that further develop into leafy shoots (**f**, **g**). Antheridia (**h**) and archegonia (**i**) form at the apex of the gametophores. Water dependent fertilization occurs in the archegonia and the fertilized egg cell gives rise to the brown diploid sporangium in which meiosis occurs and spores differentiate. Scale bars **a**, **e**, **h**, **i**: 20 μm; **b**, **d**: 50 μm; **c**, **g**, **j**: 1 mm; **f**: 0.4 cm.

Despite their relative morphological simplicity, Bryophytes share most of the biochemical, genetic and developmental processes that characterize the biology of modern plants (3–5). Mosses have proven extremely valuable systems to undertake biological research at the cellular level, while the dominance of the haploid gametophyte in the life cycle facilitates genetic approaches. Over the last decade, the moss *Physcomitrella patens* (*P. patens*) has become the reference model in moss biology: the outstanding efficiency of targeted mutagenesis (6, 7) combined with the availability of completely sequenced genome have been essential in its development (8–10).

P. patens development has been extensively reviewed (see (11) and references therein) (Fig. 1). Briefly, a haploid spore germinates to establish a juvenile developmental stage, the protonema. This two-dimensional filamentous protonema is essentially formed by two distinct cell types, the chloronema and the caulonema, that expand by tip growth and multiple divisions of apical and subapical cells (12). Protonema development is determined by a few single cell developmental transitions that can be modulated by external factors and followed *in vivo* with non-invasive technology (Table 1). It is the most characterized developmental stage of mosses (13, 14) and provides an ideal material to combine molecular, genetic and biochemical approaches to study multiple aspects of plant cell biology. Yet the morphogenetic processes underlying moss protonema development have little analogy with those controlling spermatophyte's organogenesis, except maybe for the differentiation of root hairs and pollen tubes.

The transition to three dimensional bushy growth is determined within single caulonema subapical cells in the protonema (Fig. 1). These cells enter into three cycles of asymmetric cell divisions to establish the bud, a primitive meristem formed by a single tetrahedral apical stem cell (i.e., with three cutting faces). During this developmental transition cell expansion switches from tip growth to diffuse growth. The apical stem cell then continuously divides to form a new stem cell and a basal primordia that will further differentiate into a stem internode and a leaf (15). Subsequent leafy shoot development is achieved by the reiterated differentiation of phytomers from the shoot apical stem cell, as in other land plants. Yet, there is no polar auxin transport in moss shoots (16) and the meristematic activity that generates patterning is confined to a single apical stem cell. In *P. patens*, the adult gametophyte is composed of numerous unbranched leafy shoots carrying leaves displaying phyllotaxis along the stem axis, and of filamentous rhizoids that differentiate from shoot epidermal cells. These filamentous rhizoids constitute the root apparatus of the plant (mosses have no roots) and are involved in substrate fixation and colonization, and in nutrient uptake. Despite the morphological simplicity of the plant, several organs or differentiated tissues can be recognized in leafy shoots. Transverse sections through the stem reveal the presence of an epidermal layer from which basal and mid-stem rhizoids differentiate, of cortically located parenchyma cells and of centrally located solute conducting cells or hydroids (17). Leaf primordia (phyllid) undergo a number of symmetric and asymmetric cell divisions to form a single layered photosynthetic leaf blade and a multilayered midrib. Cells specialized in solute transport differentiate within the midrib (stereids, hydroids, and deuters) but do not directly connect with the hydroids of the stem (17). Stem organogenesis is also a highly coordinated process that is so far poorly documented. In response to environmental conditions (low light and temperature (18)) a bunch of reproductive organs (male antheridia and female archegonia) differentiate at the apex of the shoot concomitantly with the arrest of shoot apical growth. Fertilization between a ciliated antherozoid and the egg cell is water dependent and the resulting zygote further differentiates into an epiphytic diploid sporophyte, the sporangium. Spore mother cells differentiate in the sporophyte, proliferate, and undergo meiosis to give rise

Table 1
Cell types and developmental characteristics of *Physcomitrella patens* protonemal cells

Cell types[a]	Cell division	Cell growth	Light requirement	Primary function	Stem cell capacity[b]	Developmental transitions[c]	Enhancing factors
Spore	24 h		Yes	Stem cell	Yes	Into chloronema apical	
Chloronema apical cell	24 h	5 μm/h	Yes	Stem cell	Yes	Into chloronema apical and subapical	
Chloronema subapical cell	24 h	5 μm/h	Yes	Photosynthesis	Yes/no	Into chloronema apical Into caulonema apical	NH_4 Auxin
Caulonema apical cell	8 h	20 μm/h	No	Stem cell	No	Into caulonema apical and subapical	
Caulonema subapical cell	8 h	20 μm/h	No	Nutrients assimilation and colonization	No	Into chloronema II (90%) Into caulonema (5%) Into buds (5%) Into skotonema	NH_4 Auxin Cytokinins Darkness
Buds			Yes	Shoot apical stem cell		Into leafy shoot	
Skotonema	8 h	20 μm/h	No		No	Into chloronema apical Into caulonema apical	Light Light
Protoplasts	24 h	5 μm/h	Yes	Stem cell	Yes	Into chloronema apical	Light

[a]In *P. patens*, subapical cells usually undergo 2–4 additional cell divisions that generate the ramified/branched pattern of the protonema, except for dark grown skotonema
[b]Refers to the ability to produce protoplasts that will reinitiate protonemal growth; this capacity decreases with aging in chloronemata. Yet all cell types are able to deprogramming and redifferentiate into a chloronemal apical cell following wounding
[c]Numbers in brackets represent the percentage of each transition in standard growth conditions which roughly corresponds to the fraction of each cell type in the protonema

to ca. 4,000 haploid spores per capsule (for reproductive organs and sporophyte organogenesis see (19)).

All the organogenetic processes described above are highly reminiscent of the basic features that control higher plant development. Yet moss organogenesis is much simpler in terms of cell types and developmental transitions and is frequently initiated within a single cell. This provides outstanding facilities to follow developmental processes at the cellular level. It also may suggest that local and/or positional cues are more important to establish the organogenetic pattern of plant growth than cell-to-cell communication mediated by polar transport of growth factors within multicellular meristems. Here, we first describe why *Physcomitrella* is a good model for cell and molecular biologists and geneticists. Then, we review a number of recent publications describing moss developmental mutants and try to show how this moss could advantageously complement currently used models to understand plant organogenesis. We believe that the moss *P. patens* provides the same advantages and limits to study plant development as those offered by *Caenorhabditis elegans* or *Drosophila melanogaster* to study vertebrate organogenesis.

2. *Physcomitrella patens*, the Green Yeast of Plant Biology

2.1. *In Vitro* Techniques

In addition to gene targeting facilities (see below), the similarity between moss and yeast extends to laboratory procedures since a *Physcomitrella* culture essentially requires microbiological techniques. For most experiments, moss tissue is cultured on modified Knop's solidified media, while growth on sterilized compost mixture is also possible (see (20) for protocols) (21). In both cases, the life cycle is completed in 2–3 months under standard conditions. Most importantly cell linage and developmental transitions can be followed visually and at the cellular level constantly. *Physcomitrella* is self-fertile but genetic crosses are possible and the recent development of fluorescent tagged lines will facilitate identification of hybrid sporophytes (22). Cultures can be initiated from spores or protonemal fragments, from a suspension of fragmented protonema or from leafy gametophores which have the ability to reinitiate protonemal growth. Thus vegetative propagation is easy at any developmental stage of the moss, which is especially important for the conservation of sterile or developmental mutants. Spores, suspension of protonema or differentiated plants in sealed containers can also be stored up to several years in dark cold rooms for medium to long term conservation. To enable the production of a large amount of homogenous tissue suitable for cellular, metabolic or biochemical analyses, 1-week-old protonemal tissue is collected, fragmented and re-inoculated in a Petri dish to maintain pure protonemal culture (yield is 1 g/week/petri). Such cultures also provide optimal material for protoplast isolation. Large numbers of protoplasts can be isolated from them (10^6/plate) and these can regenerate with high efficiencies (up to 80%). Importantly protoplast regeneration reproduces spore germination and directly gives rise to differentiated protonemal cells. This significantly reduces the risk of somaclonal variation frequently associated with protoplast regeneration in other plants. Techniques for the establishment of protonemal liquid cultures in batch and bioreactors have also been developed and are applied for the production of biopharmaceuticals (23). Recently fluorescence associated cell sorting techniques have been established for isolated protoplasts that facilitate rapid and quantitative analyses of reporter gene expression (24).

2.2. Targeted Mutagenesis and Other Molecular Genetic Tools

Genetic transformation by polyethyleneglycol-mediated DNA uptake in protoplasts was first reported 20 years ago (25) and remains the best method to transform *Physcomitrella*. Transformation was performed with vectors without sequence homology with the moss genome and the frequencies were low (ca. 2–5 clones for 10^5 regenerants). The amazing efficiency of gene targeting (GT) in *Physcomitrella* was subsequently established (6) which boosted transformation efficiencies (to 1 in 10^3) and the interest of the plant scientific community (26). Efficient GT is the Holy Grail of reverse genetics as it enables the generation of any type of mutations within a genome and *Physcomitrella* ranks No. 1 for GT efficiencies among multicellular eukaryotes (27). The subsequent publication of the complete genome of *Physcomitrella* provided the essential information for developing accurate targeted mutagenesis (8). Over the last 10 years, most GT approaches have used replacement vectors to generate specific mutations in the moss genome. Based on experimental strategies established in mouse embryonic stem cells (28), we have further developed procedures that combines GT with Cre/lox mediated site specific recombination to generate clean mutations. This enables the recycling of selectable markers and the elimination of any vector sequences that could interfere with gene expression in knock-in and complementation approaches (29). Figure 2 describes strategies combining GT with Cre/lox recom-

Fig. 2. (continued) genotyped by PCR with primers a and b to identify those carrying the expected deletion. In these deleted clones, the entire ORF is replaced by a single LoxP site. (**B**) Gene knock-in strategy. The scheme illustrates the generation of N-terminal knock-in of a fluorescent protein (FP) in YFG. The replacement cassette carries as 5′ targeting sequences 5′ UTS of YFG down to position-30 to -10 regarding the ATG and a translational fusion between an ATG-FP and as 3′ targeting sequences the coding sequence of YFG starting at amino acid 2. Selection of replacement and deletion clones is performed as described above. A single LoxP site will remain upstream of the ATG and the fluorescent fusion of YFG will be driven by its natural chromosomal environment in the final strains. (**C**) Point mutagenesis strategy. The scheme illustrates the conversion of a serine (ser) into an alanine (ala) within exon II of YFG. The 5′ targeting sequence covers exon I and half of intron I and the 3′ targeting sequence covers the second half of intron I and the genomic sequence covering exon II and III. The desired mutation is introduced within exon II in the replacement vector. After Cre mediated elimination of the M+, the final clones carry the ser to ala mutation in exon II and a single LoxP site located in intron I. With similar strategies all kinds of mutations can be generated in the moss genome including promoter exchanges or protein domain shuffling. Targeting sequences positioning and genomic location of the residual LoxP site after CRE recombination are critical but not limiting in the design of the replacement vectors.

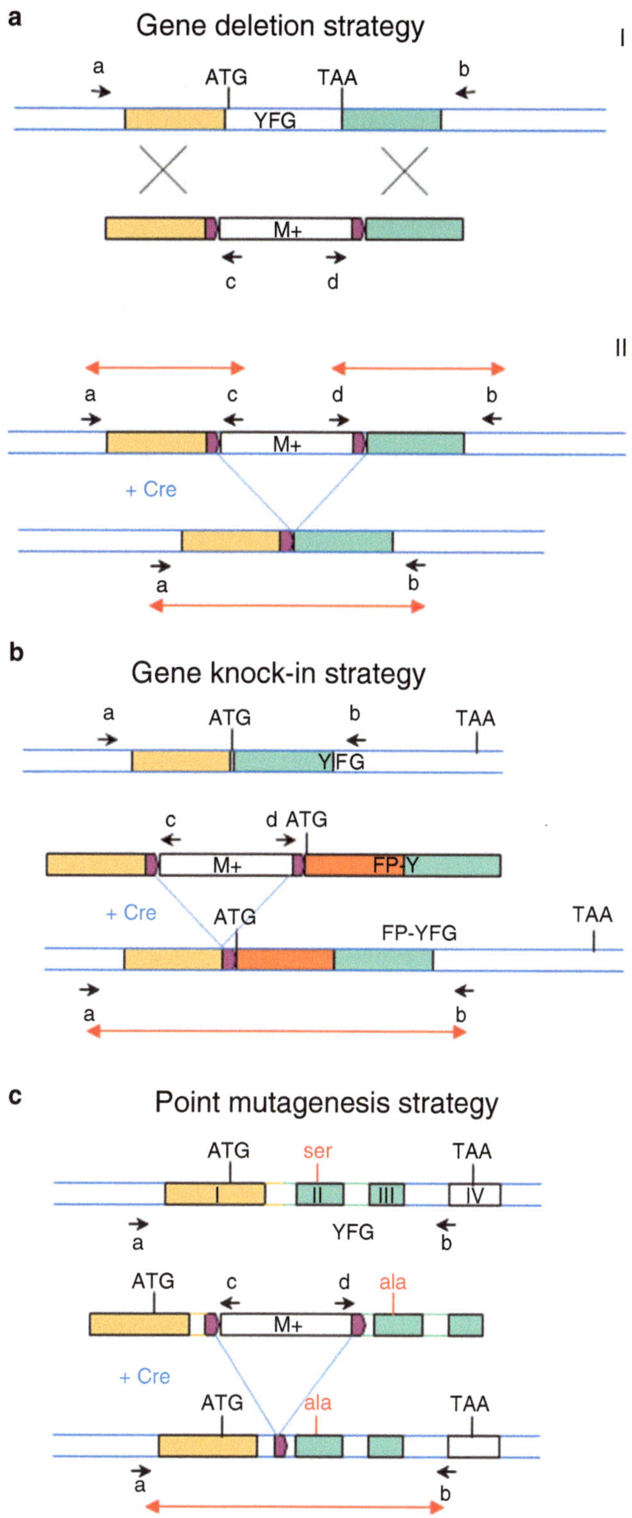

Fig. 2. (**A**) Gene deletion strategy. (I) 5′ (*yellow box*) and 3′ (*green box*) targeting sequences (600–800 bp) are cloned upstream and downstream of a positive selectable marker (M+) flanked by 2 LoxP sites (*violet box*) to build the replacement vector. The targeting sequences are homologous to the 5′and 3′ UTS of your favorite gene (YFG), respectively. Targeted integration by homologous recombination within both targeting sequences (*crosses*) will generate a gene replacement. (II) Antibiotic resistant clones are screened by PCR to identify replacement events using genomic (a, b) and vector primers (c, d). These clones are further submitted to transient expression of the Cre recombinase, single-protoplast derived colonies are regenerated on nonselective medium and then replica plated on selective medium. Antibiotic sensitive clones are further

bination to generate a gene deletion, an N-terminal tagged gene and a point mutation in the moss genome (30). Numerous examples of successful knockouts, gene deletions, knock-in or mutant complementation obtained with these strategies can be found in the articles discussed below. Thus GT facilitates reverse genetics approaches in moss at a speed and an efficiency so far only known for yeast and fungi and allows identification of developmental phenotype or marker lines within 4–6 weeks after transformation.

Efficient GT in *Physcomitrella* also promoted high throughput insertional mutagenesis approaches to establish forward genetic tools. Protoplasts were transformed using either mutagenized cDNA libraries (31) or gene-trap and enhancer-trap systems (32). Numerous developmental mutants have been identified in these collections but only a few of them have been characterized down to the molecular level due to the complexity of transgene integration patterns. They nevertheless constitute valuable resources for further developmental studies. Global gene expression analyses are also possible with the recent development of DNA-microarrays that cover most of the 38,357 predicted genes in the genome.

In addition to GT, gene knockdown can also be obtained via gene silencing techniques such as RNA interference (33), or amiRNA (34, 35). Recently, a system based on RNA interference has been described in order to screen rapidly for temperature sensitive alleles of a given gene (36). For this purpose the authors have co-expressed an RNAi that targets the gene of interest and different versions of an expression construct of the rescuing gene in which residues were mutated and tested for temperature sensitivity (36).

Extensive mutagenesis of the moss genome may require the expression of heterologous genes in overexpression or complementation assays. The introduction of a heterologous gene under the control of endogenous moss regulatory sequences is straightforward with GT and has been successfully used to complement several mutants described below. So far a number of heterologous promoters have been successfully used to drive gene expression in *Physcomitrella*, including the 35S CaMV, the rice actin-1 or the maize ubiquitin-1 gene promoters, or the ABA-responsive wheat *Em* gene and the auxin-responsive soybean GH3-1 promoter (discussed in (37)). A versatile conditional expression system based on the heat-shock-responsive promoter from soybean small HSP17.3b gene was also established and used for example to establish conditional GFP-labelling of microfilaments (38). If working on a haploid genome facilitates genetic analyses, it also impedes the identification of mutants affected in essential genes. The generation of conditional alleles is required to circumvent this situation: this has been recently achieved using the HSP promoter to drive expression of the polycomb gene *PpCLF* to study the gametophyte to sporophyte transition (39).

3. Organogenesis and Cytoskeleton

The principal function of the cytoskeleton is to spatially coordinate basic cellular processes such as cell growth, cell division, cell polarization, subcellular compartmentation, and intracellular membrane trafficking. It thus plays an essential role in organogenesis and we review below some recent studies performed in *Physcomitrella* that provided new insight in this field.

3.1. Actin Microfilaments

Filamentous actin (F-actin) is indispensable for cell viability in eukaryotes. The dynamic equilibrium between globular and F-actin is constantly modulated by regulatory proteins such as actin depolymerization factor (ADF), profilins, formins, or myosin XI. Using RNAi approaches, the groups of Vidali and Bezanilla demonstrated that these genes are essential for *P. patens* development, being required to establish apical growth and cell polarity in chloronemal cells (reviewed and references in (10) and (40)). They have also characterized a quintuple knock-out of Myosin VIII which shows that this small multigene family is required to regulate protonema patterning but is dispensable for the establishment of bushy growth (41).

The ARP2/3 (actin-related protein) complex is a downstream regulator of F-actin nucleation. Its activity is regulated by the SCAR/WAVE complex in response to Rho small GTPase signalling. This highly conserved pathway is essential in yeasts and animal cells, being required for the establishment of cell polarity, localized outgrowth and cell migration. In contrast, loss of function of either complex only affects trichome and pavement cell morphogenesis in Arabidopsis (42, 43). In *P. patens* loss of function of ARP3 or ARPc4 abolishes tip growth but not cell polarization of chloronemal cells and blocks the differentiation of caulonemata and rhizoids (44, 45). The same phenotype was observed in *brk1* mutants (BRICK1 is a component of the SCAR complex) which additionally displayed abnormal orientation of cell division during protonemal development (46). Cytological analyses demonstrate that these complexes localize at the growing tip of protonemal cells in the wild-type, where they control the formation of an F-actin array that is required to establish tip growth. Noticeably leafy shoots develop normally in these mutants, indicating that this regulatory pathway is not required to establish bushy growth. These studies have in a short time brought new insights in the essential role of actin dynamics in cell polarization, tip growth, and plant developmental processes.

3.2. Microtubules

Microtubules (MT) are essential for cell survival and mutations affecting MT dynamics are most of the time lethal in eukaryotes.

In plants, MT form four distinct networks during the cell cycle: (1) interphasic cortical arrays that are required for cell growth and the deposition of cell wall material, (2) the preprophase band (PPB) that defines at the onset of mitosis the future position of the division plane, (3) the mitotic spindle, and (4) the phragmoplast which is required for cell plate formation. In Arabidopsis, loss of function of TONNEAU1 (*AtTON1*) abolishes PPB formation at the onset of mitosis and affects the organization of interphasic MT. The resulting plants display anarchic tissue organization resulting from stochastic orientation of cell divisions and reduced cell elongation, but the differentiation pattern of the plant including phyllotaxis is not affected (47, 48). Loss of TONNEAU1 function in *P. patens* does not affect protonema development but moss gametophores phenocopy the developmental syndrome observed in Arabidopsis (49). Detailed analyses of *Ppton1* hypomorphic alleles further established that the involvement of TON1 protein in the organization of the PPB and of the interphasic MT cortical arrays could be uncoupled, accounting for the defect in the orientation of cell division and in cell elongation observed in both species. Successful reciprocal cross-complementation between Arabidopsis and *Physcomitrella* showed that the function of TON1 has been conserved and was eventually recruited from the gametophyte to the sporophyte during land plant evolution (49). These findings also implicate that proper orientation of cell division is not necessary to establish phyllotaxis in moss, a rather counter-intuitive observation if one considers that meristematic activity is confined to a single stem cell that displays a highly regular pattern of cell division. We have also generated a double *arp3/ton1* mutant to further investigate the extent of overlying function between TON1 and the ARP2/3 complex (Finka and Schaefer, unpublished data). The resulting mutant phenocopies the developmental syndromes of both mutants indicating that the function of these two regulators of the cytoskeleton do not overlap during *P.patens* development (Fig. 3). This work illustrates the plasticity of the moss system since such a mutant would probably be hard to characterize in Arabidopsis.

4. Organogenesis and Growth Factors

Auxin and cytokinin cross talk were early mentioned in moss studies as likely essential for a correct gametophytic development (50, 51). Several recent studies highlight the level of functional conservation of growth factors between *Physcomitrella* and Arabidopsis.

4.1. Auxin

In *Physcomitrella*, early studies have shown that auxin is required for chloronema to caulonema transition, for rhizoid differentiation and for normal shoot development. Auxin resistant mutants unable to

Fig. 3. Top: isolated gametophore from the wild type (WT), *Ppton1* null (TON1-1) and hypomorphic (TON1-2) alleles showing alteration of gametophore development but normal rhizoid differentiation. Bottom: isolated gametophore from *Pparp3* (ARP3) that develops normally but does not form rhizoids and of the double *Pparp3/ton1* mutants. Both the null and the hypomorphic *ton1* phenotype can be recognized in the *Pparp3* background. Scale bar: 0.3 mm.

progress beyond the chloronemal stage have also been isolated (50). Comparative analysis of the moss and Arabidopsis genomes indicates that all basic components for rapid auxin response are found in *P. patens* (52). In Arabidopsis, auxin binds to the TRANSPORT INHIBITOR RESPONSE (TIR1)/AUXIN SIGNALING F-BOX PROTEINS (AFB) which promote the degradation of AUXIN/INDOLE-3-ACETIC ACID (AUX/IAA) repressors and induce expression of auxin-responsive genes. Recent analyses of moss auxin resistant mutants identified mutations in conserved domains of *AUX/IAA* genes and demonstrated auxin-dependent interaction of moss *AUX/IAA* with *AFB* genes (53). This indicates that the molecular mechanism of auxin perception is conserved in embryophytes.

Rhizoid differentiation from shoot epidermal cells is strongly induced by auxin and gametophores grown on 1 μM NAA are leafless with numerous rhizoids (17). It was also shown that the chloronema–caulonema transition is gradual, and that auxin induction of both *PpRHD SIX-LIKE1 (PpRSL1)* and *PpRSL2* genes (transcription factors, see also below) expression is sufficient to promote this transition (54). These results suggest that the involvement of auxin in rhizoid formation may represent an ancient role, as it has been observed in many earlier-diverging streptophytes, while the more specific role of auxin in chloronema–caulonema transition may have been co-opted within the moss lineage (53).

There is no polar auxin transport in gametophore development (55) but genes homologous to the Arabidopsis auxin efflux carriers *PIN FORMED* (*PIN*) genes have been found in *P. patens* genome and their role is to be unravelled (56). The absence of polar auxin transport suggests an important role for localized auxin biosynthesis during moss development. In Arabidopsis, auxin biosynthesis is positively regulated by SHORT INTERNODE/STYLISH (SHI/STY) proteins and two *SHI* orthologues are present in the moss genome. Overexpressors and single knock-outs of *PpSHI* genes were generated, which consistently displayed increased and decreased auxin levels, respectively (57). In these strains, all the above-mentioned developmental processes were affected in a way that correlated with auxin levels and the involvement of auxin in regulating senescence was further identified. Some of these functions may be analogous in bryophytes and tracheophytes. This study further established a tight correlation between the expression profile of *PpSHI* genes and an auxin inducible reporter construct, suggesting that local auxin production could promote auxin peak formation in organogenetic processes.

4.2. Cytokinin

In *P. patens*, cytokinins (CK) are required for the differentiation of buds from caulonemal subapical and thus for the establishment of bushy growth. Bud overproduction was observed in response to exogenous CK and in *ove* mutants which also displayed altered CK metabolism (21, 50, 58 and see Fig. 4). Noticeably extracellular CK constitute the main CK responsible for bud induction in moss cultures (59). In Arabidopsis, CK signalling is transduced through a two-component system consisting of three phosphorelay signal transducers: His-kinase receptors (HK), a His-containing phosphotransmitters (HPT) and response regulators (RR) that modulate gene expression. Comparative analyses of fully sequenced plant genomes indicated that the complete set of proteins of the CK signalling pathway appeared in the moss genome (60) while initial characterization of the moss PpHK4b CK receptor showed that it exhibits His-kinase activity (61). Further studies of CK induced bud induction in moss will soon provide additional insights on the molecular network regulating CK-induced organogenetic processes.

4.3. ABA

Early physiological studies have shown that abscisic acid (ABA) induces the differentiation of swollen globular drought-tolerant cells, also called brood cells (62) and plays an important role to protect mosses against desiccation, cold or salt stresses. ABA regulated expression of the wheat Em promoter in *Physcomitrella* provided initial evidence for the conservation of ABA response pathways between mosses and seed plants (63). More recent studies

Fig. 4. Over production of buds in response to exogenous supply of CK. Protonema was grown under gravitropic conditions in darkness to enhance caulonemal differentiation. Cultures were transferred after 10 days in light on standard or CK supplemented medium. Images were taken 5 days after transfer to light. Buds differentiate from most caulonemal branch initials while chloronema formation is almost completely abolished on CK. Scale bar: 0.5 mm.

have shown that ABA signalling pathways mediated either by ABA INSENSITIVE (ABI)3-like transcription factors (64–66) or by ABI1-like type 2C protein phosphatase are functionally conserved ((67) and discussed in (68)). These studies also show that *Pp*ABI1 is required for the differentiation of brood cells in response to ABA in *Physcomitrella* (69) while ectopic expression of Arabidopsis *ABI1-1* affects moss development, suggesting a possible role of ABA in leafy shoot and reproductive organ morphogenesis (70). Finally, analysis of *Physcomitrella open stomata-1* (*ost-1*) knock-out and successful complementation of the Arabidopsis *ost1* mutant with *PpOST1* provide functional evidence for the conservation of ABA-mediated regulation of stomatal behavior in embryophytes (71).

4.4. GAs, Brassinosteroids, and Ethylene

As an exception to growth factors mentioned above, no true gibberellin (GA) pathway has been observed in *P. patens* (72, 73) and fully functional brassinosteroid signalling components are only present in vascular plants (74). Yet, *ent*-Kaurene synthase *(PpCPS/KS)* mutants, deficient in the production of *ent*-Kaurene derived GA type diterpenes, show no chloronema–caulonema transition (75). These diterpenes remain to be identified, and would shed light on GA signalling evolution in land plants (76). Finally, the role of ethylene in chloronema–caulonema transition has been early suggested, and sequences for putative ethylene-receptors have been recently found in *P. patens* genome, but clear evidence for a role of ethylene in this or other moss developmental steps are still lacking (77).

4.5. Strigolactones (SLs)

SL have prior been characterized as key signals of root proximity for both parasitic weeds seeds and symbiotic mycorrhizal fungi, then as a novel class of hormones controlling shoot branching in seed plants (for reviews: (78–80)). We obtained a *Ppccd8* mutant affected in the SL biosynthesis enzyme CCD8 (CAROTENOID CLEAVAGE DIOXYGENASE 8) that, compared to WT moss, fails in arresting plant extension after 3 weeks and shows increased caulonema branching (81). This study led us to propose that SLs or derived molecules are produced by the moss and released in the medium to control moss individual extension (81). Furthermore, the SL concentration would act as a signal for sensing neighbor individuals, an observation reminiscent of the early description of Factor H known to inhibit caulonema growth and possibly determinant in moss community structure in nature (82).

5. Organogenesis and (Nonhormonal) Signal Transduction Pathways

In addition to growth factors, other signals contribute to and/or modify the moss development, by inducing specific transduction pathways.

Light is one of these major regulators of *P. patens* morphogenesis (83): it is necessary for the very early step of spore germination, and later on for all gametophytic development, that responds to both light quality and periodicity. Early studies had shown that red-light induced membrane depolarization of the protonema (e.g., Ca^{2+} fluxes) correlated with side branch initial formation (84). Later on, both cryptochromes (cry) and phytochromes have been shown as photoreceptors contributing to protonema development (85, 86). Moreover, the study of the *Ppcry1a-cry1b* double mutants suggested that cryptochrome (blue) light signals repress auxin signals during moss development (87). Day length affects moss development (in particular short days induce sporophyte development) and several core components of the circadian clock have been found in *P. patens*. However, a recent comparison with Arabidopsis suggests a single feedback loop in moss versus a more complicated clock network in vascular plants (88).

Hexokinases (*HXK*) are enzymes that catalyze the phosphorylation of glucose and fructose, but these proteins may also play a role in sugar sensing and signalling. Since the cloning of the first *PpHXK1* gene, encoding a chloroplastic HXK and the study of the corresponding *hxk1* KO mutant affected in protonema growth (89), ten more *HXK* genes have been characterized in *Physcomitrella* (90). The predicted encoded proteins are more similar to each other than to hexokinases from vascular plants, which makes this family very interesting for evo-devo studies. Two other protein kinase genes were cloned from *Physcomitrella*, the Snf1-related protein kinase 1 genes *PpSNF1a* and *PpSNF1b*, that may redundantly contribute to moss growth (and hormone regulated?) adaptation to low energy supply conditions (91).

Tip growth regulation requires signal transduction as highlighted by both pollen tubes and root hairs studies from vascular plants. The requirement of *ARABINOGALACTAN PROTEINS* for protonema cell extension has been early demonstrated in moss (92). Recently, *P. patens* KO mutants affected in enzymes responsible for the synthesis of Phosphatidyl Inositol 4,5 diphosphate (PtdIns(4,5)P_2) (PIPK1 and PIPK2) were obtained (93). The level of PtdIns(4,5)P_2 were reduced in both single mutants, however only the *pipk1* mutant showed shorter caulonema cells and less developed rhizoids (93), suggesting a role for the PIPK1 enzyme in tip growth. The double KO *pipk1-2* did not develop any caulonema or rhizoid. In a recent survey, homologues of genes encoding proteins involved in *RAC/ROP GTPases* signalling were identified in the moss genome (94), which characterization should lead to a better knowledge of land plant tip growth control mechanisms.

6. Organogenesis and Transcription Factors

In plants many transcription factors are encoded by members of multigene families that expanded much more dramatically during land plant evolution than during the evolution of animals and fungi (95). In vascular plants, Class 1 KNOTTED1-LIKE HOMEOBOX (KNOX) transcription factors are essential to the regulation of cell division and differentiation in the shoot apical meristem (for review: (96)). The class 1 *KNOX* genes homologues (*MKN* genes), identified in the moss genome, would not regulate the leafy shoot haploid meristem (97). However, these transcription factors are found expressed in the sporophyte where they would regulate sporogenous cells divisions (97, 98). Sakakibara et al. claim that this would correspond to a situation where networks operating in vascular plants sporophyte would not be co-opted from ancestor's gametophyte network (97).

Like caulonema, rhizoid cells elongate by tip-growth. Two types of rhizoids differentiate from *P. patens* gametophore epidermal cells, namely basal and mid-stem rhizoid, and auxin induces their development (17). The *PpHB7* gene (encoding a homeodomain-leucine zipper I transcription factor) is required for rhizoids late differentiation steps, but not for their determination. Its expression is increased following both auxin and CK (6-benzylaminopurine) treatment (17). Two other transcription factors of the basic-helix–loop–helix (bHLH) type have been described as sufficient for rhizoid development, the *PpRSL1* and *PpRSL2* genes (99). Target genes of these transcription factors are yet to be discovered. Both *PpRSL1* and *PpRSL2* genes are induced by auxin, contrary to the Arabidopsis homologous genes (*AtROOT HAIR DEFECTIVE 6* and *AtRSL1*) that do regulate root hairs development (100). Therefore the mechanism of auxin action is different between moss and vascular plants for regulating rhizoid and root hair differentiation (100).

The three closely related bHLH transcription factors SPEECHLESS, MUTE, and FAMA regulate a different step in the stomatal lineage: asymmetric divisions, the acquisition of guard mother cell identity, and guard cell differentiation, respectively. Stomata were an ancient innovation of land plants predating even the evolution of leaves and roots. Cross-species complementation between Arabidopsis and *Physcomitrella* show that PpSMF1, one of the two bHLHs of the same subgroup found in *P. patens* possesses dual functionalities while MUTE and FAMA are now each specialized for a single step. A model is proposed where an ancestral multifunctional transcription factor underwent duplication followed by specialization to provide the three (now nonoverlapping) functions of the angiosperm stomatal bHLHs (101).

GAI, RGA, SCL (GRAS)-type genes have been recently reported in moss in an attempt to study the evolution of this large family of transcription factors that underwent large diversification (102); a precise role of these proteins in bryophyte development remains to be discovered.

7. Organogenesis and Reprogramming

For more than a decade now, the role of epigenetic factors on developmental processes regulation has been discovered in plants. The predominant gametophytic phase of *P. patens* development was proven very useful for demonstrating the role of chromatin modifiers protein complex (e.g., The Polycomb group complex, see below) and small RNAs.

7.1. Epigenetic Control of Moss Development

The Polycomb group (PcG) complex, involved in the epigenetic control of gene expression profiles, plays a central role in regulating the transition of the female gametophyte to the sporophyte in flowering plants. In *A. thaliana* all PcG complexes comprise the WD40 motif-containing proteins FERTILIZATION INDEPENDENT ENDOSPERM (FIE) and one of the three SET domain proteins CURLY LEAF (CLF), SWINGER (SWN) or MEDEA (MEA). Expression of a PpFIE-GUS fusion protein under the control of its native promoter indicates that the PpFIE protein accumulates only in gametophyte apical cells, and in cells that undergo fate transition. In the absence of PpFIE, gametophore meristems overproliferate, but fail to develop and to reach the reproductive phase (103). Hence, the essential FIE PcG function in regulating developmental programs along the life cycle was established early in evolution. In another work, PpCLF deletion lines were obtained, where a gametophytic vegetative cell frequently gave rise to a sporophyte-like body which did not form a sporangium but which, with continued culture, branched (104). Sporophyte branching is almost unknown among extant bryophytes demonstrating the role of the PcG complex in sustaining evolutionary innovation in land plants. In the gametophyte, PpCLF represses initiation of a sporophytic pluripotent stem cell, while in the sporophyte, it represses that stem cell activity and induces reproductive organ development.

Most eukaryotes express diverse small silencing RNAs which direct the sequence-specific repression of target RNAs. The accumulation of endogenous 24 nt short interfering RNAs (siRNAs), derived mostly from intergenic and repetitive genomic regions, requires the Dicer family member DICER-LIKE3 (DCL3). Despite the presence of a clear homolog of DCL3 in non-flowering plant, the 24 nt siRNAs are not readily apparent in the small RNA populations of several lineages. Nevertheless, "hotspots" of small RNA production were found in the genome of *Physcomitrella patens* that produced a mix of 21-24 nt siRNAs, which despite their different sizes, are reminiscent of the 24 nt siRNA loci of angiosperms (enriched in transposon content, avoid annotated genes, and densely modified with the epigenetic mark 5-methyl cytosine). *Ppdcl3* mutants failed to accumulate 22–24 nt siRNAs from repetitive regions and displayed an acceleration of leafy gametophore production, suggesting that repetitive siRNAs may play a role in moss development (105). Furthermore, KO mutants for DICER-LIKE1 (DCL1) homologs (*Ppdcl1a* and *Ppdcl1b*) show strong developmental alterations such as growth retardation and/or aberrant cell division and differentiation, and their study pinpoints specific roles for each protein in the transcriptional control of target genes expression, by MicroRNAs (106).

7.2. Moss Organogenesis Versatility

In vascular plants, dissected or wounded tissues can proliferate when treated with exogenous phytohormones to form callus, which can be fated to form shoot or root meristematic tissue bearing stem cells (107, 108). In *Physcomitrella*, when part of a gametophore leaf is excised and cultivated for a few days on culture medium without phytohormone supplementation, leaf cells facing the cut edge change into cells that are indistinguishable from the apical cells of chloronemata. The study of this reprogramming of excised leaf cells showed that cyclin-dependant kinase A (CDKA) regulates cell cycle progression and acquisition of new cell characteristics in parallel (109). Treatment with a CDK inhibitor or induction of dominant negative CDKA;1 protein inhibited not only cell cycle progression but also tip growth and protonemal gene expression, whereas a DNA synthesis inhibitor, aphidicolin, inhibited cell cycle progression but prevented neither tip growth nor protonemal gene expression. Thus CDKA concomitantly regulates cell division and cellular change in reprogramming differentiated cells to become stem cells in plants.

8. Conclusion

In the recent years the moss *Physcomitrella patens* has emerged as a powerful complementary model system to study plant organogenesis. Recent studies have addressed the relation between moss organogenesis and many regulators of plant development including growth factors, the cytoskeleton, transduction pathways, transcription factors, epigenetic control and dedifferentiation processes. These studies highlight both the similarity and the differences between *Physcomitrella* and Arabidopsis and strongly support the idea that the major developmental innovations that were associated with the colonization of land by plants can be investigated in extant moss species. The unrivalled efficiencies of GT combined with the outstanding cellular and biochemical facilities offered by *Physcomitrella* will certainly contribute to improve our understanding of the evolution of plant developmental processes.

References

1. Leyser O (2011) Auxin, self-organisation, and the colonial nature of plants. Curr Biol 21:R331–R337
2. Kenrick P, Crane PR (1997) The origin and early evolution of plants on land. Nature 389:33–39
3. Reski R, Reynolds S, Wehe M, Kleberjanke T, Kruse S (1998) Moss (Physcomitrella Patens) expressed sequence tags include several sequences which are novel for plants. Bot Acta 111:143–149
4. Rensing SA, Rombauts S, Van de Peer Y, Reski R (2002) Moss transcriptome and beyond. Trends Plant Sci 7:535–538
5. Lang D, Zimmer AD, Rensing SA, Reski R (2008) Exploring plant biodiversity: the Physcomitrella genome and beyond. Trends Plant Sci 13:542–549
6. Schaefer DG, Zryd JP (1997) Efficient gene targeting in the moss Physcomitrella patens. Plant J 11:1195–1206
7. Schaefer DG (2002) A new moss genetics: targeted mutagenesis in Physcomitrella patens. Annu Rev Plant Biol 53:477–501
8. Rensing SA, Lang D, Zimmer AD, Terry A, Salamov A, Shapiro H, Nishiyama T, Perroud PF, Lindquist EA, Kamisugi Y, Tanahashi T, Sakakibara K, Fujita T, Oishi K, Shin IT, Kuroki Y, Toyoda A, Suzuki Y, Hashimoto S, Yamaguchi K, Sugano S, Kohara Y, Fujiyama A, Anterola A, Aoki S, Ashton N, Barbazuk WB, Barker E, Bennetzen JL, Blankenship R, Cho SH, Dutcher SK, Estelle M, Fawcett JA, Gundlach H, Hanada K, Heyl A, Hicks KA, Hughes J, Lohr M, Mayer K, Melkozernov A, Murata T, Nelson DR, Pils B, Prigge M, Reiss B, Renner T, Rombauts S,

Rushton PJ, Sanderfoot A, Schween G, Shiu SH, Stueber K, Theodoulou FL, Tu H, Van de Peer Y, Verrier PJ, Waters E, Wood A, Yang L, Cove D, Cuming AC, Hasebe M, Lucas S, Mishler BD, Reski R, Grigoriev IV, Quatrano RS, Boore JL (2008) The Physcomitrella genome reveals evolutionary insights into the conquest of land by plants. Science 319:64–69

9. Quatrano RS, McDaniel SF, Khandelwal A, Perroud PF, Cove DJ (2007) Physcomitrella patens: mosses enter the genomic age. Curr Opin Plant Biol 10:182–189
10. Prigge MJ, Bezanilla M (2010) Evolutionary crossroads in developmental biology: Physcomitrella patens. Development 137: 3535–3543
11. Cove D, Bezanilla M, Harries P, Quatrano R (2006) Mosses as model systems for the study of metabolism and development. Annu Rev Plant Biol 57:497–520
12. Menand B, Calder G, Dolan L (2007) Both chloronemal and caulonemal cells expand by tip growth in the moss Physcomitrella patens. J Exp Bot 58:1843–1849
13. Schumaker KS, Dietrich MA (1998) Hormone-induced signaling during moss development. Annu Rev Plant Physiol Plant Mol Biol 49:501–523
14. Pressel S, Ligrone R, Duckett JG (2008) Cellular differentiation in moss protonemata: a morphological and experimental study. Ann Bot 102:227–245
15. Harrison CJ, Roeder AH, Meyerowitz EM, Langdale JA (2009) Local cues and asymmetric cell divisions underpin body plan transitions in the moss Physcomitrella patens. Curr Biol 19:461–471
16. Fujita T, Sakaguchi H, Hiwatashi Y, Wagstaff SJ, Ito M, Deguchi H, Sato T, Hasebe M (2008) Convergent evolution of shoots in land plants: lack of auxin polar transport in moss shoots. Evol Dev 10:176–186
17. Sakakibara K, Nishiyama T, Sumikawa N, Kofuji R, Murata T, Hasebe M (2003) Involvement of auxin and a homeodomain-leucine zipper I gene in rhizoid development of the moss Physcomitrella patens. Development 130:4835–4846
18. Hohe A, Rensing SA, Mildner M, Lang D, Reski R (2002) Day length and temperature strongly influence sexual reproduction and expression of a novel MADS-Box gene in the moss Physcomitrella patens. Plant Biol (Stuttg) 4:595–602
19. Sakakibara K, Nishiyama T, Deguchi H, Hasebe M (2008) Class 1 KNOX genes are not involved in shoot development in the moss Physcomitrella patens but do function in sporophyte development. Evol Dev 10:555–566
20. Cove DJ, Perroud PF, Charron AJ, McDaniel SF, Khandelwal A, Quatrano RS (2009) Culturing the moss Physcomitrella patens. Cold Spring Harb Protoc 2009, pdb prot5136
21. Reski R, Abel WO (1985) Induction of budding on chloronemata and caulonemata of the moss, *Physcomitrella patens* using isopentenyladenine. Planta 165:354–358
22. Perroud PF, Cove DJ, Quatrano RS, McDaniel SF (2011) An experimental method to facilitate the identification of hybrid sporophytes in the moss Physcomitrella patens using fluorescent tagged lines. New Phytol 191: 301–306
23. Decker EL, Reski R (2007) Moss bioreactors producing improved biopharmaceuticals. Curr Opin Biotechnol 18:393–398
24. Thevenin J, Dubos C, Xu W, Le Gourrierec J, Kelemen Z, Charlot F, Nogue F, Lepiniec L, Dubreucq B (2012) A new system for fast and quantitative analysis of heterologous gene expression in plants. New Phytol 193:504–512
25. Schaefer D, Zryd JP, Knight CD, Cove DJ (1991) Stable transformation of the moss Physcomitrella patens. Mol Gen Genet 226:418–424
26. Schaefer DG, Zryd JP (2001) The moss Physcomitrella patens, now and then. Plant Physiol 127:1430–1438
27. Schaefer DG (2001) Gene targeting in Physcomitrella patens. Curr Opin Plant Biol 4:143–150
28. Müller U (1999) Ten years of gene targeting: targeted mouse mutants, from vector design to phenotype analysis (review). Mech Dev 82: 3–21
29. Sauer B (1993) Manipulation of the transgene by site-specific recombination: use of *cre* recombinase. Methods Enzymol 225:890–900
30. Schaefer DG, Zrÿd J-P (2004) Principles of targeted mutagenesis in the moss Physcomitrella patens. In: Wood AJ, Oliver MJ, Cove D (eds) New frontiers in bryology. Kluwer Academic Publishers, Dordrecht, pp 37–49
31. Schween G, Egener T, Fritzowsky D, Granado J, Guitton MC, Hartmann N, Hohe A, Holtorf H, Lang D, Lucht JM, Reinhard C, Rensing SA, Schlink K, Schulte J, Reski R (2005) Large-scale analysis of 73 329 physcomitrella plants transformed with different gene dis-

ruption libraries: production parameters and mutant phenotypes. Plant Biol (Stuttg) 7:228–237
32. Hiwatashi Y, Nishiyama T, Fujita T, Hasebe M (2001) Establishment of gene-trap and enhancer-trap systems in the moss Physcomitrella patens. Plant J 28:105–116
33. Bezanilla M, Pan A, Quatrano RS (2003) RNA interference in the moss Physcomitrella patens. Plant Physiol 133:470–474
34. Khraiwesh B, Ossowski S, Weigel D, Reski R, Frank W (2008) Specific gene silencing by artificial MicroRNAs in Physcomitrella patens: an alternative to targeted gene knockouts. Plant Physiol 148:684–693
35. Khraiwesh B, Fattash I, Arif MA, Frank W (2011) Gene function analysis by artificial microRNAs in Physcomitrella patens. Methods Mol Biol 744:57–79
36. Vidali L, Augustine RC, Fay SN, Franco P, Pattavina KA, Bezanilla M (2009) Rapid screening for temperature-sensitive alleles in plants. Plant Physiol 151:506–514
37. Saidi Y, Finka A, Chakhporanian M, Zryd JP, Schaefer DG, Goloubinoff P (2005) Controlled expression of recombinant proteins in Physcomitrella patens by a conditional heat-shock promoter: a tool for plant research and biotechnology. Plant Mol Biol 59:697–711
38. Finka A, Schaefer DG, Saidi Y, Goloubinoff P, Zryd JP (2007) In vivo visualization of F-actin structures during the development of the moss Physcomitrella patens. New Phytol 174:63–76
39. Okano Y, Aono N, Hiwatashi Y, Murata T, Nishiyama T, Ishikawa T, Kubo M, Hasebe M (2009) A polycomb repressive complex 2 gene regulates apogamy and gives evolutionary insights into early land plant evolution. Proc Natl Acad Sci USA 106:16321–16326
40. Vidali L, Burkart GM, Augustine RC, Kerdavid E, Tuzel E, Bezanilla M (2010) Myosin XI is essential for tip growth in Physcomitrella patens. Plant Cell 22:1868–1882
41. Wu SZ, Ritchie JA, Pan AH, Quatrano RS, Bezanilla M (2011) Myosin VIII regulates protonemal patterning and developmental timing in the moss physcomitrella patens. Mol Plant 4:909–921
42. Mathur J (2006) Local interactions shape plant cells. Curr Opin Cell Biol 18:40–46
43. Uhrig JF, Mutondo M, Zimmermann I, Deeks MJ, Machesky LM, Thomas P, Uhrig S, Rambke C, Hussey PJ, Hulskamp M (2007) The role of Arabidopsis SCAR genes in ARP2-ARP3-dependent cell morphogenesis. Development 134:967–977
44. Perroud PF, Quatrano RS (2006) The role of ARPC4 in tip growth and alignment of the polar axis in filaments of Physcomitrella patens. Cell Motil Cytoskeleton 63:162–171
45. Finka A, Saidi Y, Goloubinoff P, Neuhaus JM, Zryd JP, Schaefer DG (2008) The knock-out of ARP3a gene affects F-actin cytoskeleton organization altering cellular tip growth, morphology and development in moss Physcomitrella patens. Cell Motil Cytoskeleton 65:769–784
46. Perroud PF, Quatrano RS (2008) BRICK1 is required for apical cell growth in filaments of the moss Physcomitrella patens but not for gametophore morphology. Plant Cell 20: 411–422
47. Traas J, Bellini C, Nacry P, Kronenberg J, Bouchez D, Caboche M (1995) Normal differentiation pattern in plants lacking microtubular preprophase band. Nature 375: 676–677
48. Azimzadeh J, Nacry P, Christodoulidou A, Drevensek S, Camilleri C, Amiour N, Parcy F, Pastuglia M, Bouchez D (2008) Arabidopsis TONNEAU1 proteins are essential for preprophase band formation and interact with centrin. Plant Cell 20:2146–2159
49. Spinner L, Pastuglia M, Belcram K, Pegoraro M, Goussot M, Bouchez D, Schaefer DG (2010) The function of TONNEAU1 in moss reveals ancient mechanisms of division plane specification and cell elongation in land plants. Development 137:2733–2742
50. Ashton NW, Grimsley NH, Cove DJ (1979) Analysis of gametophytic development in the moss, *Physcomitrella patens* using auxin and cytokinin resistant mutants. Planta 144: 427–435
51. Schumaker KS, Dietrich MA (1998) Hormone-induced signaling during moss development (review). Annu Rev Plant Physiol Plant Mol Biol 49:501–523
52. Paponov IA, Teale W, Lang D, Paponov M, Reski R, Rensing SA, Palme K (2009) The evolution of nuclear auxin signalling. BMC Evol Biol 9:126
53. Prigge MJ, Lavy M, Ashton NW, Estelle M (2010) Physcomitrella patens auxin-resistant mutants affect conserved elements of an auxin-signaling pathway. Curr Biol 20:1907–1912
54. Jang G, Dolan L (2011) Auxin promotes the transition from chloronema to caulonema in moss protonema by positively regulating PpRSL1and PpRSL2 in Physcomitrella patens. New phytol 192:319–327

55. Fujita T, Sakaguchi H, Hiwatashi Y, Wagstaff SJ, Ito M, Deguchi H, Sato T, Hasebe M (2008) Convergent evolution of shoots in land plants: lack of auxin polar transport in moss shoots. Evol Dev 10:176–186
56. Krecek P, Skupa P, Libus J, Naramoto S, Tejos R, Friml J, Zazimalova E (2009) The PIN-FORMED (PIN) protein family of auxin transporters. Genome Biol 10:249
57. Eklund DM, Thelander M, Landberg K, Staldal V, Nilsson A, Johansson M, Valsecchi I, Pederson ER, Kowalczyk M, Ljung K, Ronne H, Sundberg E (2010) Homologues of the Arabidopsis thaliana SHI/STY/LRP1 genes control auxin biosynthesis and affect growth and development in the moss Physcomitrella patens. Development 137: 1275–1284
58. Schulz PA, Hofmann AH, Russo VE, Hartmann E, Laloue M, von Schwartzenberg K (2001) Cytokinin overproducing ove mutants of Physcomitrella patens show increased riboside to base conversion. Plant Physiol 126:1224–1231
59. von Schwartzenberg K, Nunez MF, Blaschke H, Dobrev PI, Novak O, Motyka V, Strnad M (2007) Cytokinins in the bryophyte Physcomitrella patens: analyses of activity, distribution, and cytokinin oxidase/dehydrogenase overexpression reveal the role of extracellular cytokinins. Plant Physiol 145: 786–800
60. Pils B, Heyl A (2009) Unraveling the evolution of cytokinin signaling. Plant Physiol 151:782–791
61. Ishida K, Yamashino T, Nakanishi H, Mizuno T (2010) Classification of the genes involved in the two-component system of the moss Physcomitrella patens. Biosci Biotechnol Biochem 74:2542–2545
62. Goode JA, Stead AD, Duckett JG (1993) Redifferentiation of moss Protonemata—an experimental and immunofluorescence study of brood cell formation. Can J Bot-Rev Can Bot 71:1510–1519
63. Knight CD, Sehgal A, Atwal K, Wallace JC, Cove DJ, Coates D, Quatrano RS, Bahadur S, Stockley PG, Cuming AC (1995) Molecular responses to abscisic acid and stress are conserved between moss and cereals. Plant Cell 7:499–506
64. Marella HH, Sakata Y, Quatrano RS (2006) Characterization and functional analysis of ABSCISIC ACID INSENSITIVE3-like genes from Physcomitrella patens. Plant J 46:1032–1044
65. Khandelwal A, Cho SH, Marella H, Sakata Y, Perroud PF, Pan A, Quatrano RS (2010) Role of ABA and ABI3 in desiccation tolerance. Science 327:546
66. Sakata Y, Nakamura I, Taji T, Tanaka S, Quatrano RS (2010) Regulation of the ABA-responsive Em promoter by ABI3 in the moss Physcomitrella patens: role of the ABA response element and the RY element. Plant Signal Behav 5:1061–1066
67. Komatsu K, Nishikawa Y, Ohtsuka T, Taji T, Quatrano RS, Tanaka S, Sakata Y (2009) Functional analyses of the ABI1-related protein phosphatase type 2C reveal evolutionarily conserved regulation of abscisic acid signaling between Arabidopsis and the moss Physcomitrella patens. Plant Mol Biol 70: 327–340
68. Takezawa D, Komatsu K, Sakata Y (2011) ABA in bryophytes: how a universal growth regulator in life became a plant hormone? J Plant Res 124:437–453
69. Tougane K, Komatsu K, Bhyan SB, Sakata Y, Ishizaki K, Yamato KT, Kohchi T, Takezawa D (2010) Evolutionarily conserved regulatory mechanisms of abscisic acid signaling in land plants: characterization of ABSCISIC ACID INSENSITIVE1-like type 2C protein phosphatase in the liverwort Marchantia polymorpha. Plant Physiol 152:1529–1543
70. Sakata Y, Komatsu K, Taji T, Tanaka S (2009) Role of PP2C-mediated ABA signaling in the moss Physcomitrella patens. Plant Signal Behav 4:887–889
71. Chater C, Kamisugi Y, Movahedi M, Fleming A, Cuming AC, Gray JE, Beerling DJ (2011) Regulatory mechanism controlling stomatal behavior conserved across 400 million years of land plant evolution. Curr Biol 21:1025–1029
72. Hirano K, Nakajima M, Asano K, Nishiyama T, Sakakibara H, Kojima M, Katoh E, Xiang H, Tanahashi T, Hasebe M, Banks JA, Ashikari M, Kitano H, Ueguchi-Tanaka M, Matsuoka M (2007) The GID1-mediated gibberellin perception mechanism is conserved in the Lycophyte Selaginella moellendorffii but not in the Bryophyte Physcomitrella patens. Plant Cell 19:3058–3079
73. Yasumura Y, Crumpton-Taylor M, Fuentes S, Harberd NP (2007) Step-by-step acquisition of the gibberellin-DELLA growth-regulatory mechanism during land-plant evolution. Curr Biol 17:1225–1230
74. Depuydt S, Hardtke CS (2011) Hormone signalling crosstalk in plant growth regulation. Curr Biol 21:R365–R373
75. Hayashi K, Horie K, Hiwatashi Y, Kawaide H, Yamaguchi S, Hanada A, Nakashima T, Nakajima M, Mander LN, Yamane H, Hasebe M,

Nozaki H (2010) Endogenous diterpenes derived from ent-kaurene, a common gibberellin precursor, regulate protonema differentiation of the moss Physcomitrella patens. Plant Physiol 153:1085–1097
76. Sun TP (2011) The molecular mechanism and evolution of the GA-GID1-DELLA signaling module in plants. Curr Biol 21: R338–R345
77. Ishida K, Yamashino T, Nakanishi H, Mizuno T (2010) Classification of the genes involved in the two-component system of the moss Physcomitrella patens. Biosci Biotechnol Biochem 74:2542–2545
78. Dun EA, Brewer PB, Beveridge CA (2009) Strigolactones: discovery of the elusive shoot branching hormone. Trends Plant Sci 14:364–372
79. Rameau C (2010) Strigolactones, a novel class of plant hormone controlling shoot branching. C R Biol 333:344–349
80. Xie X, Yoneyama K (2010) The strigolactone story. Annu Rev Phytopathol 48:93–117
81. Proust H, Hoffmann B, Xie X, Yoneyama K, Schaefer DG, Nogue F, Rameau C (2011) Strigolactones regulate protonema branching and act as a quorum sensing-like signal in the moss Physcomitrella patens. Development 138:1531–1539
82. Watson MA (1981) Chemically mediated interactions among juvenile mosses as possible determinants of their community structure. J Chem Ecol 7:367–376
83. Schaefer DG, Zryd JP (2001) The moss Physcomitrella patens, now and then. Plant Physiol 127:1430–1438
84. Ermolayeva E, Sanders D, Johannes E (1997) Ionic mechanism and role of phytochrome-mediated membrane depolarisation in caulonemal side branch initial formation in the moss *Physcomitrella patens*. Planta 201:109–118
85. Mittmann F, Brucker G, Zeidler M, Repp A, Abts T, Hartmann E, Hughes J (2004) Targeted knockout in Physcomitrella reveals direct actions of phytochrome in the cytoplasm. Proc Natl Acad Sci USA 101:13939–13944
86. Uenaka H, Wada M, Kadota A (2005) Four distinct photoreceptors contribute to light-induced side branch formation in the moss Physcomitrella patens. Planta 222:623–631
87. Imaizumi T, Kadota A, Hasebe M, Wada M (2002) Cryptochrome light signals control development to suppress auxin sensitivity in the moss Physcomitrella patens. Plant Cell 14:373–386
88. Holm K, Kallman T, Gyllenstrand N, Hedman H, Lagercrantz U (2010) Does the core circadian clock in the moss Physcomitrella patens (Bryophyta) comprise a single loop? BMC Plant Biol 10:109
89. Olsson T, Thelander M, Ronne H (2003) A novel type of chloroplast stromal hexokinase is the major glucose-phosphorylating enzyme in the moss Physcomitrella patens. J Biol Chem 278:44439–44447
90. Nilsson A, Olsson T, Ulfstedt M, Thelander M, Ronne H (2011) Two novel types of hexokinases in the moss Physcomitrella patens. BMC Plant Biol 11:32
91. Thelander M, Olsson T, Ronne H (2004) Snf1-related protein kinase 1 is needed for growth in a normal day-night light cycle. EMBO J 23:1900–1910
92. Lee KJ, Sakata Y, Mau SL, Pettolino F, Bacic A, Quatrano RS, Knight CD, Knox JP (2005) Arabinogalactan proteins are required for apical cell extension in the moss Physcomitrella patens. Plant Cell 17:3051–3065
93. Saavedra L, Balbi V, Lerche J, Mikami K, Heilmann I, Sommarin M (2011) PIPKs are essential for rhizoid elongation and caulonemal cell development in the moss Physcomitrella patens. Plant J 67:635–647
94. Eklund DM, Svensson EM, Kost B (2010) Physcomitrella patens: a model to investigate the role of RAC/ROP GTPase signalling in tip growth. J Exp Bot 61:1917–1937
95. Melzer R, Theissen G (2011) MADS and more: transcription factors that shape the plant. Methods Mol Biol 754:3–18
96. Hamant O, Pautot V (2010) Plant development: a TALE story. C R Biol 333:371–381
97. Sakakibara K, Nishiyama T, Deguchi H, Hasebe M (2008) Class 1 KNOX genes are not involved in shoot development in the moss Physcomitrella patens but do function in sporophyte development. Evol Dev 10:555–566
98. Singer SD, Ashton NW (2007) Revelation of ancestral roles of KNOX genes by a functional analysis of Physcomitrella homologues. Plant Cell Rep 26:2039–2054
99. Menand B, Yi K, Jouannic S, Hoffmann L, Ryan E, Linstead P, Schaefer DG, Dolan L (2007) An ancient mechanism controls the development of cells with a rooting function in land plants. Science 316:1477–1480
100. Jang G, Yi K, Pires ND, Menand B, Dolan L (2011) RSL genes are sufficient for rhizoid system development in early diverging land plants. Development 138:2273–2281

101. MacAlister CA, Bergmann DC (2011) Sequence and function of basic helix-loop-helix proteins required for stomatal development in Arabidopsis are deeply conserved in land plants. Evol Dev 13:182–192
102. Engstrom EM (2011) Phylogenetic analysis of GRAS proteins from moss, lycophyte and vascular plant lineages reveals that GRAS genes arose and underwent substantial diversification in the ancestral lineage common to bryophytes and vascular plants. Plant Signal Behav 6:850–854
103. Mosquna A, Katz A, Decker EL, Rensing SA, Reski R, Ohad N (2009) Regulation of stem cell maintenance by the Polycomb protein FIE has been conserved during land plant evolution. Development 136:2433–2444
104. Okano Y, Aono N, Hiwatashi Y, Murata T, Nishiyama T, Ishikawa T, Kubo M, Hasebe M (2009) A polycomb repressive complex 2 gene regulates apogamy and gives evolutionary insights into early land plant evolution. Proc Natl Acad Sci USA 106: 16321–16326
105. Cho SH, Addo-Quaye C, Coruh C, Arif MA, Ma Z, Frank W, Axtell MJ (2008) Physcomitrella patens DCL3 is required for 22-24 nt siRNA accumulation, suppression of retrotransposon-derived transcripts, and normal development. PLoS Genet 4:e1000314
106. Khraiwesh B, Arif MA, Seumel GI, Ossowski S, Weigel D, Reski R, Frank W (2010) Transcriptional control of gene expression by microRNAs. Cell 140:111–122
107. Skoog F, Miller CO (1957) Chemical regulation of growth and organ formation in plant tissues cultured in vitro. Symp Soc Exp Biol 11:118–130
108. Raghavan V (1989) Developmental biology of fern gametophytes. Cambridge University Press, Cambridge
109. Ishikawa M, Murata T, Sato Y, Nishiyama T, Hiwatashi Y, Imai A, Kimura M, Sugimoto N, Akita A, Oguri Y, Friedman WE, Hasebe M, Kubo M (2011) Physcomitrella cyclin-dependent kinase a links cell cycle reactivation to other cellular changes during reprogramming of leaf cells. Plant Cell 23:2924–2938

Chapter 3

The Dicot Root as a Model System for Studying Organogenesis

Julien Lavenus, Mikaël Lucas, Laurent Laplaze, and Soazig Guyomarc'h

Abstract

Organogenesis is the developmental process for producing new organs from undifferentiated cells. In plants, most organs are formed during postembryonic development. Shoot lateral organs are generated in the shoot apical meristem whereas lateral roots develop outside the root apical meristem. While lateral organ formation at the shoot and root might seem quite different, recent genetic studies have highlighted numerous parallels between these processes. In particular, the dynamic accumulation of auxin has been shown to play a crucial role both as a "morphogenetic trigger" and as a morphogen in both phenomena. This suggests that a unique model system could be adopted to study organogenesis in plants. In this chapter we describe the conceptual and technical advantages that support lateral root development as a good model system for studying organogenesis in plants.

Key words: Lateral root, Organogenesis, Genetic regulation, Hormonal regulation, Development, Auxin

1. Introduction

Development is the process by which multicellular eukaryotes form a complete organism from a single totipotent cell, the zygote. The developmental process of producing new organs from undifferentiated cells is defined as organogenesis. Unlike most animals, plants are able to develop new organs continuously, and therefore their development ends only with their death. The embryo of dicotyledonous plants basically contains only two storage leaves, the cotyledons, and two pools of stem cells, namely, the shoot apical meristem (SAM) and the root apical meristem (RAM). These apical meristems are responsible for the growth in length of the stem and the root, respectively, and their activity makes it possible for all the other organs of the plant to be produced through postembryonic

Ive De Smet (ed.), *Plant Organogenesis: Methods and Protocols*, Methods in Molecular Biology, vol. 959, DOI 10.1007/978-1-62703-221-6_3, © Springer Science+Business Media New York 2013 *de novo* organogenesis. On the one hand the shoot lateral organs, such as leaves or flowers, are initiated in a regular pattern in the shoot apical meristem. On the other hand, the root lateral organs such as lateral roots (LR) or symbiotic nitrogen-fixing nodules develop from inner tissues outside the root apical meristem.

Despite the apparent differences in the processes of shoot and root organogenesis, recent research tends to bring them together. Most notably, many studies have highlighted the central role played by the phytohormone auxin and its accumulation pattern in plant organogenesis (1–3). Furthermore, genetic approaches have identified a number of regulators of plant development that are involved in both root and shoot organogenesis (4–6). In this chapter, we discuss why lateral root development is a very good model system for studying organogenesis in plants.

2. The Organogenesis of Lateral Organs in *Arabidopsis*

2.1. The Formation of Lateral Roots in Arabidopsis

Most research on root development in dicots has been carried out on the model plant *Arabidopsis thaliana* (*Arabidopsis*) because it displays a transparent root with a very simple radial organization (7). The *Arabidopsis* root is composed of a central vasculature containing two xylem poles and two phloem poles, surrounded by a single-layered pericycle and only three outer layers, namely, the endodermis, the cortex, and the epidermis. This organization is built by the root apical meristem, which itself originates from the basal part of the embryo (Fig. 1A). These two processes—root meristem organogenesis and root histogenesis—involve precise regulation of cell divisions and identity patterning (8, 9). Although all pericycle cells are morphologically similar, electron microscopy analyses showed that there are actually two types of pericycle cell. The pericycle cells located in front of the phloem poles are highly vacuolated, differentiated cells, whereas the xylem pole pericycle cells (XPP) display meristematic features, such as a dense cytoplasm and small fragmented vacuoles (10, 11).

In *Arabidopsis*, the first divisions leading to lateral root formation occur in the mature part of the root, several millimeters above the RAM. This event, called initiation, involves three pairs of adjacent XPP founder cells and is followed by a stereotyped pattern of development (2, 12, 13). The lateral root founder cells are selected from all the other XPP cells very early on in the basal meristem by an endogenous "clock-like" process called priming (14). The first visible event following priming is the migration of the nuclei of the pairs of primed founder cells towards the common cell walls (14, 15). This is immediately followed by anticlinal asymmetric division that gives birth to two neighboring small daughter cells surrounded by two bigger ones. This corresponds to

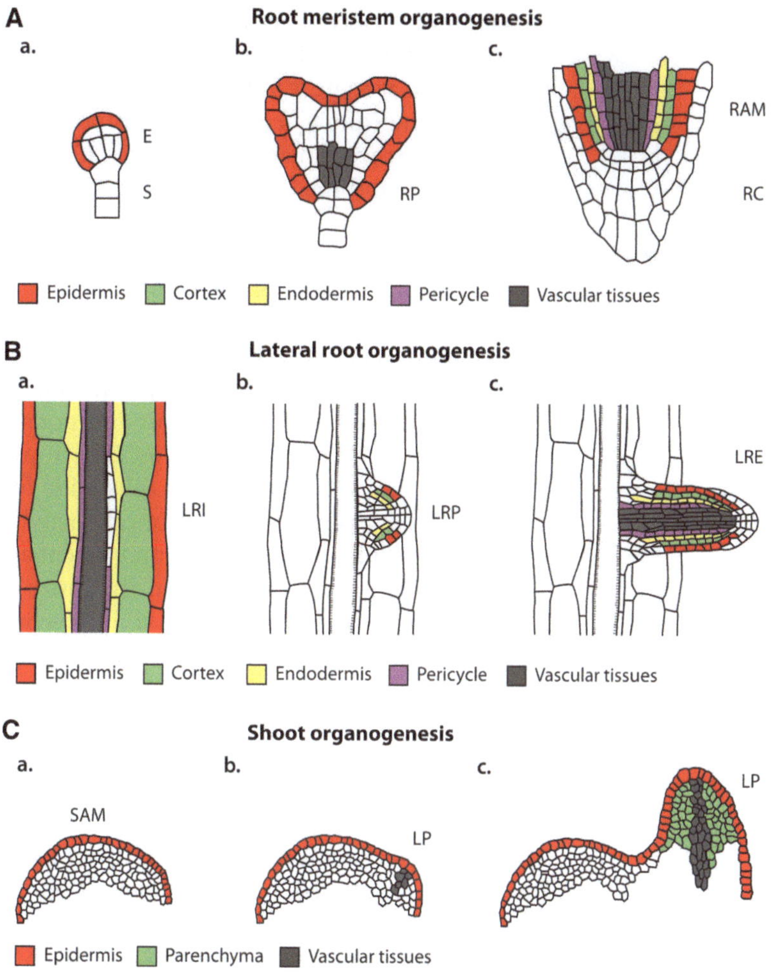

Fig. 1. Organogenesis of root meristems, lateral roots, and shoot organs in *Arabidopsis thaliana*, a dicotyledonous model plant. (**A**) Root meristem organogenesis: (a) globular embryo (E, embryo proper; S, suspensor); (b) heart-stage embryo with a defined root pole (RP); (c) postembryonic root apical meristem (RAM) and root cap (RC). (**B**) Lateral root organogenesis: (a) lateral root initiation (LRI) in the pericycle of the primary root; (b) development of a lateral root primordium (LRP; stage VI) and beginning of cell differentiation; (c) a lateral root (LR) after emergence and meristem activation. (**C**) Shoot organogenesis: (a) initiation of a new lateral organ at the flank of the shoot apical meristem (SAM) is not readily visible; (b) bulge of a new leaf primordium (LP); (c) outgrowth of the leaf primordium (LP) and progressive differentiation of leaf tissues.

the initiation event. Additional rounds of asymmetric divisions in the two biggest cells lead to a stage I LR primordium composed of 4–10 adjacent small cells (Fig. 1B-a). Those cells then divide periclinally to form a two-layered primordium (Stage II). A sequence of other anti- and periclinal divisions forms a multilayered dome-shaped primordium (Stage III–VII; Fig. 1B-b) that protrudes into the external tissues (endodermis, cortex, and epidermis). The cells in the primordium start to differentiate and acquire different identities depending on their position in the primordium (2, 12), leading to recognizable root meristem organization at stage VI. This process corresponds to the patterning of the primordium. Finally the primordium emerges (Stage VIII) through the primary root tissues (2, 16). The meristem is then activated and the lateral root begins to elongate (Fig. 1B-c).

2.2. The Formation of Lateral Shoot Organs in Arabidopsis

The formation of the aerial lateral organs such as leaves and flowers occurs in the flank of the SAM following a robust repetitive phyllotactic pattern. The SAM is a dome-shaped structure divided into several layers and zones (17). The two outermost layers (L1 and L2) in which cells always divide anticlinally form the tunica. Below the tunica, cells divide in various directions and form a less organized cellular mass called the corpus. The top of the dome, called the central zone, contains stem cells with low mitotic activity (18). These cells maintain their mitotically active neighboring cells located in the peripheral zone in an undifferentiated state (18, 19). Shoot lateral organs are produced in this latter zone following a regular phyllotactic pattern. The first visible events of leaf primordium formation are the periclinal divisions that generate a bulge at a precise location in the peripheral zone of the SAM (Fig. 1C-b) (17). Additional divisions in the bulge allow the primordium to grow out while it exits the meristem (Fig. 1C-c) (17). The polarity and morphology of the organ are then progressively set up (Fig. 1C-c) (20).

Shoot lateral organs are initiated by the SAM, whereas lateral roots are initiated in the mature part of the root. The development of shoot and root lateral organs were thus long supposed to involve different mechanisms to define the sites of organ initiation. However, the recent discovery that pericycle cells are primed in the root basal meristem suggests that these two developmental programs are more similar than previously thought.

3. Many Features Are Common to Both Root and Shoot Organ Formation

Two key questions surround organogenesis of the leaf and lateral root. First, how is the initiation site of the new organ determined? Second, how does the cellular population deriving from a very limited number of dividing founder cells organize itself into a well-patterned organ? Comparing the organogenesis processes in shoots and roots has shown that in many ways these two different processes are actually very similar. Most remarkably, the phytohormone auxin has been shown to play a key role in controlling organogenesis in both the root and the shoot. Auxin has been shown to be able to trigger the formation of new shoot and root organs, to form morphogenetic gradients and to activate the expression of master regulators both in the shoot and the root (see below). Furthermore, genetic approaches have unravelled the downstream genetic network controlling the postembryonic initiation and patterning of plant organs. Once again, they highlighted many similarities at the molecular level between the organogenesis of shoot and root organs.

3.1. The Accumulation of Auxin at the Organ Initiation Sites Depends on Its Polar Transport and Triggers Organogenesis

Many studies have shown that the phytohormone auxin plays a key role in shoot lateral organ formation in plants. Studies on *Epilobium* showed that exogenous auxin application can modify the phyllotactic pattern and therefore controls organogenesis (21). However, the instructive role of auxin in the organogenesis of shoot lateral organs was demonstrated only a decade ago (22). In their study of the role of auxin in phyllotaxis, Reinhardt et al. showed that a mutation (*pin-formed1*, *pin1*) or pharmacological treatments (using naphthyl phthalamic acid, NPA) impairing auxin transport blocked shoot organogenesis in *Arabidopsis* or tomato, respectively. In addition, local application of auxin on the shoot apical meristem of these plants was sufficient to restore organogenesis. Auxin accumulation in some meristematic cells of the peripheral zone is therefore necessary and enough to trigger organ initiation. Importantly, the amount of auxin and site of application in the peripheral zone of the meristem influenced the size of the primordium and its precise location, respectively (22–24). This demonstrates the role of auxin as a trigger and as an instructive signal for shoot organogenesis. In the wild-type plants, non uniform auxin distribution in the shoot apical meristem and especially auxin accumulation at sites of future initiation in the peripheral zone depend on PIN1 auxin transporter activity (25–27). Several studies explored the dynamics of *PIN1* expression and polar localization in the SAM, controlling rhythmic auxin accumulation (monitored by auxin-responsive *DR5::GFP* expression) at sites in the peripheral zone where this triggers changes in gene expression and organ initiation (24, 26, 27). Several observations of auxin feedback on *PIN1* expression and polarization suggest the existence of feedback loops between auxin transport, auxin distribution and auxin signalling cascades that could generate the self-organizing properties of organogenesis in the shoot meristem (28, 29). Based on these observations, the mathematical models generated could effectively simulate the phyllotactic pattern (25, 26, 30). In addition, supplemental hypotheses about the regulation of PIN1 polarization allowed these models to recapitulate both organ generation and organ patterning with respect to mid-vein formation and leaf serration (30–35).

The current model for the functioning of the SAM is that auxin triggers the formation of the primordium and generates the phyllotactic pattern by controlling its own distribution through the dynamic polar localization of its transporters. Two models for the mechanism governing PIN1 relocalization are proposed and they successfully explain the data (3).

In parallel to these studies showing the central role played by auxin in the organogenesis of shoot lateral organs, similar studies have also highlighted the role of auxin and of polar transport in the organogenesis of lateral roots and the histogenesis of the primary root by the RAM. Large amounts of exogenous auxin promote the termination of the RAM and the formation of lateral roots (36), suggesting a role for auxin in controlling lateral root initiation and the maintenance of the RAM. As in the shoot, blocking polar auxin transport has been shown to block lateral root initiation (37). This shows that, as in shoot organogenesis, polar auxin transport is crucial for LR organogenesis. Analyzing the expression profile of *PIN* genes and subcellular localization of PIN proteins has consistently showed that several of the *PIN* genes are up-regulated at the lateral root initiation sites, in particular *PIN1*, and that PIN proteins are polarized toward the center of the primordium during early stages and toward the tip of the primordium at later stages (1). Analyses of *DR5* promoter activity and auxin immunolocalization experiments suggested that auxin accumulates at the sites of LR initiation and later in the tip of LR primordia (1, 38, 39). This auxin accumulation pattern was found to be modified or even abolished in the *pin1* mutant or in NPA-treated seedlings as shown by the enlarged staining profile of *DR5::GUS* (1, 39). This demonstrates that as in the shoot, the accumulation of auxin at the organ initiation sites is strongly dependent on its polar transport.

More recently, Dubrovsky et al. (40) have demonstrated that the increase in auxin response monitored by *DR5* in two adjacent XPP cells was a very early marker distinguishing lateral founder cells among the other XPP cells. Using the heat shock inducible Cre-Lox system, the authors showed that any XPP cell could be converted into a LR founder cell in response to local artificial production of auxin. Auxin accumulation in XPP cells is therefore necessary and enough to trigger LR initiation.

Though these data show that auxin accumulation plays a crucial role in lateral root initiation, it does not give any mechanistic explanation for the fact that only a small proportion of XPP pericycle cells starts to accumulate auxin and initiate a LR, whereas the others do not. It also does not explain the regular "rhizotactic" spacing pattern of LR along the primary root (41) and between the opposing xylem poles (42). The mechanism explaining these striking features has recently been found to depend on the auxin response in a zone located just above the RAM and called the basal meristem, therefore linking histogenesis of the primary root and organogenesis of the lateral roots. De Smet et al. (14) showed by monitoring activity of the *DR5* promoter that the auxin response in the young protoxylem in the basal meristem oscillates over time with a frequency that corresponds to the frequency of LR initiation.

These authors showed that auxin maxima in the protoxylem correspond later to LR initiation sites. The *aux1* mutant shows a decreased number of initiation sites and fails to initiate LRs in a regular left-right pattern. The authors showed that this defect correlates with impaired oscillation of the auxin response in the basal meristem. Recently, Moreno-Risueno et al. (43), using a *DR5::LUC* reporter to monitor the change in DR5 activity over time in individual plants, have confirmed that the auxin response oscillates in a zone encompassing the basal meristem and the adjacent proximal elongation zone. Using this tool, the authors also showed that the spots of *DR5* activity are established in the oscillating zone every 6 h by an endogenous mechanism that may itself be controlled by auxin. Interestingly, Moreno-Risueno et al. also identified by micro-array analysis several sets of transcription factors that are potentially involved in the generation of these oscillations. Once the spots of high *DR5* activity are generated, they retain this activity and later systematically colocalize with LR initiation sites. Thus, these recent studies strongly indicate that the periodic increases in auxin response around and in the basal meristem are responsible for priming the lateral root founder cells. Therefore, as in the shoot, patterns of auxin accumulation in the meristem determine the sites of lateral organ initiation in the root.

Interestingly, auxin also acts later during lateral root formation in the outer tissues to locally reprogram cell identity. Swarup et al. (16) showed that expression of *LIKE-AUX1 3* (*LAX3*) encoding an auxin influx carrier is induced in the outer tissues in front of the lateral root primordia by auxin diffusing from the primordia. This generates a positive feedback loop leading to highly localized accumulation of auxin in the cortical and epidermal cells overlaying the primordia, which subsequently makes possible spatially restricted expression of cell wall remodelling enzymes. The local action of degrading enzymes is necessary for the primordia to emerge through the outer cell layers without damaging the parental root. The emergence process requires complex local cell-to-cell communication mechanisms that are not yet fully understood.

To sum up, research on shoot and root lateral organ development has shown that in both cases auxin plays a crucial role in the organogenesis of plant organs. Auxin has been found to accumulate at the sites of lateral organ initiation and to act as a "morphogenetic trigger" for organogenesis (44) (Fig. 2). That is to say that local accumulation of auxin is able to alter the fate of cells by changing cell identity. Biological and modelling approaches have highlighted the essential role played by polar auxin transport and its control by auxin itself in the generation of robust dynamic auxin accumulation patterns. However, the mechanisms generating certain accumulation patterns (such as the oscillating auxin response in the basal meristem) are still unclear and will need further research efforts to be understood (14, 43). Finally, many other endogenous and

a

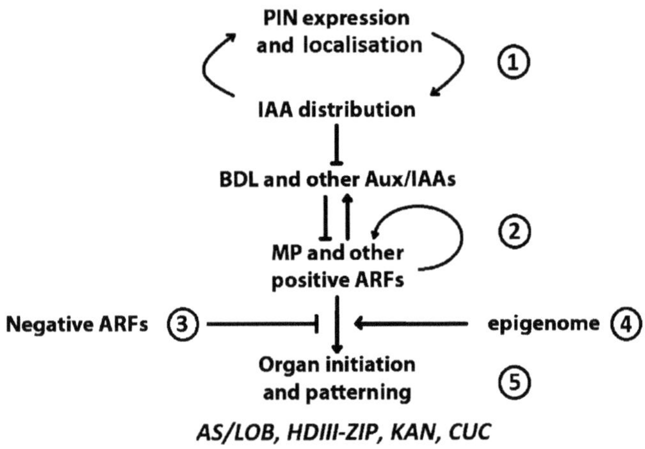

b

	Root organogenesis	Shoot organogenesis
1. Self-organizing auxin distribution pattern	References: ***(1, 14, 37, 112)***	References: ***(1, 23, 25, 27, 113)***
2. Positive ARFs and Aux/IAAs network	*IAA14* ***(72, 101)***; *BDL* ***(5)***, *MP* ***(5)***; *ARF7/19* ***(71, 96)***; *ARF6/8* ***(75)***	*MP* ***(23)***; *ARF7/19* ***(71)***; *ARF7/MP* ***(105)***; *ARF6/8* ***(114)***
3. Negative ARFs	*ARF17* ***(75, 115)***; *ARF2/3/4* ***(76)***	*ARF17* **(115)**; *ARF2/3/4* ***(116, 117)***
4. Importance of epigenomic marks	*PKL* ***(108, 111)***; *GCN5* ***(107)***; *CLF* ***(108)***	*AS1* ***(118)***; *TPL* ***(110)***
5. Downstream pathway	*HDIII-ZIP* and *KAN* ***(4)***; *CUC3* ***(1)***, *NAC1* ***(119)***; *AS/LOB* ***(96, 120)***; *PUCHI* ***(90)***	*AS2* ***(121)***; *CUC* ***(31, 122, 123)***; *KAN* ***(116, 124)***; *HDIII-ZIP* ***(125)***; *PUCHI* ***(126)***

Fig. 2. (**a**) The genetic pathways controlling different organogenetic events in plants share many of their components. Organ initiation sites are determined by self-regulating auxin distribution dependent on auxin polar transport by PIN proteins (1). Auxin signals through positive ARF-Aux/IAA modules (2). Activity of the positive ARFs is modulated by negative ARFs (3) and by the epigenome (4). Positive ARFs activate genetic cascades that lead to organ initiation and patterning. Many of the genes involved are shared between the different organogenesis pathways (5). (**b**) Examples of components of the same gene family shared between the shoot and root organogenesis processes.

environmental cues have been shown to influence auxin biosynthesis, auxin transport or auxin signalling, providing mechanisms by which various signals could be integrated into the regulation of shoot and root organogenesis (3, 13, 45–47).

3.2. Auxin Acts as a Morphogen During Organogenesis

Besides its role in triggering organogenesis in plants, it has also been suggested that auxin plays an important role in the patterning of newly formed organs by influencing cell fate in a concentration-dependent manner. For this reason, auxin could be compared to animal morphogens (44). Morphogens are moving signals that influence cell fate and act at low concentrations, and whose effect on competent cells (i.e., cells able to perceive the morphogen) depends on their concentration.

Direct and indirect mapping of auxin concentration in developing organs have shown that during shoot lateral organ formation, root histogenesis at the RAM and lateral root formation, auxin is not evenly distributed. Auxin accumulates at the organ tips and forms concentration gradients (1, 48, 49). Disrupting these gradients has been shown to affect polarity as well as morphological and histological patterning of the organs. For example, *pin1* and *35S::PIN1* plants fail to restrict the auxin maxima at the tip of LR primordia (1). As a consequence, division patterns in the primordia are affected in these lines. This leads to the formation of flattened lateral root primordia with badly defined tips. This phenotype is intensified in the *pin3pin7* and *pin1pin3pin4* multiple mutants (1). The authors showed that this phenotype is accompanied by a misexpression of several cell identity markers and this thus indicates a clear link between auxin concentration patterns and cell identity. The very same morphogenetic mechanisms were also found to occur during the organogenesis of various shoot derived organs such as leaves, flower organs and ovules (1).

An additional example suggesting that auxin acts as a morphogen has been provided by the work of Xu et al. (50) on the process of regeneration in the RAM. The authors showed that disruption of auxin fluxes in the RAM by means of laser-induced wounding (ablation of the quiescent center, QC) caused the auxin accumulation pattern to change in the root tip and triggered the formation of a new, more basipetal, auxin maximum. This was followed by reprogramming of cells in the root tip. In particular, expression of the QC marker *WUSCHEL-RELATED HOMEOBOX 5* (*WOX5*) (6) shifted to the new auxin maximum and subsequently the new auxin distribution pattern was stabilized by changes in *PIN* gene expression and polarity. This showed that, in the RAM, cell fate is influenced by local auxin concentration. This is in full accordance with the hypothesis that auxin acts as a morphogen.

More recently, Ding and Friml (51) have studied the impact of auxin on columella stem cells (CSC). These cells, located below and in contact with the QC, divide asymmetrically to regenerate themselves and produce daughter cells that differentiate into statocytes, as shown by the accumulation of starch granules. Laser ablation experiments of the QC showed that auxin plays an active role in maintaining CSC in an undifferentiated state (52). However, our knowledge about the signal(s) involved in this process remains fragmented. The *WOX5* gene was found to be expressed in the QC and to have a non-cell autonomous effect on CSC differentiation (6). *WOX5* expression is controlled by the SHORT ROOT (SHR)/SCARECROW (SCR) pathway on the one hand (6) and the ARABIDOPSIS CRINKLY4 (ACR4) receptor-like kinase-CLAVATA3/EMBRYO SURROUNDING REGION-RELATED 40 (CLE40) peptide pathway on the other hand (53). Ding and Friml (51) observed that the *pin3* and *pin4* mutants, as well as auxin biosynthesis mutants, showed decreased auxin accumulation in columella cells as shown by reduced activity of the *DR5* promoter. This decrease in auxin concentration is accompanied by an abnormally elevated number of CSC. On the other hand, increased auxin accumulation at the root tip in plants treated with the auxin transport inhibitor NPA or the auxin analogue naphthalene acetic acid (NAA) caused the CSC to differentiate, as shown by the presence of starch granules. This suggested that at high concentrations auxin promotes the differentiation of the columella stem cells. This effect of auxin was subsequently linked with an auxin-mediated decrease in the expression of *WOX5* in the QC, therefore linking the fate of CSC to the auxin concentration in the QC (51).

The clearest example of the morphogenic action of auxin was brought by studies carried out on the *PLETHORA* (*PLT*) gene family (54, 55). These genes encode auxin-inducible transcription factors that are expressed in the RAM. Interestingly, their expression levels follow the auxin gradient and their impact on the cell depends on their dose. At high doses, *PLT*s were shown to promote stem cell identities. At lower concentrations, these genes promote cell proliferation. Finally, cells can undergo differentiation only once the level of *PLT* gene expression is very low (55). Therefore auxin can influence cell fates in a concentration-dependent manner in the RAM by modulating the expression of the *PLT* genes.

In developing organs, auxin concentration forms gradients that act as instructive morphogenetic patterns, providing positional cues that dictate cell identities. As discussed below, auxin was shown to control organ patterning by activating various developmental programs. In many ways, the genetic networks controlling the organogenesis of shoot and root lateral organs in plants are very comparable.

3.3. A Gene Regulatory Network Downstream of Auxin Controlling Organ Initiation and Patterning

Auxin probably influences cell activity by different mechanisms, but the best-known auxin signalling pathway influencing cell identity involves the auxin receptor TRANSPORT INHIBITOR RESPONSE 1 (TIR1) F-box protein (56) and its homologs AUXIN RECEPTOR F-BOX PROTEINS (AFB1-5) (57). These proteins are part of E3-ubiquitin ligase complexes (complexes $SCF^{TIR/AFB}$) and are involved in the recognition of short-lived nuclear proteins of the Aux/IAA family (38, 58, 59). At low auxin concentrations, Aux/IAA proteins can form heterodimers with a subgroup of the transcriptional regulators called AUXIN RESPONSE FACTORS (ARFs), repressing their action on their target genes (60–63). When auxin is present, $SCF^{TIR/AFB}$ complexes interact with Aux/IAAs leading to their poly-ubiquitinylation and to their targeting to the 26S proteasome-dependent degradation pathway (64). High auxin concentrations thus release positive ARF inhibition by Aux/IAAs and enable these ARFs to induce the expression of their target genes (60). The other ARFs, known as negative ARFs, do not strongly interact with the repressors Aux/IAA and have been shown to be negative regulators of transcription (60, 63, 65). These ARFs can therefore modulate the expression of auxin response genes in an auxin-independent manner.

Not surprisingly, genetic screens for mutants affected in organogenesis led to the identification of several *ARFs* and *Aux/IAAs* as central regulators of plant development. In particular, *ARF5/MONOPTEROS* (*MP*) and *IAA12/BODENLOS* (*BDL*) were identified very early on for the very strong defects in embryogenesis they yield when mutated (66–69). Full loss-of-function *mp* alleles and dominant negative *bdl* alleles were found to result in rootless embryos (70), emphasizing the crucial role played by these genes in primary root organogenesis. Furthermore, these genes were shown to be crucial not only for embryogenesis but also for postembryonic organogenesis. In particular *mp* mutants are not able to initiate flower primordia (23) and present strong defects in lateral root initiation and patterning as well (5), which suggests a general role for *MP* in organogenesis. The *arf7arf19* double loss-of-function mutant and the gain-of-function *iaa14/solitary root* (*slr*) mutant show almost no lateral roots nor lateral root primordia (71–73), demonstrating the role played by the ARF7-ARF19-IAA14 module in controlling lateral root organogenesis. Besides its strong root phenotype, the *arf7arf19* mutants also show a defect in leaf expansion, suggesting that this module is involved in both root and shoot organogenesis (71).

Recent findings in the field of lateral root development have highlighted the multi-modularity of the auxin network controlling LR organogenesis. In addition to the *ARF7-SLR* and *MP-BDL* modules that act successively during initiation and patterning of lateral root primordia, another part of the network has been characterized. This module, involving *IAA28*, has been shown to be essential for priming lateral root founder cells in the basal meristem (15, 74). Recently, a chemical genetics approach has led to the identification of the GATA23 transcription factor as one of its targets (15). It can be reasonably hypothesized that the multi-modularity of the auxin signalling network may be another feature shared with other organogenesis processes.

Many *ARF* genes are posttranscriptionally controlled by miRNA, which are themselves under the transcriptional control of the ARF gene. This yields intricate networks with positive and negative feedback loops, the roles of which have recently been highlighted in the control of organogenesis. For example, Gutierrez et al. (75) unravelled a network between *ARF6*, *ARF8* (positive ARFs), *ARF17* (a negative ARF), and two *miRNA* (*miR160* regulating *ARF17* and *miR167* regulating *ARF6* and *ARF8*). The authors showed that this network yields a subtle light-responsive balance between positive and negative ARF activity in the hypocotyl. In turn, this controls the organogenesis of adventitious roots. Similarly, Marin et al. (76) identified another *ARF/miRNA* transcriptional network involving *ARF2*, *ARF3*, *ARF4* (negative ARFs), *miR390*, and *TAS3a* controlling LR development. When cleaved by *miR390*, *TAS3a* generates trans-acting siRNAs (tasiRNAs) that target *ARF2*, *ARF3*, and *ARF4* transcripts. In turn, the targeted ARFs regulate *miR390* expression providing positive and negative feedback loops. In *TAS3* overexpression lines, lateral root primordia develop faster and lateral roots grow faster than in wild-type plants, whereas the opposite phenotype is observed in *tas3a-1* mutants. Together, these data show the importance of the balance between Aux/IAAs, positive ARFs and negative ARFs in controlling organogenesis in plants. Interestingly, these genes were also shown to be critical for controlling the transition from the vegetative to reproductive phase and the histological patterning of shoot organs (77, 78), once again demonstrating that similar genetic pathways are recruited for root and shoot organogenesis.

The research carried out on plant organogenesis over the last few decades has shown that in many ways, the organogenesis of root and shoot organs are very similar (Fig. 2). In particular, the plant hormone auxin plays an essential role in organogenesis, first as a morphogenetic trigger initiating the organogenesis process, and later as a morphogen during the patterning of the organ (29, 44). Furthermore, when one member of a transcription factor encoding gene family is involved in the postembryonic organogenesis of a certain organ, it is often also involved in embryogenesis or in the organogenesis of other types of organs (Fig. 2B). If not, another closely related member of the family usually is.

Together, these data suggest that the genetic programs controlling shoot and root organogenesis are evolutionary related and therefore a unique model system could be adopted to study organogenesis in plant. In addition, recent data have enhanced the functional significance of the lateral root developmental pathway in plant development by revealing striking similarities between lateral root formation and plant regeneration via callogenesis. It has indeed been shown that the first steps in callus formation are very similar to early lateral root development both at a genetic and morphological level, even if the callus is generated from shoot explants (Fig. 3). Interestingly, the authors showed that a mutation blocking lateral root initiation such as the *alf4-1* mutation also impairs callogenesis (79).

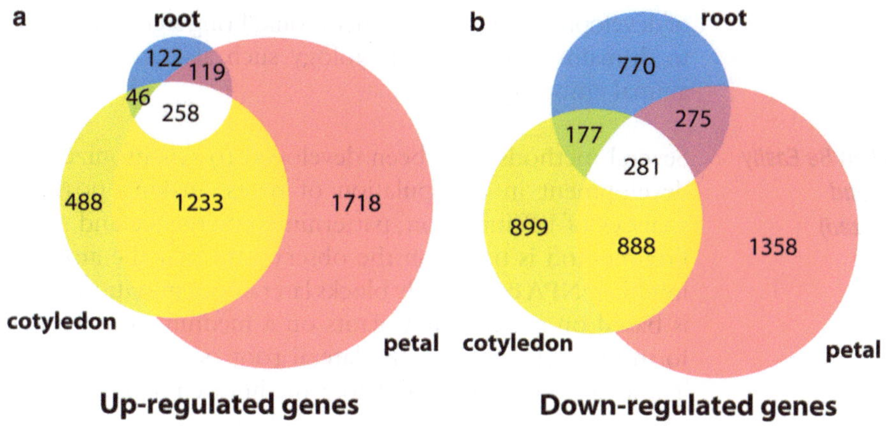

Fig. 3. Gene expression during callus regeneration highlights genetic similarities in organogenesis pathways. (**a**, **b**) Venn diagrams of the differentially expressed genes that exhibit significant upregulation (**a**) or downregulation (**b**) during callus regeneration compared to the tissue of origin of the callus (p value <0.0001, fold change >3 or <−3). Reproduced from ref. (79), with permission from Cell Press.

4. LR Formation as a Model System for studying Organogenesis

Generally speaking, a model system should comply with several criteria. First, studying the model system should allow one to address as many general issues as possible and therefore should allow as far as possible generalization to non-model systems. Second, studying the system should be technically easy (and ideally cheap) in comparison to the other systems thanks to its intrinsic qualities. Third, the system should benefit from as many tools as possible to manipulate and study it. We believe that lateral root formation in *Arabidopsis* is a good model for organogenesis in plants.

4.1. LRs Formation Addresses Many General Relevant Issues

Several questions currently surround the formation of lateral roots and focus research in the field of lateral root development. First, what makes the xylem pole pericycle cells *competent* to respond to the auxin signal? How are the lateral root initiation sites determined (i.e., what is priming and what is the priming signal in the basal meristem)? How is the *polarity* of founder cells determined prior to the first asymmetric division? What are the mechanisms that control the *structural patterning* (shaping) of the primordium? How is the primordium *histologically patterned*? What are the mechanisms that allow the primordium to *emerge* through three layers of outer tissues without causing damages? How does lateral root formation respond to *exogenous cues*, such as soil nutrient availability or soil moisture? Importantly, many of these issues have their counterpart in embryogenesis or other organogenesis programs. All are biologically relevant for our general understanding of development and solving them could bring significant contribution to domains outside plant biology such as stem cell biology and animal embryogenesis.

4.2. LRs Can Be Easily Induced (and Synchronized)

Several methods have been developed to synchronize lateral root development in a population of roots, making possible kinetic analyses of LR initiation, patterning, emergence and growth. The first method is based on the observation that the auxin transport inhibitor NPA completely blocks lateral root initiation. The method is based on germinating plants on a medium containing NPA to form plantlets without any lateral root or lateral root primordia. These are then transferred to a medium enriched with the auxin analogue NAA. This efficiently yields synchronous activation of all the xylem pole pericycle cells to form lateral roots. Lack of specificity is the major drawback of this pharmacological method. This disadvantage is nevertheless somewhat limited by the fact that in the mature part of the root, the xylem pole pericycle has been shown to be particularly responsive to auxin (10, 80). The second drawback of this lateral root induction method is the formation of a continuous series of adjacent primordia, since the whole pericycle is activated. Therefore this method is probably not very well suited to investigating the genes involved in steps occurring much after lateral root initiation. Despite these drawbacks, this method has been found to be very well suited to identifying key genes involved in LR initiation (81, 82).

Several groups have recently published another synchronization method based on the observation that when the *Arabidopsis* root bends in response to a gravistimulus or mechanical stimulus, a lateral root is almost systematically formed at the bend (83–85). Thus, by gravistimulating a population of primary roots (by simply turning the Petri dishes containing the plants), one can synchronize the formation of lateral root primordia and predict their position. The main advantage of this method is that it does not involve the use of any drugs.

4.3. LRs Can Be Easily Visualized and Imaged

One of the principle advantages of studying lateral root formation in the commonly used plant model *Arabidopsis* is the relatively easy phenotyping of root architecture. In *Arabidopsis* the primary root grows first, indefinitely, and lateral secondary roots are initiated quickly after germination. By 10 days, the primary root of Col-0 seedlings can show around 20 emerged laterals. A reliable estimate of the lateral root density—the number of lateral roots per centimeter of primary root—can thus be obtained quickly by counting the number of emerged lateral roots directly on the plate under a binocular microscope. The average primary root length can then be precisely measured by using informatics tools such as ImageJ. Efforts for automation of this kind of large-scale phenotyping methods are currently being made in different laboratories.

For instance, the *Rootrace* software (86) was extended to measure lateral root density. Despite the difficulty of this task, the software performs quite efficiently (87). Other software, such as *EZ-rhizo*, can perform this task too and provide an interesting series of various root architecture parameters (88).

The choice of the *Arabidopsis* root as a model system is of high interest when it comes to more detailed lateral root phenotyping. First of all, the *Arabidopsis* root is thin and almost transparent, which makes lateral root primordia very accessible. After clearing, lateral root primordia can be easily spotted along the primary root under a microscope using DIC optics with no need for any further preparation. Furthermore, thanks to the presence of secondary walls in the xylem vessels and tracheids, the xylem poles can be identified without any need for additional markers. The transparency and thinness of the *Arabidopsis* root also make it very convenient for confocal analysis (15, 84, 89) and defects in LR primordium size, structure or shape are therefore easy to spot (90).The absence of chlorophyll and the scarcity of strongly lignified or suberified tissues limit the background noise in microscopy.

These qualities of the root system also facilitate the use of numerous reporter constructs, expressing the GUS enzyme or fluorescent markers under the control of informative promoters. For example, ARF-dependent auxin signalling activity can be readily assessed in time and space using *DR5::GUS*, *DR5::GFP* or other Aux-IAA—ARF-related artificial constructs (81, 90, 91). Cell mitotic activity can be monitored using *CycB1;1::GUS* or *GFP* lines (81, 90). Cell fate can be studied using cell-specific promoters fused to fluorescent proteins or to the GUS enzyme (12), or enhancer trap lines showing expression of GFP in specific subpopulations of root cells (92). Cell membranes can be highlighted by plasma membrane fluorescent markers. Cell walls in the outer tissues of the root and in the RAM can be stained by readily absorbed fluorochromes such as propidium iodide (93).

The permeability of the root, at least until the differentiated endodermis, makes possible simple staining and whole mount immunolocalization without tissue embedding or sectioning (94). Still, simple and efficient methods for performing root cross-sections have also been developed. This is particularly useful for precisely identifying the tissues in which a gene of interest is expressed during lateral root development (15, 95).

In addition, the root system is readily accessible for dissection, and the ease with which specific organ structures can be visualized and sampled makes possible specific analyses (e.g., transcriptomics) of selected portions of the organ.

The formation of a lateral root is rather stereotyped in *Arabidopsis*, which allowed Malamy and Benfey (12) to define eight developmental stages. Therefore, the detailed quantification of each of these stages along the primary root can readily provide information for studying putative changes in the initiation or emergence rate or in the speed of development (15, 76). This kind of defect can also be assessed more accurately by synchronizing LR formation and following the development kinetics (83).

4.4. Abundant Transcriptomic Data

Over the last few decades, several transcriptomic analyses have been performed to unravel the genetic network controlling LR formation. At the moment, the published transcriptomic datasets for LR formation have focused on the early steps surrounding initiation. They provided and still provide very valuable information for ongoing research in the field.

Vanneste et al. (81) made use of the auxin-mediated LR synchronization method and of transcriptomic analyses with ATH1 Affymetrix chips to identify genes differentially regulated during initiation. The authors compared the response of Col-0 and *slr-1* (dominant negative allele of *IAA14* that blocks lateral root initiation) at three time points following the transfer onto NAA. Their analysis not only shows that auxin activates the cell cycle but also that it is necessary for specifying different cell identities during the first asymmetric divisions. Using the same chip, Okushima et al. (73) compared the auxin response of Col-0, *arf7*, *arf19*, and *arf7arf19* (whole seedlings). This dataset led to the identification of several LOB domain transcription factors as major direct targets of ARF7 and ARF19 during lateral root formation (96). In order to monitor the transcriptome variations that occur specifically in the xylem pole pericycle cells during LR initiation, De Smet et al. (82) combined the auxin-mediated LR induction with a fluorescence-activated cell sorting approach using the J0121 enhancer trap line (*see* Subheading 4.5 for further details). By combining their list of differentially expressed genes with several other datasets available, the authors identified a receptor-like kinase regulating asymmetric division during lateral root initiation (82).

Lately, Parizot et al. (97) have integrated these various datasets together with cell cycle (98) and spatial expression data (99) to create a user-friendly tool, VisuaLRTC that greatly facilitates the use of all these available data for studying lateral root initiation.

4.5. Tools for Manipulating Gene Expression During LR Development

Another useful tool for studying lateral root development is the collection of well-characterized GAL4 enhancer trap lines. These lines have been generated by genomic insertions of a T-DNA cassette containing the *GAL4* yeast gene—which encodes a transcription factor—driven by a minimal *35S* promoter, and the *GFP* gene under the synthetic Upstream Activating Sequence (*UAS*) promoter. GAL4 strongly binds the *UAS* promoter and can therefore activate the expression of the *UAS::GFP* construct when expressed. The *GAL4* expression pattern depends on recruitment by *GAL4* of local enhancers around the insertion site. Laplaze et al. (100)

screened a population of GAL4 enhancer trap lines and identified two particularly useful lines for studying lateral root development, namely, J0121—which displays a strong GFP signal in the xylem pole pericycle—and J0192—in which *GFP* is expressed specifically in young lateral root primordia. Interestingly, these enhancer trap lines can be used to drive the expression of any gene in the *GAL4* expression domain by simply introducing into the appropriate line a construct containing the gene of interest fused to the *UAS* promoter. This is particularly convenient for performing tissue-specific mutant complementation or tissue-specific RNAi expression for example (15), or for modifying hormonal responses in only a specific tissue type (101).

4.6. Modelling Tools

New integrative approaches have recently proven to be successful for understanding the mechanism controlling lateral root development (41, 83). For instance, using stochastic modelling, Lucas et al. (41) looked for probabilistic rules in the distribution of lateral primordia along the primary root. Their analysis showed a trade-off between initiation and emergence processes that could be successfully predicted by a mechanistic model based on auxin fluxes. Another model, based on the impact of curvature on cell size, was also proposed by Laskowski et al. to explain the pattern of lateral root distribution along the primary root (102). Szymanowska-Pulka and Nakielski (103) modelled cell division and growth using a tensor-based model. Interestingly, their purely mechanical model was able to generate a lateral root with a correct cellular organization, suggesting that the patterning of lateral root primordia could be based on cellular perception of mechanical forces. More generally speaking, a large variety of system approaches can be used to model lateral root development at the tissue or whole root scale, as has recently been reviewed by Lucas et al. (104).

5. Discussion–Conclusion

How do organs develop? This question has attracted considerable research in the field of multicellular eukaryote biology and much progress has been made over the last two decades thanks to technical advances. The research performed on plant organogenesis has notably shown that although the formation of the various organs of the plant during embryogenesis and postembryonic development may seem at first sight quite different, many genetic pathways are actually common to all these organogenesis processes.

This suggests that a model system could be adopted for studying organogenesis in plants. We believe that lateral root formation in *Arabidopsis* could be a very well-suited model system for several reasons. First it addresses many relevant questions that transcend the field of root and plant development. Second, it is easily accessible, can be synchronized and therefore can be easily studied. Third, it benefits from a number of useful tools for manipulating gene expression, such as the GAL4 enhancer trap lines as well as several publicly available transcriptomic datasets.

One current research question in the field of plant development concerns the way auxin differentially affects cells depending on its concentration and on cell type. The discovery of partially overlapping auxin signalling modules—notably during lateral root development—have improved our understanding of this intriguing aspect of auxin signalling (5, 105, 106). However, this has also raised many questions: how do these signalling modules regulate each other? Could different doses of the various signalling modules determine the fate of cells? Identifying the downstream components of the different auxin signalling modules will be of great help in addressing these crucial issues, as will mathematical modelling of the corresponding genetic networks. Because of its numerous experimental advantages, LR development could definitely help solve these questions.

In addition to the existence of several auxin signalling modules, the role of epigenomic marks in the perception of auxin at the transcriptional level also seems to be extremely important. The *PLT* genes are known to be dose-dependent master regulators of root development downstream of auxin (55). Recently, Kornet and Scheres (107) have showed that mutations in the genes that encode the GCN5 histone acetyltransferase and its associated cofactor caused a strong decrease in the expression of *PLT*s at the root apex, though the auxin gradient remained undisturbed in these lines. Similarly, mutations in *PICKLE* and *CURLY LEAF*, two chromatin remodelling factors, have recently been shown to affect RAM activity and the expression of marker genes such as *PLT*s and *WOX5* without altering the auxin distribution pattern (108). Taken together this shows that modulation of expression of key patterning genes such as *PLT* by auxin is dependent on the epigenome.

Like the auxin signal, the epigenome is also known to be crucial for determining cell identity. How do these two mechanisms interact with each other? It is already known that at low auxin concentrations, the primary auxin-responsive genes are epigenomically silenced by a mechanism dependent on the recruitment by Aux/IAA of the co-repressor TOPLESS (109, 110), which itself recruits the histone deacetylase HDA19. But together these data do not fully explain the differentiation process. Once again, studying the role of epigenomic changes during lateral root formation could provide significant knowledge on this subject (111).

Acknowledgments

Work in our laboratory is supported by the IRD, the Université Montpellier 2 (grant from the Conseil Scientifique to SG and LL), the Région Languedoc-Roussillon (grant "Chercheur d'Avenir" to LL and SG), and the Agropolis Fondation (Rhizopolis federative grant).

References

1. Benková E, Michniewicz M, Sauer M et al (2003) Local, efflux-dependent auxin gradients as a common module for plant organ formation. Cell 115:591–602
2. Péret B, De Rybel B, Casimiro I et al (2009) Arabidopsis lateral root development: an emerging story. Trends Plant Sci 14:399–408
3. Vernoux T, Besnard F, Traas J (2010) Auxin at the shoot apical meristem. Cold Spring Harb Perspect Biol 2:a001487
4. Hawker NP, Bowman JL (2004) Roles for class III HD-Zip and KANADI genes in Arabidopsis root development. Plant Physiol 135:2261–2270
5. De Smet I, Lau S, Voss U et al (2010) Bimodular auxin response controls organogenesis in Arabidopsis. Proc Natl Acad Sci USA 107:2705–2710
6. Sarkar AK, Luijten M, Miyashima S et al (2007) Conserved factors regulate signalling in *Arabidopsis thaliana* shoot and root stem cell organizers. Nature 446:811–814
7. Dolan L, Janmaat K, Willemsen V et al (1993) Cellular organisation of the *Arabidopsis thaliana* root. Development 119:71
8. De Smet I, Lau S, Mayer U et al (2010) Embryogenesis – the humble beginnings of plant life. Plant J 61:959–970
9. Peris CIL, Rademacher EH, Weijers D (2010) Green beginnings – pattern formation in the early plant embryo. Curr Top Dev Biol 91:1–27
10. Himanen K, Vuylsteke M, Vanneste S et al (2004) Transcript profiling of early lateral root initiation. Proc Natl Acad Sci USA 101:5146–5151
11. Parizot B, Laplaze L, Ricaud L et al (2008) Diarch symmetry of the vascular bundle in Arabidopsis root encompasses the pericycle and is reflected in distich lateral root initiation. Plant Physiol 146:140–148
12. Malamy JE, Benfey PN (1997) Organization and cell differentiation in lateral roots of *Arabidopsis thaliana*. Development 124:33
13. Guyomarc'h S, Lucas M, Laplaze L (2010) Lateral/secondary roots. In: Encyclopedia of life sciences (ELS). John Wiley & Sons, Ltd, Chichester, UK
14. De Smet I, Tetsumura T, De Rybel B et al (2007) Auxin-dependent regulation of lateral root positioning in the basal meristem of Arabidopsis. Development 134:681–690
15. De Rybel B, Vassileva V, Parizot B et al (2010) A novel aux/IAA28 signaling cascade activates GATA23-dependent specification of lateral root founder cell identity. Curr Biol 20:1697–1706
16. Swarup K, Benková E, Swarup R et al (2008) The auxin influx carrier LAX3 promotes lateral root emergence. Nat Cell Biol 10:946–954
17. Satina S, Blakeslee AF, Avery AG (1940) Demonstration of the three germ layers in the shoot apex of Datura by means of induced polyploidy in periclinal chimeras. Am J Bot 27:895–905
18. Laufs P, Grandjean O, Jonak C et al (1998) Cellular parameters of the shoot apical meristem in Arabidopsis. Plant Cell 10:1375–1390
19. Mayer KF, Schoof H, Haecker A et al (1998) Role of WUSCHEL in regulating stem cell fate in the Arabidopsis shoot meristem. Cell 95:805–815
20. Husbands AY, Chitwood DH, Plavskin Y et al (2009) Signals and prepatterns: new insights into organ polarity in plants. Genes Dev 23:1986–1997
21. Meicenheimer RD (1981) Changes in Epilobium phyllotaxy induced by N-1-naphthylphthalamic acid and alpha-4-chlorophenoxyisobutyric acid. Am J Bot 68:1139–1154
22. Reinhardt D, Mandel T, Kuhlemeier C (2000) Auxin regulates the initiation and radial position of plant lateral organs. Plant Cell 12:507–518
23. Reinhardt D, Pesce E-R, Stieger P et al (2003) Regulation of phyllotaxis by polar auxin transport. Nature 426:255–260

24. Bohn-Courseau I (2010) Auxin: a major regulator of organogenesis. C R Biol 333: 290–296
25. Smith RS, Guyomarc'h S, Mandel T et al (2006) A plausible model of phyllotaxis. Proc Natl Acad Sci USA 103:1301–1306
26. Heisler MG, Ohno C, Das P et al (2005) Patterns of auxin transport and gene expression during primordium development revealed by live imaging of the Arabidopsis inflorescence meristem. Curr Biol 15:1899–1911
27. de Reuille PB, Bohn-Courseau I, Ljung K et al (2006) Computer simulations reveal properties of the cell–cell signaling network at the shoot apex in Arabidopsis. Proc Natl Acad Sci USA 103:1627–1632
28. Paciorek T, Zazimalova E, Ruthardt N et al (2005) Auxin inhibits endocytosis and promotes its own efflux from cells. Nature 435:1251–1256
29. Vanneste S, Friml J (2009) Auxin: a trigger for change in plant development. Cell 136:1005–1016
30. Stoma S, Lucas M, Chopard J et al (2008) Flux-based transport enhancement as a plausible unifying mechanism for auxin transport in meristem development. PLoS Comput Biol 4:e1000207
31. Bilsborough GD, Runions A, Barkoulas M et al (2011) Model for the regulation of *Arabidopsis thaliana* leaf margin development. Proc Natl Acad Sci USA 108:3424–3429
32. Bayer EM, Smith RS, Mandel T et al (2009) Integration of transport-based models for phyllotaxis and midvein formation. Genes Dev 23:373
33. Wabnik K, Kleine-Vehn J, Balla J et al (2010) Emergence of tissue polarization from synergy of intracellular and extracellular auxin signaling. Mol Syst Biol 6:447
34. Sauer M, Balla J, Luschnig C et al (2006) Canalization of auxin flow by Aux/IAA-ARF-dependent feedback regulation of PIN polarity. Genes Dev 20:2902–2911
35. Scarpella E, Marcos D, Friml J et al (2006) Control of leaf vascular patterning by polar auxin transport. Genes Dev 20:1015–1027
36. Torrey JG (1986) Endogenous and exogenous influences on the regulation of lateral root formation. In: Jackson MB (ed) New root formation in plants and cuttings. Martinus Nijhoff, Hingham, MA
37. Casimiro I, Marchant A, Bhalerao RP et al (2001) Auxin transport promotes Arabidopsis lateral root initiation. Plant Cell 13:843–852
38. Ulmasov T, Murfett J, Hagen G et al (1997) Aux/IAA proteins repress expression of reporter genes containing natural and highly active synthetic auxin response elements. Plant Cell 9:1963–1971
39. Sabatini S, Beis D, Wolkenfelt H et al (1999) An auxin-dependent distal organizer of pattern and polarity in the Arabidopsis root. Cell 99:463–472
40. Dubrovsky JG, Sauer M, Napsucialy-Mendivil S et al (2008) Auxin acts as a local morphogenetic trigger to specify lateral root founder cells. Proc Natl Acad Sci USA 105:8790–8794
41. Lucas M, Guédon Y, Jay-Allemand C et al (2008) An auxin transport-based model of root branching in *Arabidopsis thaliana*. PLoS One 3:e3673
42. Dubrovsky JG, Gambetta GA, Hernández-Barrera A et al (2006) Lateral root initiation in Arabidopsis: developmental window, spatial patterning, density and predictability. Ann Bot 97:903–915
43. Moreno-Risueno MA, Van Norman JM, Moreno A et al (2010) Oscillating gene expression determines competence for periodic Arabidopsis root branching. Science 329:1306–1311
44. Benková E, Ivanchenko MG, Friml J et al (2009) A morphogenetic trigger: is there an emerging concept in plant developmental biology? Trends Plant Sci 14:189–193
45. Malamy JE (2005) Intrinsic and environmental response pathways that regulate root system architecture. Plant Cell Environ 28:67–77
46. Fukaki H, Tasaka M (2009) Hormone interactions during lateral root formation. Plant Mol Biol 69:437–449
47. Shani E, Yanai O, Ori N (2006) The role of hormones in shoot apical meristem function. Curr Opin Plant Biol 9:484–489
48. Petersson SV, Johansson AI, Kowalczyk M et al (2009) An auxin gradient and maximum in the Arabidopsis root apex shown by high-resolution cell-specific analysis of IAA distribution and synthesis. Plant Cell 21: 1659–1668
49. Ljung K, Bhalerao RP, Sandberg G (2001) Sites and homeostatic control of auxin biosynthesis in Arabidopsis during vegetative growth. Plant J 28:465–474
50. Xu J, Hofhuis H, Heidstra R et al (2006) A molecular framework for plant regeneration. Science 311:385–388
51. Ding Z, Friml J (2010) Auxin regulates distal stem cell differentiation in Arabidopsis roots. Proc Natl Acad Sci USA 107: 12046–12051

52. van den Berg C, Willemsen V, Hendriks G et al (1997) Short-range control of cell differentiation in the Arabidopsis root meristem. Nature 390:287–289
53. Stahl Y, Wink RH, Ingram GC et al (2009) A signaling module controlling the stem cell niche in Arabidopsis root meristems. Curr Biol 19:909–914
54. Aida M, Beis D, Heidstra R et al (2004) The PLETHORA genes mediate patterning of the Arabidopsis root stem cell niche. Cell 119: 109–120
55. Galinha C, Hofhuis H, Luijten M et al (2007) PLETHORA proteins as dose-dependent master regulators of Arabidopsis root development. Nature 449:1053–1057
56. Dharmasiri N, Dharmasiri S, Estelle M (2005) The F-box protein TIR1 is an auxin receptor. Nature 435:441–445
57. Dharmasiri N, Dharmasiri S, Weijers D et al (2005) Plant development is regulated by a family of auxin receptor F box proteins. Dev Cell 9:109–119
58. Tiwari SB, Hagen G, Guilfoyle TJ (2004) Aux/IAA proteins contain a potent transcriptional repression domain. Plant Cell 16: 533–543
59. Tiwari SB, Wang XJ, Hagen G et al (2001) AUX/IAA proteins are active repressors, and their stability and activity are modulated by auxin. Plant Cell 13:2809–2822
60. Ulmasov T, Hagen G, Guilfoyle TJ (1999) Activation and repression of transcription by auxin-response factors. Proc Natl Acad Sci USA 96:5844–5849
61. Ulmasov T, Hagen G, Guilfoyle TJ (1999) Dimerization and DNA binding of auxin response factors. Plant J 19:309–319
62. Guilfoyle T, Hagen G, Ulmasov T et al (1998) How does auxin turn on genes? Plant Physiol 118:341–347
63. Guilfoyle TJ, Hagen G (2007) Auxin response factors. Curr Opin Plant Biol 10:453–460
64. Gray WM, Kepinski S, Rouse D et al (2001) Auxin regulates SCFTIR1-dependent degradation of AUX/IAA proteins. Nature 414:271–276
65. Tiwari SB, Hagen G, Guilfoyle T (2003) The roles of auxin response factor domains in auxin-responsive transcription. Plant Cell 15:533–543
66. Berleth T, Jurgens G (1993) The role of the monopteros gene in organising the basal body region of the Arabidopsis embryo. Development 118:575
67. Hardtke CS, Berleth T (1998) The Arabidopsis gene MONOPTEROS encodes a transcription factor mediating embryo axis formation and vascular development. EMBO J 17:1405–1411
68. Hamann T, Mayer U, Jurgens G (1999) The auxin-insensitive bodenlos mutation affects primary root formation and apical-basal patterning in the Arabidopsis embryo. Development 126:1387–1395
69. Hamann T, Benkova E, Baurle I et al (2002) The Arabidopsis BODENLOS gene encodes an auxin response protein inhibiting MONOPTEROS-mediated embryo patterning. Genes Dev 16:1610–1615
70. Mayer U, Ruiz RAT, Berleth T et al (1991) Mutations affecting body organization in the Arabidopsis embryo. Nature 353:402–407
71. Wilmoth JC, Wang S, Tiwari SB et al (2005) NPH4/ARF7 and ARF19 promote leaf expansion and auxin-induced lateral root formation. Plant J 43:118–130
72. Fukaki H, Tameda S, Masuda H et al (2002) Lateral root formation is blocked by a gain-of-function mutation in the SOLITARY-ROOT/IAA14 gene of Arabidopsis. Plant J 29:153–168
73. Okushima Y, Overvoorde PJ, Arima K et al (2005) Functional genomic analysis of the AUXIN RESPONSE FACTOR gene family members in *Arabidopsis thaliana*: unique and overlapping functions of ARF7 and ARF19. Plant Cell 17:444–463
74. Rogg LE, Lasswell J, Bartel B (2001) A gain-of-function mutation in IAA28 suppresses lateral root development. Plant Cell 13:465–480
75. Gutierrez L, Bussell JD, Pacurar DI et al (2009) Phenotypic plasticity of adventitious rooting in Arabidopsis is controlled by complex regulation of AUXIN RESPONSE FACTOR transcripts and microRNA abundance. Plant Cell 21:3119–3132
76. Marin E, Jouannet V, Herz A et al (2010) miR390, Arabidopsis TAS3 tasiRNAs, and their AUXIN RESPONSE FACTOR targets define an autoregulatory network quantitatively regulating lateral root growth. Plant Cell 22:1104–1117
77. Fahlgren N, Montgomery TA, Howell MD et al (2006) Regulation of AUXIN RESPONSE FACTOR3 by TAS3 ta-siRNA affects developmental timing and patterning in Arabidopsis. Curr Biol 16:939–944
78. Garcia D, Collier SA, Byrne ME et al (2006) Specification of leaf polarity in Arabidopsis via the trans-acting siRNA pathway. Curr Biol 16:933–938
79. Sugimoto K, Jiao Y, Meyerowitz EM (2010) Arabidopsis regeneration from multiple tissues

occurs via a root development pathway. Dev Cell 18:463–471
80. Himanen K, Boucheron E, Vanneste S et al (2002) Auxin-mediated cell cycle activation during early lateral root initiation. Plant Cell 14:2339–2351
81. Vanneste S, De Rybel B, Beemster GTS et al (2005) Cell cycle progression in the pericycle is not sufficient for SOLITARY ROOT/IAA14-mediated lateral root initiation in *Arabidopsis thaliana*. Plant Cell 17:3035–3050
82. De Smet I, Vassileva V, De Rybel B et al (2008) Receptor-like kinase ACR4 restricts formative cell divisions in the Arabidopsis root. Science 322:594–597
83. Lucas M, Godin C, Jay-Allemand C et al (2008) Auxin fluxes in the root apex co-regulate gravitropism and lateral root initiation. J Exp Bot 59:55–66
84. Ditengou FA, Teale WD, Kochersperger P et al (2008) Mechanical induction of lateral root initiation in *Arabidopsis thaliana*. Proc Natl Acad Sci USA 105: 18818–18823
85. Richter GL, Monshausen GB, Krol A et al (2009) Mechanical stimuli modulate lateral root organogenesis. Plant Physiol 151: 1855–1866
86. French A, Ubeda-Tomás S, Holman TJ et al (2009) High-throughput quantification of root growth using a novel image-analysis tool. Plant Physiol 150:1784–1795
87. Naeem A, French AP, Wells DM et al (2011) High-throughput feature counting and measurement of roots. Bioinformatics 27:1337–1338
88. Armengaud P, Zambaux K, Hills A et al (2009) EZ Rhizo: integrated software for the fast and accurate measurement of root system architecture. Plant J 57:945–956
89. Lucas M, Swarup R, Paponov IA et al (2011) Short-root regulates primary, lateral, and adventitious root development in Arabidopsis. Plant Physiol 155:384–398
90. Hirota A, Kato T, Fukaki H et al (2007) The auxin-regulated AP2/EREBP gene PUCHI is required for morphogenesis in the early lateral root primordium of Arabidopsis. Plant Cell 19:2156–2168
91. Vernoux T, Brunoud G, Farcot E et al (2011) The auxin signalling network translates dynamic input into robust patterning at the shoot apex. Molecular System Biology 7:508
92. Haseloff J (1999) GFP variants for multispectral imaging of living cells. Methods Cell Biol 58:139–151
93. Geldner N, Denervaud□Tendon V, Hyman DL et al (2009) Rapid, combinatorial analysis of membrane compartments in intact plants with a multicolor marker set. Plant J 59:169–178
94. De Smet I, Chaerle P, Vanneste S et al (2004) An easy and versatile embedding method for transverse sections. J Microsc 213:76–80
95. Sauer M, Paciorek T, Benkova E et al (2006) Immunocytochemical techniques for whole-mount in situ protein localization in plants. Nat Protoc 1:98–103
96. Okushima Y, Fukaki H, Onoda M et al (2007) ARF7 and ARF19 regulate lateral root formation via direct activation of LBD/ASL genes in Arabidopsis. Plant Cell 19:118–130
97. Parizot B, De Rybel B, Beeckman T (2010) VisuaLRTC: a new view on lateral root initiation by combining specific transcriptome data sets. Plant Physiol 153:34–40
98. Menges M, Hennig L, Gruissem W et al (2003) Genome-wide gene expression in an Arabidopsis cell suspension. Plant Mol Biol 53:423–442
99. Brady SM, Orlando DA, Lee J-Y et al (2007) A high-resolution root spatiotemporal map reveals dominant expression patterns. Science 318:801–806
100. Laplaze L, Parizot B, Baker A et al (2005) GAL4-GFP enhancer trap lines for genetic manipulation of lateral root development in *Arabidopsis thaliana*. J Exp Bot 56:2433–2442
101. Fukaki H, Nakao Y, Okushima Y et al (2005) Tissue-specific expression of stabilized SOLITARY-ROOT/IAA14 alters lateral root development in Arabidopsis. Plant J 44:382–395
102. Laskowski M, Grieneisen VA, Hofhuis H et al (2008) Root system architecture from coupling cell shape to auxin transport. PLoS Biol 6:e307
103. Szymanowska-Pułka J, Nakielski J (2010) The tensor-based model for growth and cell divisions of the root apex. II. Lateral root formation. Planta 232:1207–1218
104. Lucas M, Laplaze L, Bennett MJ (2011) Plant systems biology: network matters. Plant Cell Environ 34:535–553
105. Hardtke CS, Ckurshumova W, Vidaurre DP et al (2004) Overlapping and non-redundant functions of the Arabidopsis auxin response factors MONOPTEROS and NONPHOTOTROPIC HYPOCOTYL 4. Development 131:1089–1100
106. Weijers D, Benkova E, Jäger KE et al (2005) Developmental specificity of auxin response

by pairs of ARF and Aux/IAA transcriptional regulators. EMBO J 24:1874–1885

107. Kornet N, Scheres B (2009) Members of the GCN5 histone acetyltransferase complex regulate PLETHORA-mediated root stem cell niche maintenance and transit amplifying cell proliferation in Arabidopsis. Plant Cell 21:1070–1079
108. Aichinger E, Villar CBR, Di Mambro R et al (2011) The CHD3 chromatin remodeler PICKLE and polycomb group proteins antagonistically regulate meristem activity in the Arabidopsis root. Plant Cell 23:1047–1060
109. Szemenyei H, Hannon M, Long JA (2008) TOPLESS mediates auxin-dependent transcriptional repression during Arabidopsis embryogenesis. Science 319:1384–1386
110. Long JA, Ohno C, Smith ZR, Meyerowitz EM (2006) TOPLESS regulates apical embryonic fate in Arabidopsis. Science 312:1520–1523
111. Fukaki H, Taniguchi N, Tasaka M (2006) PICKLE is required for SOLITARY-ROOT/IAA14-mediated repression of ARF7 and ARF19 activity during Arabidopsis lateral root initiation. Plant J 48:380–389
112. Blilou I, Xu J, Wildwater M, Willemsen V et al (2005) The PIN auxin efflux facilitator network controls growth and patterning in Arabidopsis roots. Nature 433:39–44
113. Jönsson H, Heisler MG, Shapiro BE et al (2006) An auxin-driven polarized transport model for phyllotaxis. Proc Natl Acad Sci USA 103:1633–1638
114. Wu M-F, Tian Q, Reed JW (2006) Arabidopsis microRNA167 controls patterns of ARF6 and ARF8 expression, and regulates both female and male reproduction. Development 133:4211–4218
115. Mallory AC, Bartel DP, Bartel B (2005) MicroRNA-directed regulation of Arabidopsis AUXIN RESPONSE FACTOR17 is essential for proper development and modulates expression of early auxin response genes. Plant Cell 17:1360–1375
116. Pekker I, Alvarez JP, Eshed Y (2005) Auxin response factors mediate Arabidopsis organ asymmetry via modulation of KANADI activity. Plant Cell 17:2899–2910
117. Okushima Y, Mitina I, Quach HL et al (2005) AUXIN RESPONSE FACTOR 2 (ARF2): a pleiotropic developmental regulator. Plant J 43:29–46
118. Guo M, Thomas J, Collins G et al (2008) Direct repression of KNOX loci by the ASYMMETRIC LEAVES1 complex of Arabidopsis. Plant Cell 20:48–58
119. Xie Q, Frugis G, Colgan D et al (2000) Arabidopsis NAC1 transduces auxin signal downstream of TIR1 to promote lateral root development. Genes Dev 14:3024–3036
120. Lee HW, Kim NY, Lee DJ et al (2009) LBD18/ASL20 regulates lateral root formation in combination with LBD16/ASL18 downstream of ARF7 and ARF19 in Arabidopsis. Plant Physiol 151:1377–1389
121. Lin WC, Shuai B, Springer PS (2003) The Arabidopsis LATERAL ORGAN BOUNDARIES-domain gene ASYMMETRIC LEAVES2 functions in the repression of KNOX gene expression and in adaxial-abaxial patterning. Plant Cell 15:2241–2252
122. Aida M, Vernoux T, Furutani M et al (2002) Roles of PIN-FORMED1 and MONOPTEROS in pattern formation of the apical region of the Arabidopsis embryo. Development 129:3965–3974
123. Hibara K, Karim MR, Takada S et al (2006) Arabidopsis CUP-SHAPED COTYLEDON3 regulates postembryonic shoot meristem and organ boundary formation. Plant Cell 18:2946–2957
124. Kerstetter RA, Bollman K, Taylor RA et al (2001) KANADI regulates organ polarity in Arabidopsis. Nature 411:706–709
125. McConnell JR, Emery J, Eshed Y et al (2001) Role of PHABULOSA and PHAVOLUTA in determining radial patterning in shoots. Nature 411:709–713
126. Karim MR, Hirota A, Kwiatkowska D et al (2009) A role for Arabidopsis PUCHI in floral meristem identity and bract suppression. Plant Cell 21:1360–1372

Chapter 4

Genetic Control of Root Organogenesis in Cereals

Caroline Marcon, Anja Paschold, and Frank Hochholdinger

Abstract

Monocot cereals develop a complex root system comprising embryonic roots at an early seedling stage and postembryonic roots which make up the fibrous root system of adult crops. In the model cereals maize, rice, and barley a number of mutants affecting root development have been identified in the past and a subset of the affected genes have been recently cloned and functionally characterized. The present review summarizes genetic and molecular data of cereal root mutants impaired in the elongation or initiation of embryonic and postembryonic roots and the elongation of root hairs for which the affected genes have been recently cloned.

Key words: (Lateral) root, Cereal, Maize, Rice

1. Introduction

Cereals such as rice, maize, wheat, and barley are of outstanding importance as source for human food or livestock feed. Currently, approximately 75% of the human energy demand is provided by cereal starch (1). For rice, it is estimated that by 2030, a 40% increase in production will be required to feed the large and growing number of people mainly living on this crop (2). Despite the outstanding agronomic relevance of cereals, a systematic genetic analysis of their root system development has only recently been initiated (3). This chapter covers the current status of the molecular and genetic dissection of cereal root organogenesis with a focus on recently cloned genes that are known to determine root system architecture as illustrated by their mutant phenotypes.

Ive De Smet (ed.), *Plant Organogenesis: Methods and Protocols*, Methods in Molecular Biology, vol. 959,
DOI 10.1007/978-1-62703-221-6_4, © Springer Science+Business Media New York 2013

2. Morphology of the Cereal Root Architecture

Root architecture of different monocot cereal species is similar and characterized by root type that emerge during embryogenesis and roots that are formed during postembryonic development (4–6). Nevertheless, different cereal species display differences in early seedling root system development already a few days after germination. In Fig. 1a 5-day-old rice, maize, and barley seedling root systems are displayed from the left. At this developmental stage only embryonic roots, i.e., roots that are laid down during embryogenesis, emerged. Rice forms only a single embryonic primary root from the basal pole of the seedling. In contrast, maize forms a primary root and a variable number of seminal roots from an embryonic structure called scutellar node. Similarly, barley develops a primary root and a variable number of seminal roots. However,

Fig. 1. Organization of cereal root systems. (**a**) Embryonic primary (PR) and seminal (SR) roots of 5-day-old rice, maize, and barley seedlings (from *left*). (**b**) Detail of a 10-day-old maize, seedling illustrating shoot-borne crown roots (CR). (**c**) Aboveground brace roots (BR) of adult rice plants.

since primary and seminal roots emerge simultaneously in barley it is not possible to visually distinguish between these different types of embryonic roots as in maize. While embryonic roots of cereals are characteristic for the early seedling root stock, later in development they are often replaced by postembryonic shoot-borne roots that make up the dense fibrous root architecture of crop plants (5, 7). The shoot-borne root system is subdivided into below-ground crown roots (maize as an example in Fig. 1b) and above-ground brace roots (rice as an example in Fig. 1c) (8).

All root types of cereals form lateral roots and root hairs. Lateral root primordia are initiated from pericycle and endodermal cells of all underground roots (9). Aboveground brace roots only form lateral roots after they penetrated the soil (10). Due to their branching capacity lateral roots expand the absorbing root surface and are therefore important for water and nutrient uptake of the plant from the soil (11). The complex root architecture of monocot cereals with a fibrous root system made up by different root types differs from dicot plants such as *Arabidopsis thaliana* that establishes an embryonic primary root with its lateral roots and adventitious roots later in development (12–14). A more detailed comparison of Arabidopsis versus cereal roots is provided in (15).

On the histological level, the longitudinal organization of cereal roots is characterized by a root cap at the terminal end, followed by the meristematic, elongation, and maturation or root hair zone (16). The root apex localized proximal to the root cap, consists of the quiescent center (QC) which is a mitotically inactive region of 800–1,200 cells (17). The QC is surrounded by distal and proximal meristems, which contain several hundred cells (10). The distal meristematic cells initiate new root cap cells, while the proximal meristematic cells provide cells for the elongation zone, in which the new cells stop to divide and start to elongate. The radial patterning of cereal roots is characterized by a cortical parenchyma that consists of several cell layers and a single endodermal cell layer (6, 18).

3. Phenotyping Cereal Root Systems as a Prerequisite for Genetic Analyses

Gene function in root organogenesis can be studied by analyzing altered root phenotypes conditioned by natural variation or by induced mutations. Natural variation of root system architecture is typically manifested by the quantitative variation of traits controlled by multiple genes (19). Recently, several root-related quantitative trait loci (QTL) have been mapped in maize (20–22) and rice (23, 24). In addition to natural variation, alterations of maize root system architecture can be induced by mutagenesis with chemical mutagens such as ethyl methanesulfonate (EMS) (25), transposon tagging (26), or T-DNA insertions (27). Such mutagens often affect only the function of a single gene and lead in many instances to the loss of specific root types (28–30). Large-scale phenotyping of mutagenized populations with respect to root traits is challenging because roots in soil are not directly accessible for phenotypic inspection. Moreover, cereal root systems have a considerable size and complexity and root architecture can be significantly influenced by environmental interactions (31). Therefore, various high-throughput screening procedures have been recently developed to screen large populations of cereal seedlings at different developmental stages under laboratory or field conditions (32).

Phenotyping root system architecture at the seedling stage on germination paper allows screening of large plant populations at a short time period and requires only little space. Moreover, environmental conditions can be tightly controlled in an approach in which surface sterilized kernels are arrayed on germination paper sheets, which are then rolled up and put into beakers with distilled water (28). Such paper rolls can be opened and inspected for aberrant root phenotypes after 10–14 days. Moreover, these seedlings can be subsequently transplanted to soil for propagation. In a similar system maize roots are grown on the surface of blotting paper in pouches and root phenotypes are subsequently captured by a digital scanner and analyzed by image analysis programs (33). For the observation of qualitative changes in root architecture these 2D approaches on germination paper are often sufficient. To study later developmental stages, roots can be grown in soil or gel-based systems. While soil grown roots can be imaged in 3D by X-ray computer tomography (CT) or magnetic resonance imaging (MRI) (32), gel-grown roots can be captured by 3D laser scanning devices (34) or digital cameras (35).

On the field scale root system architecture can be evaluated by excavating and visually inspecting the top portion of the shoot-borne root system (36). In rare instances changes in root system architecture can also be indirectly identified in large-scale field tests of segregating F_2-populations based on their aboveground phenotype as for instance reduced lodging resistance (28). A comprehensive account of currently available tools to phenotype root system architecture is provided by (32).

As a result of such large-scale screening efforts, mutants affected in different aspects of root architecture were identified in cereals. In the following section an account of genes involved in cereal root formation as demonstrated by mutant phenotypes is provided (Table 1). A list of additional cereal mutants affected in root development with not yet cloned genes is provided elsewhere (8).

Table 1
Summary of cloned genes conferring a root type-specific mutant phenotype in maize, rice, and barley

Gene	Encoded protein	Affected developmental process	References
Zea mays			
Rth1	SEC3 homolog	Elongation of root hairs	(28)
Rtcs	LBD protein	Initiation of seminal and shoot-borne roots	(37)
Rth3	COBRA-like protein	Elongation of root hairs	(63)
Rum1	Aux/IAA protein	Initiation of lateral roots at the primary root and seminal roots	(56)
Oryza sativa			
Arl1	LBD protein	Initiation of crown roots, reduced number of lateral roots	(40)
Crl1	LBD protein	Initiation of crown roots, reduced number of lateral roots	(39)
Crl4/Gnom1	ARF-GEF protein	Elongation of crown and lateral roots	(46, 47)
Spr1	Mitochondrial protein	Elongation of crown and lateral roots	(57)
Cand1	Cullin-associated and neddylation-dissociated protein 1	Crown root emergence	(51)
Rth1	Apyrase	Elongation of root hairs, seminal, and crown roots	(66)
Expa17	α-Expansin	Elongation of root hairs	(70)
Hordeum vulgare			
rhl1.a	β-Expansin	Initiation of root hairs	(68)

4. Genetic Control of Cereal Root Organogenesis

4.1. Genes Controlling Embryonic, Lateral, and Shoot-Borne Roots

Embryonic roots are instrumental during the early stages of seedling development. Later in development an extensive shoot-borne root system dominates the adult root stock. The maize mutant *rtcs* (*rootless concerning crown and seminal roots*) is impaired in the initiation of embryonic seminal and the postembryonic shoot-borne roots. The *Rtcs* gene was identified by map-based cloning and encodes an auxin-inducible lateral organ boundary domain (LBD) transcription factor (37). LBDs are DNA-binding regulators of plant lateral organ development (summarized in (38)). *Rtcs* is expressed in emerging crown roots (37). The rice ortholog of *Rtcs* is encoded by the allelic genes *Arl1* (*Adventitious rootless 1*) and *Crl1* (*Crownrootless 1*). Both rice mutants do not form crown roots and display in contrast to the maize mutant *rtcs* fewer lateral roots (39, 40). The reduced number of lateral roots in *arl1/crl1* may be explained by the observation that the *Rtcs* gene has been duplicated in evolution in contrast to its rice ortholog and that the paralogous *Rtcl* might functionally replace *Rtcs* in lateral root development (37). The phytohormone auxin plays an important role in root development (41, 42). The identification of an auxin-responsive element (ARE) in their promotors, their inducibility by auxin, and the binding capacity of auxin response factors (ARF) suggest that *Arl1/Crl1* and *Rtcs* are early auxin response genes (37, 39). For Arabidopsis it was shown that ARF7 and ARF19 directly activate LBD16 and LBD29, both regulators of the lateral root initiation (43). Both genes are closely related to maize *Rtcs* and rice *Arl1/Crl1* (38, 43, 44). Exogenous application of auxin activates cell division in Arabidopsis pericycle cells resulting in the stimulation of lateral root formation (12). Moreover, auxin increases the number of crown roots in rice (39). In contrast, inhibition of auxin transport suppresses the formation of lateral roots in Arabidopsis (45) and reduces the number of crown roots in rice (39).

Further studies on rice root mutants underscored the important role of auxin in crown root formation. In the *crl4/gnom1* mutants, which are affected in the same gene, crown and lateral root emergence are impaired (46, 47). *CRL4/GNOM1* encode a 133 kDa ADP-ribosylation factor (ARF) which is a homolog of the Arabidopsis GNOM protein, a guanine nucleotide exchange factor for ADP-ribosylation factor (ARF-GEF). GNOM is involved in polar auxin transport in Arabidopsis (48). It is suggested that the ARF-GEF protein GNOM regulates the distribution and trafficking of the auxin efflux carrier PIN1 in Arabidopsis cells (48, 49). This notion is supported by mislocation of PIN1 in *gnom* mutants (50). The rice *cand1* mutant displays normal crown root primordia, whereas the emergence of this root type is impaired (51). It was demonstrated that *cand1* mutants are blocked in the G2/M transition of the cell cycle in crown root meristems, thus avoiding crown root emergence. The Arabidopsis *CAND1* homolog encodes a CULLIN-ASSOCIATED AND NEDDYLATION-DISSOCIATED PROTEIN 1 (51) that is required for the activity of the SCF^{TIR1} ubiquitin ligase (52, 53). The SCF^{TIR1} complex is a positive regulator of auxin signalling mediating the ubiquitination of Aux/IAA proteins (54).

The maize mutant *rum1* (*rootless with undetectable meristems 1*) does not initiate seminal roots and lateral roots at the primary root (55). The primary root of *rum1* shows a delayed gravitropic response and the mutant is affected in acropetal auxin transport (55). The *Rum1* gene encodes an unstable auxin-inducible Aux/IAA protein that binds to the auxin-responsive transcription factors ARF25 and ARF34 (56). Aux/IAA proteins are known to interact with auxin-response factors at moderate auxin levels leading to repressed expression of auxin-inducible genes (54). Increase of auxin concentrations result in binding of Aux/IAA proteins to the F-box receptor TIR1 and the subsequent fast proteasomal degradation of

Aux/IAAs. As a consequence, auxin response factors (ARFs) bind to auxin response elements of auxin-inducible genes which activate their expression (54). It was suggested that the maize RUM1 protein inhibits downstream auxin-inducible genes at low auxin levels by interaction with ARF25 and ARF34 (56). At high auxin levels RUM1 is degraded in lateral root precursor pericycle cells leading to an activation of auxin response genes via ARFs and the initiation of lateral roots. Time course experiments with GFP-tagged wild-type and mutant proteins demonstrated that the mutant protein rum1-R is very stable, whereas wild-type RUM1 has a very short half-life time that is typical for Aux/IAA proteins (56). The stability of rum1-R was explained by the lack of the degron sequence which is instrumental for degradation. It was suggested that auxin-responsive genes are repressed as a result of the presence of the stable rum1-R in homozygous mutants (56).

The rice mutant *short post-embryonic roots 1* (*spr1*) displays shorter postembryonic crown and lateral roots compared to wild-type roots as a consequence of decreased root cell elongation (57). In addition, mature mutant plants show less tillers than the wild-type and the iron content of mutant shoots is altered. Interestingly, in the *spr1* root elongation zone, higher concentrations of the reactive oxygen species (ROS) H_2O_2 were measured, that caused increased cell death. As a result, *spr1* roots are very sensitive and easy to damage. Rice *SPR1* encodes a 109 kDa protein containing a mitochondrial targeting signal (57).

4.2. Genes Regulating Root Hair Formation

Root hairs are unicellular, tubular extensions of epidermis cells formed in the root maturation zone that play an important role in facilitating the absorption of water and nutrients (58). Elongation of root hair cells is mediated by exocytosis of secretory vesicles in the root tip, resulting in localized tip growth (59). In maize, a transposon tagging approach enabled the isolation of three monogenic recessive mutants *rth1* to *rth3* (*roothairless 1* to *3*) (60). Compared to wild-type plants, *rth1* and *rth2* show normal root hair primordia formation but a reduced elongation of root hair cells. Especially *rth1* root hairs display a severe reduction of one-twentieth to one-tenth of the wild-type root hair length. The mutant *rth3* shows very short root hairs comparable to that of *rth1*. Scanning electron microscopy revealed abnormal root hair primordia in *rth3* mutants probably leading to impaired root hair elongation.

The severe phenotype of *rth1* plants is characterized by a dwarfed growth, the lack of ear formation and only rare production of tassels probably caused by nutrient deficiency (60). *Rth1* encodes a 100 kDa homolog of *SEC3* (28) that is involved in polar exocytosis of the root hair tip. *SEC3* is a subunit of the exocyst complex in yeast and mammals that acts as tethering factor to provide the initial interaction of vesicles and their target membrane prior to the exocytotic fusion (summarized in (61)). Recently, experiments in Arabidopsis and tobacco have supported the hypothesis that exocyst complexes are present in plants and play a pivotal role in plant cell expansion (62). The delayed development of *rth1* could be in part explained by the preferential accumulation of a prohibitin in the primary root of the mutant as demonstrated by proteome analyses (60).

Compared to *rth1* the mutants *rth2* and *rth3* grow vigorously (60). *Rth3* encodes a 71 kDa monocot specific COBRA-like protein that shows structural similarities to a glycosylphosphatidylinositol (GPI)-anchor protein (63). The function of COBRA-like genes is not yet fully understood but they appear to be involved in cell expansion and cell wall biosynthesis (64). In situ hybridization studies revealed that *Rth3* is expressed in root hair-forming trichoblasts of the primary root. In replicated field trials *rth3* mutants showed significant grain yield reduction suggesting that COBRA-like proteins are required for root hair elongation as well as for proper production of seeds (63).

Recently, the rice *Roothairless 1 (Rth1)* gene has been characterized (65) and identified as an apyrase (*APY*) by positional cloning (66). Scanning electron microscopy revealed that root hair elongation is arrested in the mutant after bulge formation of trichoblasts. Furthermore, *rth1* mutants of rice show a reduced height and shorter seminal and crown roots at the seedling stage (66). The enzyme apyrase hydrolyzes NTPs/NDPs to nucleoside monophosphates (67). In *rth1* roots, twice as much ATP per mg fresh weight was measured compared to wild-type roots suggesting a reduced apyrase activity in the mutant. It is suggested that the short root phenotype and impaired root hair elongation are a consequence of less energy provided by ATP hydrolysis (66).

By transcriptome subtractive hybridization (cDNA representational difference analysis) differentially expressed genes were identified in wild-type versus mutant *rhl1.a* (*roothairless1.a*) barley (*Hordeum vulgare*) roots (68). A root-specific β-expansin (*HvEXPB1*) was identified to be exclusively expressed in wild-type roots and the expression cosegregated with the root hair phenotype in the F_2-generation of crosses between wild-type and *rhl1.a*. EXPA and EXPB are the major subgroups of expansin proteins (EXPs) and they potentially mediate cell wall-loosening which is required for cell enlargement (summarized in (69)), as for instance during root hair growth. The important role of expansins in root hair elongation was further supported by a recent study of the rice *expA17* mutant showing reduced root hair length (70). Cloning of the affected gene revealed that the mutant phenotype is conferred by suppression of *OsEXPA17* which encodes an EXPA expansin. It was demonstrated that suppression of *EXPA17* by RNA interference resulted in short root hairs confirming the crucial role of EXPA17 in root hair elongation (70). In addition, the mutant phenotype could be at least partly complemented by OsEXPA and Arabidopsis EXPA7. These results suggested that class A and B expansins are required for root hair elongation in grasses and at least for EXPAs there might be a functional conservation between monocots and dicots.

More recently, ten new candidate genes involved in root hair formation have been identified in barley using a microarray approach to compare root gene expression profiles of wild-type versus the root hair-deficient mutant *rhl1.a* (71). One common feature of all these genes revealing reduced expression in *rhl1.a* is that they are associated with membranes and cell walls.

5. Conclusions and Outlook

Cereal root systems are characterized by embryonic and postembryonic root types that emerge at different stages of development. Currently, a growing number of root mutants are isolated in the monocot model species maize, rice, and barley, and several of these genes have already been cloned, thus providing novel insights into the molecular basis of root type-specific organogenesis in cereals. Some of the genes discussed here control the initiation or emergence of multiple root types (37, 39, 40, 46, 47, 56, 57, 66), whereas other genes only control the shaping of one single root cell type, the trichoblast (28, 51, 63, 68, 70).

Further mapping of genes involved in root development and the identification of the molecular networks in which these genes are interwoven will deepen the understanding of monocot specific root organogenesis. The availability of the maize and rice genome sequences (72–74) and the anticipated draft sequence of barley (75) facilitate the systematic analysis of related genes and gene families and might therefore help to determine a detailed framework of root developmental processes. To achieve these goals, functional genomic tools in cereals need to be further refined to enable analyses currently only feasible in the dicot model plant Arabidopsis. Among others, urgently required tools in cereal genomics include improved genome annotation of sequenced genomes, sequence-indexed reverse genetic resources, centralized expression databases, progress in transformation technology, and comprehensive collections of marker lines.

Acknowlegments

Root research in the Hochholdinger lab is supported by the DFG (Deutsche Forschungsgemeinschaft). We thank Dr. Dierk Wanke (University of Tübingen, Germany) for providing rice seeds to generate Fig. 1a and Dr. Caroline Gutjahr (LMU Munich, Germany) for the picture displayed in Fig. 1c.

References

1. Chandler VL, Brendel V (2002) The maize genome sequencing project. Plant Physiol 130:1594–1597
2. Khush GS (2005) What it will take to feed 5.0 billion rice consumers in 2030. Plant Mol Biol 59:1–6
3. Hochholdinger F, Park WJ, Sauer M, Woll K (2004) From weeds to crops: genetic analysis of root development in cereals. Trends Plant Sci 9:42–48
4. Coudert Y, Perin C, Courtois B, Khong NG, Gantet P (2010) Genetic control of root development in rice, the model cereal. Trends Plant Sci 15:219–226
5. Hochholdinger F (2009) The maize root system: morphology, anatomy and genetics. In: Bennetzen J, Hake S (eds) Handbook of maize. Springer, New York, pp 145–160
6. Moriata S, Nemoto K (1995) Morphology and anatomy of rice roots with special reference to coordination in organo- and histogenesis. In: Baluska T et al (eds) Structure and function of roots. Kluwer academic, Dordrecht
7. Kawata S, Yamazaki K, Ishihara K, Shibayama H, Lai KL (1963) Studies on root system formation in rice plants in a paddy. Proc Crop Sci Soc Japan 32:163
8. Hochholdinger F, Zimmermann R (2009) Molecular and genetic dissection of cereal root system development. In: Beeckman T (ed) Annual plant reviews root development. Wiley, New York, pp 175–191
9. Esau K (1965) Plant anatomy, 2nd edn. Wiley, New York
10. Feldman L (1994) The maize root. In: Freeling M, Walbot V (eds) The maize handbook. Springer, New York, pp 29–37
11. Varney GT, Canny MJ (1993) Rates of water uptake into the mature root system of maize plants. New Phytol 123:775–786
12. Laskowski MJ, Williams ME, Nusbaum HC, Sussex IM (1995) Formation of lateral root meristems is a two-stage process. Development 121:3303–3310
13. Malamy JE, Benfey PN (1997) Organization and cell differentiation in lateral roots of Arabidopsis thaliana. Development 124:33–44
14. Schiefelbein JW, Masucci JD, Wang H (1997) Building a root: the control of patterning and morphogenesis during root development. Plant Cell 9:1089–1098
15. Hochholdinger F, Zimmermann R (2008) Conserved and diverse mechanisms in root development. Curr Opin Plant Biol 11:70–74
16. Ishikawa H, Evans ML (1995) Specialized zones of development in roots. Plant Physiol 109:725–727
17. Jiang K, Meng YL, Feldman LJ (2003) Quiescent center formation in maize roots is associated with an auxin-regulated oxidizing environment. Development 130:1429–1438
18. Feix G, Hochholdinger F, Park WJ (2002) Maize root system and genetic analysis of its formation. In: Waisel Y et al (eds) Plant roots: the hidden half, 4th edn. Marcel Dekker, New York, pp 239–248
19. Gerald JNF, Lehti-Shiu MD, Ingram PA, Deak KI, Biesiada T, Malamy JE (2006) Identification of quantitative trait loci that regulate Arabidopsis root system size and plasticity. Genetics 172:485–498
20. Trachsel S, Messmer R, Stamp P, Hund A (2009) Mapping of QTLs for lateral and axile root growth of tropical maize. Theor Appl Genet 119:1413–1424
21. Ruta N, Liedgens M, Fracheboud Y, Stamp P, Hund A (2010) QTLs for the elongation of axile and lateral roots of maize in response to low water potential. Theor Appl Genet 120:621–631
22. Landi P, Giuliani S, Salvi S, Ferri M, Tuberosa R, Sanguineti MC (2010) Characterization of root-yield-1.06, a major constitutive QTL for root and agronomic traits in maize across water regimes. J Exp Bot 61:3553–3562
23. Norton GJ, Price AH (2009) Mapping of quantitative trait loci for seminal root morphology and gravitropic response in rice. Euphytica 166:229–237
24. Obara M, Tamura W, Ebitani T, Yano M, Sato T, Yamaya T (2010) Fine-mapping of qRL6.1, a major QTL for root length of rice seedlings grown under a wide range of NH_4^+ concentrations in hydroponic conditions. Theor Appl Genet 121:535–547
25. Neuffer MG (1994) Mutagenesis. In: Freeling M, Walbot V (eds) The maize handbook, Springer, New York 212–219
26. Walbot V (2000) Saturation mutagenesis using maize transposons. Curr Opin Plant Biol 3:103–107
27. Jeong DH, An S, Park S, Kang HG, Park GG, Kim SR, Sim J, Kim YO, Kim MK, Kim J, Shin M, Jung M, An G (2006) Generation of a flanking sequence-tag database for activation-tagging lines in japonica rice. Plant J 45:123–132
28. Hetz W, Hochholdinger F, Schwall M, Feix G (1996) Isolation and characterization of *rtcs*, a maize mutant deficient in the formation of nodal roots. Plant J 10:845–857

29. Hochholdinger F, Feix G (1998) Early post-embryonic root formation is specifically affected in the maize mutant *lrt1*. Plant J 16:247–255
30. Wen TJ, Hochholdinger F, Sauer M, Bruce W, Schnable PS (2005) The *roothairless1* gene of maize encodes a homolog of *sec3*, which is involved in polar exocytosis. Plant Physiol 138:1637–1643
31. Hochholdinger F, Woll K, Sauer M, Feix G (2005) Functional genomic tools in support of the genetic analysis of root development in maize (*Zea mays* L.). Maydica 50:437–442
32. Zhu J, Ingram PA, Benfey PN, Elich T (2011) From lab to field, new approaches to phenotyping root system architecture. Curr Opin Plant Biol 14:1–8
33. Hund A, Trachsel S, Stamp P (2009) Growth of axile and lateral roots of maize: I development of a phenotying platform. Plant Soil 325:335–349
34. Fang SQ, Yan XL, Liao H (2009) 3D reconstruction and dynamic modeling of root architecture *in situ* and its application to crop phosphorus research. Plant J 60:1096–1108
35. Iyer-Pascuzzi AS, Symonova O, Mileyko Y, Hao YL, Belcher H, Harer J, Weitz JS, Benfey PN (2010) Imaging and analysis platform for automatic phenotyping and trait ranking of plant root systems. Plant Physiol 152: 1148–1157
36. Trachsel S, Kaeppler SM, Brown KM, Lynch JP (2011) Shovelomics: high throughput phenotyping of maize (*Zea mays* L.) root architecture in the field. Plant Soil 341:75–87
37. Taramino G, Sauer M, Stauffer JL Jr, Multani D, Niu X, Sakai H, Hochholdinger F (2007) The maize (*Zea mays* L.) *RTCS* gene encodes a LOB domain protein that is a key regulator of embryonic seminal and post-embryonic shoot-borne root initiation. Plant J 50:649–659
38. Majer C, Hochholdinger F (2011) Defining the boundaries: structure and function of LOB domain proteins. Trends Plant Sci 16:47–52
39. Inukai Y, Sakamoto T, Ueguchi-Tanaka M, Shibata Y, Gomi K, Umemura I, Hasegawa Y, Ashikari M, Kitano H, Matsuoka M (2005) *Crown rootless1*, which is essential for crown root formation in rice, is a target of an auxin response factor in auxin signaling. Plant Cell 17:1387–1396
40. Liu H, Wang S, Yu X, Yu J, He X, Zhang S, Shou H, Wu P (2005) ARL1, a LOB-domain protein required for adventitious root formation in rice. Plant J 43:47–56
41. Friml J (2003) Auxin transport—shaping the plant. Curr Opin Plant Biol 6:7–12
42. Peret B, De Rybel B, Casimiro I, Benkova E, Swarup R, Laplaze L, Beeckman T, Bennett MJ (2009) Arabidopsis lateral root development: an emerging story. Trends Plant Sci 14:399–408
43. Okushima Y, Fukaki H, Onoda M, Theologis A, Tasaka M (2007) ARF7 and ARF19 regulate lateral root formation via direct activation of *LBD/ASL* genes in Arabidopsis. Plant Cell 19:118–130
44. Shuai B, Reynaga-Pena CG, Springer PS (2002) The LATERAL ORGAN BOUNDARIES gene defines a novel, plant-specific gene family. Plant Physiol 129: 747–761
45. Casimiro I, Marchant A, Bhalerao RP, Beeckman T, Dhooge S, Swarup R, Graham N, Inze D, Sandberg G, Casero PJ, Bennett M (2001) Auxin transport promotes Arabidopsis lateral root initiation. Plant Cell 13:843–852
46. Kitomi Y, Ogawa A, Kitano H, Inukai Y (2008) CRL4 regulates crown root formation through auxin transport in rice. Plant Root 2:19–28
47. Liu S, Wang J, Wang L, Wang X, Xue Y, Wu P, Shou H (2009) Adventitious root formation in rice requires OsGNOM1 and is mediated by the OsPINs family. Cell Res 19:1110–1119
48. Geldner N, Anders N, Wolters H, Keicher J, Kornberger W, Muller P, Delbarre A, Ueda T, Nakano A, Jurgens G (2003) The Arabidopsis GNOM ARF-GEF mediates endosomal recycling, auxin transport, and auxin-dependent plant growth. Cell 112:219–230
49. Steinmann T, Geldner N, Grebe M, Mangold S, Jackson CL, Paris S, Galweiler L, Palme K, Jurgens G (1999) Coordinated polar localization of auxin efflux carrier PIN1 by GNOM ARF GEF. Science 286:316–318
50. Friml J, Vieten A, Sauer M, Weijers D, Schwarz H, Hamann T, Offringa R, Jurgens G (2003) Efflux-dependent auxin gradients establish the apical-basal axis of Arabidopsis. Nature 426: 147–153
51. Wang XF, He FF, Ma XX, Mao CZ, Hodgman C, Lu CG, Wu P (2011) OsCAND1 is required for crown root emergence in rice. Mol Plant 4:289–299
52. Chuang HW, Zhang W, Gray WM (2004) Arabidopsis ETA2, an apparent ortholog of the human cullin-interacting protein CAND1, is required for auxin responses mediated by the SCF(TIR1) ubiquitin ligase. Plant Cell 16:1883–1897
53. Feng S, Shen Y, Sullivan JA, Rubio V, Xiong Y, Sun TP, Deng XW (2004) Arabidopsis CAND1, an unmodified CUL1-interacting protein, is involved in multiple developmental

pathways controlled by ubiquitin/proteasome-mediated protein degradation. Plant Cell 16: 1870–1882

54. Quint M, Gray WM (2006) Auxin signaling. Curr Opin Plant Biol 9:448–453
55. Woll K, Borsuk LA, Stransky H, Nettleton D, Schnable PS, Hochholdinger F (2005) Isolation, characterization, and pericycle-specific transcriptome analyses of the novel maize lateral and seminal root initiation mutant *rum1*. Plant Physiol 139:1255–1267
56. von Behrens I, Komatsu M, Zhang Y, Berendzen KW, Niu X, Sakai H, Taramino G, Hochholdinger F (2011) *Rootless with undetectable meristem 1* encodes a monocot-specific AUX/IAA protein that controls embryonic seminal and postembryonic lateral root initiation in maize. Plant J 66:341–353
57. Jia L, Wu Z, Hao X, Carrie C, Zheng L, Whelan J, Wu Y, Wang S, Wu P, Mao C (2011) Identification of a novel mitochondrial protein, short postembryonic roots 1 (SPR1), involved in root development and iron homeostasis in *Oryza sativa*. New Phytol 189:843–855
58. Gilroy S, Jones DL (2000) Through form to function: root hair development and nutrient uptake. Trends Plant Sci 5:56–60
59. Sieberer BJ, Ketelaar T, Esseling JJ, Emons AM (2005) Microtubules guide root hair tip growth. New Phytol 167:711–719
60. Wen TJ, Schnable PS (1994) Analyses of mutants of three genes that influence root hair development in *Zea mays* (Gramineae) suggest that root hairs are dispensable. Am J Bot 81:833–843
61. He B, Guo W (2009) The exocyst complex in polarized exocytosis. Curr Opin Cell Biol 21:537–542
62. Hala M, Cole R, Synek L, Drdova E, Pecenkova T, Nordheim A, Lamkemeyer T, Madlung J, Hochholdinger F, Fowler JE, Zarsky V (2008) An exocyst complex functions in plant cell growth in Arabidopsis and tobacco. Plant Cell 20:1330–1345
63. Hochholdinger F, Wen TJ, Zimmermann R, Chimot-Marolle P, Da Costa e Silva O, Bruce W, Lamkey KR, Wienand U, Schnable PS (2008) The maize (*Zea mays* L.) *roothairless3* gene encodes a putative GPI-anchored, monocot-specific, COBRA-like protein that significantly affects grain yield. Plant J 54:888–898
64. Brady SM, Song S, Dhugga KS, Rafalski JA, Benfey PN (2007) Combining expression and comparative evolutionary analysis. The *COBRA* gene family. Plant Physiol 143:172–187
65. Suzuki N, Taketa S, Ichii M (2003) Morphological and phiosiological characteristics of a root-hairless mutant in rice. Plant Soil 255:9–17
66. Yuo T, Toyota S, Ichii M, Taketa S (2009) Molecular cloning of a root hairless gene *rth1* in rice. Breed Sci 59:13–20
67. Shibata K, Morita Y, Abe S, Stankovic B, Davies E (1999) Apyrase from pea stems: isolation, purification, characterization and identification of a NTPase from the cytoskeleton fraction of pea stem tissue. Plant Physiol Biochem 37:881–888
68. Kwasniewski M, Szarejko I (2006) Molecular cloning and characterization of β-expansin gene related to root hair formation in barley. Plant Physiol 141:1149–1158
69. Sampedro J, Cosgrove DJ (2005) The expansin superfamily. Genome Biol 6:242
70. Zhiming Y, Bo K, Xiaowei H, Shaolei L, Youhuang B, Wona D, Ming C, Hyung-Taeg C, Ping W (2011) Root hair-specific EXPANSINs modulate root hair elongation in rice. Plant J 66:725–734
71. Kwasniewski M, Janiak A, Mueller-Roeber B, Szarejko I (2010) Global analysis of the root hair morphogenesis transcriptome reveals new candidate genes involved in root hair formation in barley. J Plant Physiol 167:1076–1083
72. Goff SA, Ricke D, Lan TH, Presting G, Wang R, Dunn M, Glazebrook J, Sessions A, Oeller P, Varma H, Hadley D, Hutchison D, Martin C, Katagiri F, Lange BM, Moughamer T, Xia Y, Budworth P, Zhong J, Miguel T, Paszkowski U, Zhang S, Colbert M, Sun WL, Chen L, Cooper B, Park S, Wood TC, Mao L, Quail P, Wing R, Dean R, Yu Y, Zharkikh A, Shen R, Sahasrabudhe S, Thomas A, Cannings R, Gutin A, Pruss D, Reid J, Tavtigian S, Mitchell J, Eldredge G, Scholl T, Miller RM, Bhatnagar S, Adey N, Rubano T, Tusneem N, Robinson R, Feldhaus J, Macalma T, Oliphant A, Briggs S (2002) A draft sequence of the rice genome (*Oryza sativa* L. *ssp. japonica*). Science 296:92–100
73. Schnable PS, Ware D, Fulton RS, Stein JC, Wei F, Pasternak S, Liang C, Zhang J, Fulton L, Graves TA, Minx P, Reily AD, Courtney L, Kruchowski SS, Tomlinson C, Strong C, Delehaunty K, Fronick C, Courtney B, Rock SM, Belter E, Du F, Kim K, Abbott RM, Cotton M, Levy A, Marchetto P, Ochoa K, Jackson SM, Gillam B, Chen W, Yan L, Higginbotham J, Cardenas M, Waligorski J, Applebaum E, Phelps L, Falcone J, Kanchi K, Thane T, Scimone A, Thane N, Henke J, Wang T, Ruppert J, Shah N, Rotter K, Hodges J, Ingenthron E, Cordes M, Kohlberg S, Sgro J, Delgado B, Mead K, Chinwalla A, Leonard S, Crouse K, Collura K, Kudrna D, Currie J, He R, Angelova A, Rajasekar S, Mueller T, Lomeli R, Scara G, Ko A, Delaney K, Wissotski

M, Lopez G, Campos D, Braidotti M, Ashley E, Golser W, Kim H, Lee S, Lin J, Dujmic Z, Kim W, Talag J, Zuccolo A, Fan C, Sebastian A, Kramer M, Spiegel L, Nascimento L, Zutavern T, Miller B, Ambroise C, Muller S, Spooner W, Narechania A, Ren L, Wei S, Kumari S, Faga B, Levy MJ, McMahan L, Van Buren P, Vaughn MW et al (2009) The B73 maize genome: complexity, diversity, and dynamics. Science 326:1112–1115

74. Yu J, Hu S, Wang J, Wong GK, Li S, Liu B, Deng Y, Dai L, Zhou Y, Zhang X, Cao M, Liu J, Sun J, Tang J, Chen Y, Huang X, Lin W, Ye C, Tong W, Cong L, Geng J, Han Y, Li L, Li W, Hu G, Li J, Liu Z, Qi Q, Li T, Wang X, Lu H, Wu T, Zhu M, Ni P, Han H, Dong W, Ren X, Feng X, Cui P, Li X, Wang H, Xu X, Zhai W, Xu Z, Zhang J, He S, Xu J, Zhang K, Zheng X, Dong J, Zeng W, Tao L, Ye J, Tan J, Chen X, He J, Liu D, Tian W, Tian C, Xia H, Bao Q, Li G, Gao H, Cao T, Zhao W, Li P, Chen W, Zhang Y, Hu J, Liu S, Yang J, Zhang G, Xiong Y, Li Z, Mao L, Zhou C, Zhu Z, Chen R, Hao B, Zheng W, Chen S, Guo W, Tao M, Zhu L, Yuan L, Yang H (2002) A draft sequence of the rice genome (*Oryza sativa* L. *ssp. indica*). Science 296:79–92

75. Schulte D, Close TJ, Graner A, Langridge P, Matsumoto T, Muehlbauer G, Sato K, Schulman AH, Waugh R, Wise RP, Stein N (2009) The international barley sequencing consortium—at the threshold of efficient access to the barley genome. Plant Physiol 149:142–147

Chapter 5

Gene Expression Analysis of Aquatic Angiosperms Podostemaceae to Gain Insight into the Evolution of Their Enigmatic Morphology

Satoshi Koi and Natsu Katayama

Abstract

Podostemaceae is a family of aquatic angiosperms growing submerged on rocks in fast-flowing water and called moss-like or alga-like riverweeds. It evolved remarkable innovations to adapt to such an extreme environment, one of which is reduced shoots borne on roots adhering to rock surface. Recent observations revealed that the basal subfamily Tristichoideae, like most other angiosperms, has typical shoot apical meristems (SAMs). In species of the subfamily Podostemoideae, however, shoot apical meristems (SAMs) are not formed during development and new leaves arise from the meristematic basal region of preexisting leaves. The genetic basis of this shoot organogenesis process, e.g., the expression patterns of genes homologous to transcription factors regulating shoot development, is essential to better understand the evolution of Podostemaceae. A gene expression analysis found that the SAM-less Podostemoideae leaf has mixed identity of SAM and leaf, and provided insight into the evolution of the shoot in Podostemaceae.

Key words: Aquatic angiosperms, Body plan, Evolution, In situ hybridization, Leaf, Podostemaceae, Shoot apical meristem

1. Introduction

The aim of morphology is to find the rule governing body plans of organisms. Body plans have been investigated with developmental and evolutionary approaches. In general, vascular plants are considered to consist of roots and stems with leaves, which are derived from root apical meristems (RAMs) and shoot apical meristems (SAMs), respectively. Developmental processes of organ formation are basically common across vascular plants (1). However, there are exceptions that are difficult to explain by the basic rule of development. One such is the morphology of Podostemaceae, in which leaf formation seems to occur in absence of SAM. A recent comparative analysis on the underlying gene expression patterns in the

Ive De Smet (ed.), *Plant Organogenesis: Methods and Protocols*, Methods in Molecular Biology, vol. 959,
DOI 10.1007/978-1-62703-221-6_5, © Springer Science+Business Media New York 2013

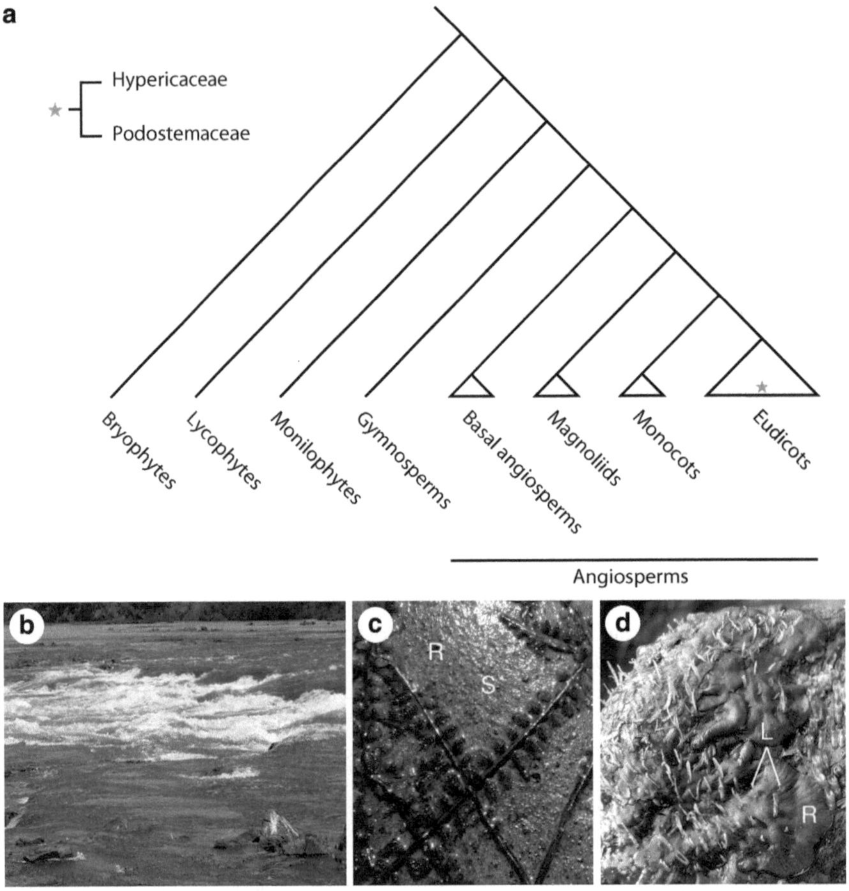

Fig. 1. Phylogeny of land plants, and habitat and morphology of Podostemaceae. (**a**) Podostemaceae is sister to Hypericaceae. (**b**) Plants (*green*) growing on rocks in rapid. (**c**) *Terniopsis* sp. (Tristichoideae). Elongated roots with short shoots on lateral sides adhering to rock surface. (**d**) *Hydrobryum japonicum* (Podostemoideae). Foliose root with filiform leaves scattered on dorsal surface. L, leaf. R, root. S, shoot

SAM-less and SAM-mediated species of the family offered an implication on how such an enigmatic shoot evolved (2).

Podostemaceae is one of the ecologically and morphologically most enigmatic families in angiosperms (3, 4). This family grows submerged on rocks in rapids and waterfalls in tropical, subtropical and some temperate regions in marked rainy and dry seasons (Fig. 1b). The plants are attached to the rock surfaces exposed to torrential pressure during the rainy season. In the dry season when the water level becomes lower, they are exposed to the air, bear flowers, set fruits, and eventually blight. Among the angiosperms, Podostemaceae are dominant in such habitats (5). Only Hydrostachyaceae grow in similar habitats (6).

Podostemaceae is the most closely related to the terricolous Hypericaceae within the eurosids I Malpighiales of eudicots (Fig. 1a) (7–9). The family has diversified into ca. 300 species assigned to ca. 50 genera, though all species occur in a uniform ecological niche (10). Most species have roots creeping on rock surfaces and adventitious shoots (root-borne shoots) on the lateral or dorsal sides of the root (Fig. 1c, d). A few species are devoid of roots, and have crustose shoots or elongate shoots floating in the water. Therefore, Podostemaceae, called moss-like or alga-like river-weeds, seems to be a good target for plant evolutionary studies, with particular focus on ecological and morphological innovations.

2. Evolution of Body Plan in Podostemaceae

In Podostemaceae, drastic changes in the body plan occurred through the invasion of the extreme aquatic environment. In most angiosperms, through embryogenesis, the radicle with an embryonic root apical meristem and the plumule with an embryonic SAM are established at both ends of the hypocotyl, giving rise to the vertical body axis (1, 11). In contrast, Podostemaceae acquired remarkable body plans with a novel horizontal axis. In most species, the radicle is not formed and an adventitious root arises from the hypocotyl, and only a few or several leaves are produced in the plumular site between the cotyledons (12–15). *Hydrobryum*, which is one of the most derived genera, is devoid of both plumule and radicle (15). In some other species, such as *Dalzellia zeylanica* and *Hydrodiscus koyamae*, no plumule, radicle nor adventitious root are formed in their seedlings, and the plants consist of only shoots formed in the axils of cotyledons or from the base of hypocotyl (16, 17). Exceptionally, *Terniopsis brevis*, a member of the basal genus *Terniopsis* in Podostemaceae, has a typical radicle and a plumule in the seedling, suggesting that the species retains the ancestral vertical body plan and reduction of vertical axis occurred step-by-step in Podostemaceae (18). In this species an adventitious root also forms from the hypocotyl.

The root morphology of Podostemaceae underwent evolutionary modifications. The roots contain chloroplasts for photosynthesis and are flattened ranging from subcylindrical to ribbon-like to foliose (Fig. 1c, d). The subcylindrical root has an apical meristem at the tip, while the ribbon-like and foliose roots have band-like meristems at the margin (19, 20).

3. Shoot Organogenesis of Podostemaceae

The shoot organogenesis had been controversial. Typical angiosperms have dome-shaped SAMs at their stem tips (1, 11). Cells in the central zone (CZ) of the SAM exhibit low rates of cell division and are considered to be multipotent. The CZ cells divide to produce themselves and supply extra cells in the peripheral zone (PZ) where the rate of cell division increases. The leaves arise exogenously in the PZ of the SAM. In some species of Podostemaceae, however, there is no obvious SAM. Hammond (21) showed that a leaf primordium arises at the base of the second-youngest leaf in *Podostemum ceratophyllum*, a member of the subfamily Podostemoideae. In some other species of Podostemoideae, the presence of SAM and its role for leaf formation have been doubted (22–24). On the other hand, Jäger-Zürn (25–28) claimed that SAMs exist in the shoots of the subfamily, but her interpretation of data obtained with a traditional anatomical technique was criticized. Imaichi et al. (29) and Koi et al. (30) performed close observations using serial semi-thin sectioning and transmission electron microscopy, and concluded that leaf formation occurs in the absence of the SAM in Podostemoideae. In *Cladopus queenslandicus* and *Zeylanidium subulatum*, the SAM is never recognizable and a new leaf primordium arises endogenously from the base of the second youngest leaf (Fig. 2a–c) (29, 30). The leaf primordium protrudes from the second youngest leaf (L2), accompanied by separation of vacuolated cells in the L2 epidermis. This repeated process makes a pile of leaves, i.e., a "stem-less shoot." The same developmental pattern was observed in leaves and bracts in *Hydrobryum japonicum* (31). In the other two subfamilies, Tristichoideae and Weddellinoideae, which together are paraphyletic to Podostemoideae (32), the SAM is apparent at the shoot apex and leaves arise exogenously in the peripheral zone of the SAM in acropetal order (Fig. 2d–g) (12, 22, 33–41). The two subfamilies show sympodial branching that a new branch arises extra-axillarily at the base of the youngest branch (Fig. 2d–g) (40, 41). Based on the similarity between the leaf organogenesis of Podostemoideae and the shoot branching of Tristichoideae and Weddellinoideae in that both organs arise from the basal meristematic region of older organs, Koi and Kato (40) and Fujinami and Imaichi (41) proposed a hypothesis that the "leaf" of Podostemoideae and the "branch" of Tristichoideae and Weddellinoideae are comparable. A comparison of the underlying gene expression patterns between the "leaf" of Podostemoideae and the "branch" of Tristichoideae and Weddellinoideae was expected to provide informative evidence on organ homology and the evolution of body plan.

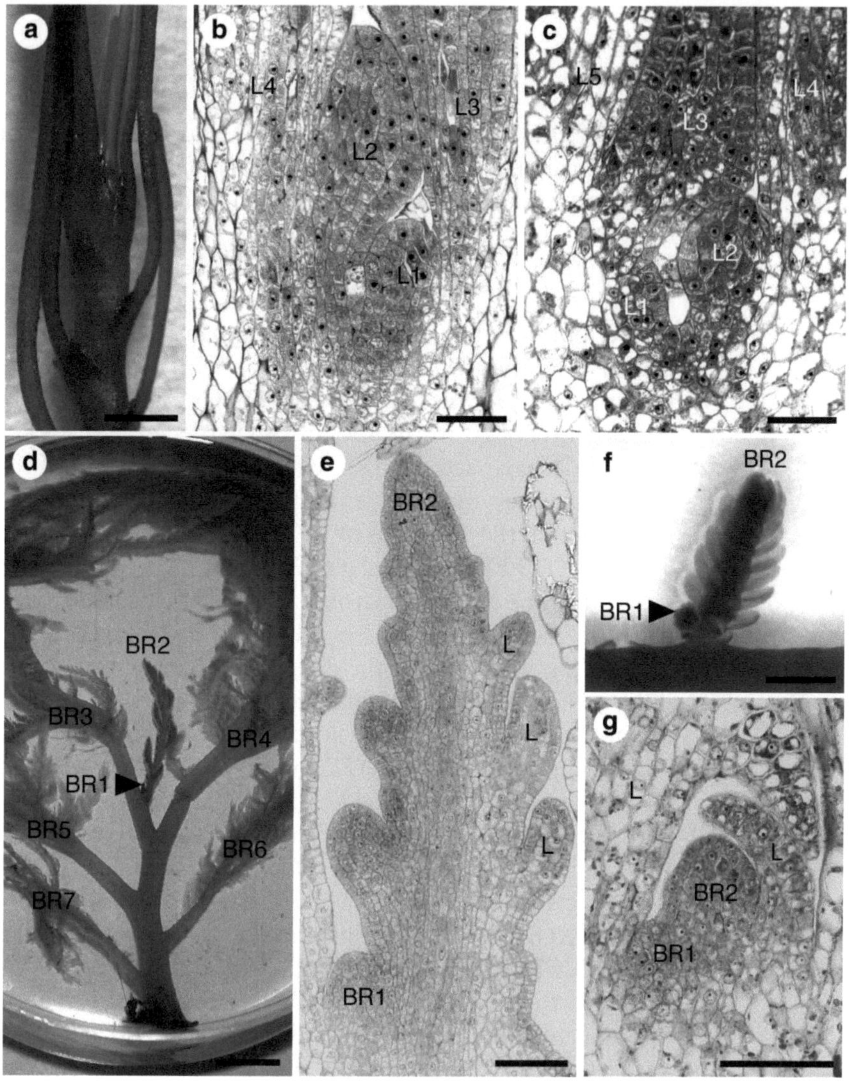

Fig. 2. Shoot development in Podostemaceae. (**a**–**c**) SAM-less shoot of *Cladopus queenslandicus* (Podostemoideae). (**a**) Shoot tip with leaves arranged in two orthostichies. (**b**) Longitudinal section of shoot, parallel to orthostichies. No SAM is visible between L1 and L2. Note that one dermal cell of L2 is vacuolated and detached cells above L1. (**c**) New leaf primordium (L1) arises just below vacuolated dermal cells of L3. (**d**, **e**) SAM-mediated shoot of *Weddellina squamulosa* (Weddellinoideae). (**d**) Sympodially branching shoot. (**e**) Longitudinal section of shoot, parallel to branches. New branch (BR1) arises at base of BR2. (**f**, **g**) SAM-mediated shoot of *Tristicha trifaria* (Tristichoideae). (**f**) Sympodially branching shoot. (**g**) Longitudinal section of shoot, parallel to branches. New branch (BR1) arises at base of BR2. L1–L5, youngest to fifth youngest leaf primordia. BR1–BR7, youngest to seventh youngest shoot branches. Scale bars = 1 mm in A, D, F; 50 μm in B, C, E, G. (**a**) Reproduced from (30) (http://www.jstor.org/stable/10.1086/427482). © 2005 by The University of Chicago. (**d**, **e**) Reproduced from (40) with permission from Oxford University Press.

4. Gene Expression Patterns of Shoot Organogenesis in Podostemaceae

Critical regulatory pathways for SAM and leaf development have been revealed in the model plant, *Arabidopsis thaliana*. A feedback loop between *CLAVATA3* (*CLV3*) and *WUSCHEL* (*WUS*) regulates the size of a stem cell population in the SAM. *CLV3*, encoding a glycopeptide secreted from undifferentiated stem cells in the central zone (CZ), limits the expression of *WUS*, a homeobox transcription factor, within the organizing center (OC) beneath the CZ (42–44). *WUS* is required for *CLV3* expression in the CZ to regulate the size of the stem cell population (44). *WUS* orthologs are also expressed in the OC of other species, e.g., *Antirrhinum majus* (Plantaginaceae) (45), *Medicago truncatula* and *Pisum sativum* (Fabaceae) (46, 47), and *Petunia hybrida* (Solanaceae) (48), implying that the functions of *WUS* genes are conserved in eudicots. The *KNOTTED1-LIKE HOMEOBOX* (*KNOX*) gene family is one of the critical factors for SAM regulation. *SHOOT MERISTEMLESS* (*STM*), encoding a transcription factor of Class I KNOX, is expressed in the entire SAM, and functions to prevent differentiation of cells within the SAM, but is down-regulated at the initiation site of leaf and floral organs (49–51). An antagonistic relationship between *KNOX* genes and *ARP* (*ASYMMETRIC LEAVES1/ROUGH SHEATH2/PHANTASTICA*) genes, encoding MYB-like transcription factors, regulates leaf development (52–55). Down-regulation of *STM* expression allows *ARP* expression at leaf initiation sites, while *ARP* represses the expression of other *KNOX* genes in leaf primordia, resulting in leaf differentiation (56, 57).

With the background of the knowledge of the genetic control of the plant shoot, Katayama et al. (2) performed a gene expression analysis of the Podostemaceae shoot, using *STM*, *WUS* and *ARP* orthologous genes as markers for cell identity in shoot organogenesis. The genes isolated from Podostemaceae possess conserved characteristic domains for each gene family. In the SAM-less shoot of *Hydrobryum japonicum*, *STM* and *WUS* orthologs (*HjSTM* and *HjWUS*) are expressed in such unusual sites as the entire incipient leaf/bract primordium and in its inner body, respectively (Fig. 3a, b). This expression pattern is similar to the one in the SAMs of other angiosperms including the model plant *A. thaliana* (Fig. 4a) (42, 44, 50, 51). In the developing reproductive shoot, the floral meristem (FM) arises following the bracts and produces floral organs exogenously in the peripheral zone in acropetal order (31). The expression patterns of *HjSTM* and *HjWUS* in the FM are the same as those in the SAMs and FMs of other angiosperms, implying that the functions of *HjSTM* and *HjWUS* may be conserved in the SAM-less *H. japonicum* (Fig. 3c, d). The similar expression pat-

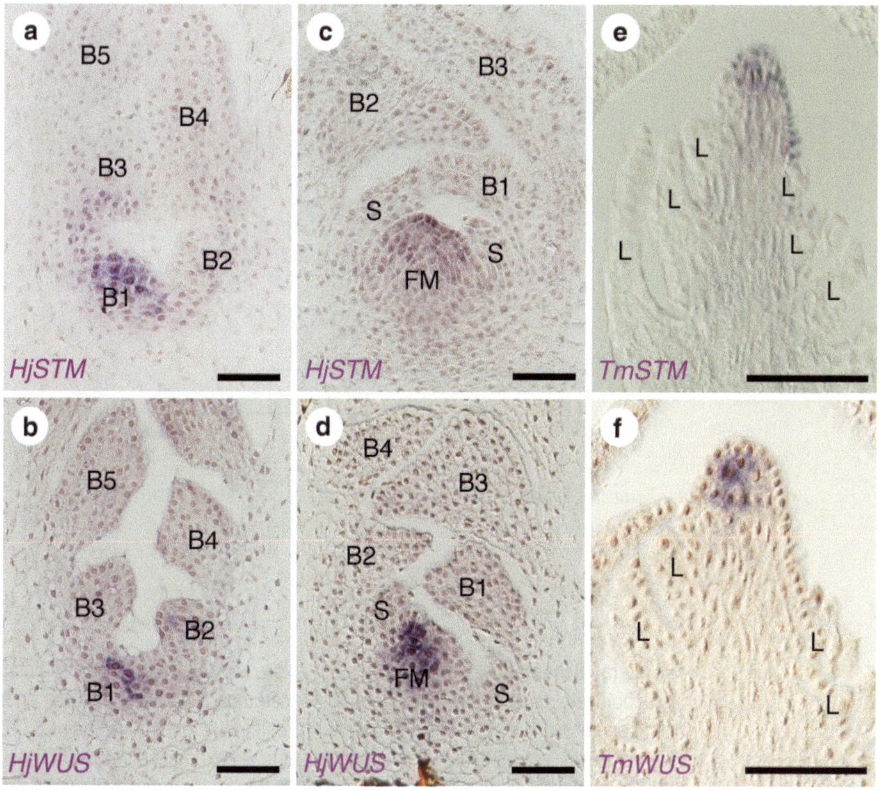

Fig. 3. Gene expression patterns in shoots of Podostemaceae. (**a–d**) SAM-less reproductive shoots and floral meristems of *Hydrobryum japonicum* (Podostemoideae). (**a**) *HjSTM* is expressed in entire incipient bract primordium. (**b**) *HjWUS* is expressed in inner cells of incipient bract primordium. (**c**) *HjSTM* is expressed in entire floral meristem. (**d**) *HjWUS* is expressed in inner cells of FM. (**e, f**) SAM-mediated vegetative shoot of *Terniopsis minor* (Tristichoideae). (**e**) *TmSTM* is expressed in entire shoot apex. (**f**) *TmWUS* is expressed in inner body of shoot apex. B1–B5, youngest to fifth youngest bracts. FM, floral meristem. L, leaf. S, spathella (cupular envelope). Reproduced from (2) with permission from the American Society of Plant Biologists.

terns of *HjSTM* and *HjWUS* in the incipient leaf and the FM suggest that the initiating leaf/bract is comparable to the SAM. As the leaf/bract primordium develops, *HjWUS* expression disappears and *HjSTM* expression is restricted to the basal part of the primordium. An *ARP* ortholog (*HjPHAN*) becomes expressed in the distal part of a primordium in a complementary pattern with *HjSTM* (Fig. 4a) (2). Based on these results, Katayama et al. (2) suggested that the "leaf" initiates as a SAM and then differentiates into a single apical leaf in *H. japonicum*. The expression pattern of *STM* and *WUS* orthologs in *Cladopus doianus* (Podostemoideae) is quite similar to that of *H. japonicum*, and it suggests that the pattern is common in Asian Podostemoideae and possibly other clades (2).

In *Terniopsis minor* of the basal subfamily Tristichoideae with the SAM-mediated shoot, *STM* and *WUS* orthologs (*TmSTM* and *TmWUS*) are expressed in the apex of branching shoots in the same

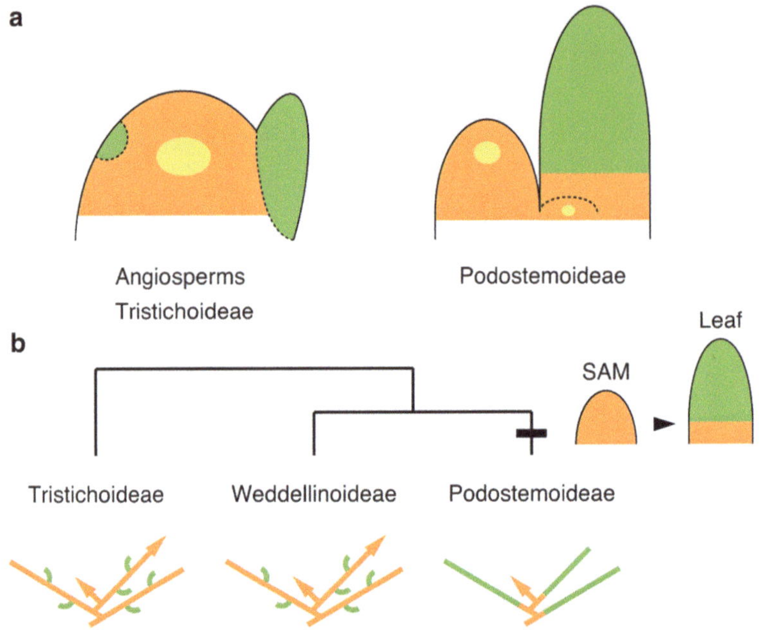

Fig. 4. (**a**) Schematic illustration of gene expression. *Left*, SAM of angiosperms and Tristichoideae. *Right*, SAM-less shoot of Podostemoideae. Leaf initiates endogenously from base of older leaf. *Orange, yellow* and *green* indicate expressions of *STM*, *WUS* and *ARP* homologs, respectively. *ARP* homolog is actually expressed in third or older youngest leaf. (**b**) Model for evolution of shoot in Podostemaceae, under phylogeny of three subfamilies. New organs (shoot in Tristichoideae and Weddellinoideae, leaf in Podostemoideae) initiate from existing organs. Developmental program of from-SAM-to-leaf transformation was acquired in Podostemoideae diverged from Weddellinoideae. *Orange bar* with arrowhead and *green bar* indicate developing shoot with SAM and leaf, respectively.

pattern as that of other angiosperms (Fig. 3e, f) (2). It is suggested that Tristichoideae (and perhaps Weddellinoideae) retain typical SAMs at the gene expression level.

5. Evolution of Unusual Shoot in Podostemaceae

Katayama et al. (2) put forward a model of the evolution of novel shoot organogenesis in Podostemaceae: the mode of the SAM-less shoot organogenesis of Podostemoideae evolved by the acquisition of the developmental program that the SAM transforms into a single apical leaf, and it was derived from the mode of the sympodial shoot branching of Tristichoideae and Weddellinoideae (Fig. 4b). In this model, down-regulation of the genetic program of SAM and ectopic expression of the genetic program of a leaf in the entire SAM contribute to the shoot-to-leaf transformation. Further analysis of the function and expression of genes will verify this hypothesis. As noted by Fujinami and Imaichi (41), there is a gap in the initiation sites of new organs between Tristichoideae and Weddelinoideae, and Podostemoideae. A new branch arises from the youngest branch in the SAM-mediated Tristichoideae and Weddelinoideae, while a new leaf arises from the second youngest leaf in the SAM-less Podostemoideae. Taking into account the uniqueness of the extra-axillary sympodial branching of Tristichoideae and Weddellinoideae among angiosperms with axillary shoot branching (40, 41), the branching of the two subfamilies may be a prototype for the leaf organogenesis of Podostemoideae. Developmental anatomical analysis for American and African Podostemoideae, which are placed phylogenetically at the base of Podostemoideae, will give useful evidence.

The reduction of the vertical shoot in Podostemaceae is caused by immediate differentiation of the SAM into a leaf. The resulting body plan seems to be an adaptation to their habitat with fast water currents to which the plants are exposed. It is likely that such a shoot was selected against the elongate vertical shoot with a SAM. Lemnaceae (duckweeds) show another case of unusual morphology. The family is a free-floating aquatic angiosperm that grows on the surface of still water where vertical growth is also restricted (58). The plant consists of leaves (which are interpreted as a mixed organ of leaf and shoot) floating on the water surface and roots (if present) immersed in water. Like in Podostemaceae, no SAM is recognizable during shoot development and a new leaf arises from the base of a young leaf (59). Similarity in leaf development between Lemnaceae and Podostemaceae implies that the leaf of Lemnaceae may have mixed identity of SAM and leaf, like the leaf of Podostemoideae. Analysis at the gene expression level will help one understand the mechanisms underlying novel shoot organogenesis in Lemnaceae and give their evolutionary perspectives.

6. Conclusions and Perspectives

The evolutionary developmental study of the unusual shoots of three species of Podostemaceae by Kayatama et al. (2) gave an insight into the evolution of their enigmatic morphology, by providing evidence on a genetic basis for transformation from shoot to leaf. Podostemaceae have other unusual morphological features in shoot organogenesis, together representing a misfit morphology (4). For instance, in some American species such as *Apinagia multibranchiata* and Asian *Zeylanidium subulatum* with adaxial and abaxial leaf-sheaths subtending young leaves, a supernumerary leaf (or a chain of leaves) arises from the abaxial base of a leaf, in addition to an ordinary leaf formed from the adaxial base (24, 25, 29). Imaichi et al. (29) showed that in *Z. subulatum* a new abaxial leaf arises adventitiously from meristematic cells of a mother leaf. Many species of Podostemoideae, mostly American species, have ensiform (unifacial-leaf-like), bifacial leaves that are produced in the absence of SAM (61, 62). In Tristichoideae, *Dalzellia* has flattened dorsiventral SAM, resulting in a foliose shoot instead of a cylindrical shoot (16). Furthermore, deviated features are seen in root organogenesis. *Hydrobryum* and related genera have foliose roots with marginal root meristem instead of root apical meristem (RAM) (19). Asian and some African Podostemoideae exhibit dichotomous root branching (i.e., exogenous branching) instead of endogenous branching (63). Molecular phylogenetic analyses suggest that the evolution of foliose root morphology is associated with the evolution of dichotomous root branching (32). How such modification occurred on SAM and RAM is one of the most intriguing subjects for the evolutionary study of Podostemaceae morphology.

Reduction of the vertical axis in the seedling (i.e. loss of plumule and radicle) occurred through the evolution of Podostemaceae. Such reduction of the principal vertical axis of the body plan may be triggered by acquisition of a horizontal new axis, i.e. a shoot-producing root and is probably an adaptation to the extreme habitats. On the other hand, the body plan is diversified in uniform habitats. *Hydrodiscus*, *Mourera* and *Rhyncholacis* have no conventional stems (SAM-less shoots) and no roots, and their plant consists of leaves (and flowers) (3, 17, 23), and *Dalzellia* is devoid of roots (16, 60). Such diversification of body plan is in contrast to the stability of seedling morphologies seen in terricolous angiosperms with a ventral axis formed by the plumule and radicle.

Comparative genomic and molecular genetic approaches will provide insights into the evolution of Podostemaceae. In comparison to other plants, however, the non-model plant Podostemaceae is difficult to be analyzed on molecular and genetic levels so far, primarily because of the difficulty to culture Podostemaceae in a laboratory. In the case of in situ hybridization by Katayama et al. (2), fresh materials were fixed in the field (in situ) immediately after they were collected. Establishments of cultivation in a laboratory and applications of molecular tools on Podostemaceae will contribute to better understand the morphological evolution in Podostemaceae.

Acknowledgments

We thank M. Kato for his encouragement for the preparation of this paper and useful suggestions on the manuscript. We also thank H. Hirano, T. Suzaki, and T. Yamada for their technical advices, and T. Wongprasert, R. Imaichi, and R. Fujinami who helped us collect materials used by the studies cited in this paper. This study was supported by a Grant-in Aid for Scientific Research from the Japan Society for the Promotion of Science.

References

1. Steeves TA, Sussex IM (1989) Patterns in plant development, 2nd edn. Cambridge University Press, Cambridge
2. Katayama N, Koi S, Kato M (2010) Expression of *SHOOT MERISTEMLESS*, *WUSCHEL*, and *ASYMMETRIC LEAVES 1* homologs in the shoots of Podostemaceae: implications for the evolution of novel shoot organogenesis. Plant Cell 22:2131–2140
3. Rutishauser R (1997) Structural and developmental diversity in Podostemaceae (river-weeds). Aquat Bot 57:29–70
4. Bell AD (2008) Plant form: an illustrated guide to flowering plant morphology. Timber Press, Portland
5. van Steenis CGGJ (1981) Rheophytes of the world. Sijthoff & Noordhoff, Alphen aan den Rijn
6. Jäger-Zürn I (1998) Anatomy of the Hydrostachyaceae. In: Landolt E, Jäger-Zürn I, Schnell RAA (eds) Extreme adaptations in angiospermous hydrophytes. Gebrüder Borntraeger, Berlin
7. Wurdack KJ, Davis CC (2009) Malpighiales phylogenetics: gaining ground on one of the most recalcitrant clades in the angiosperm tree of life. Am J Bot 96:1551–1570
8. Ruhfel BR et al (2011) Phylogeny of the clusioid clade (Malpighiales): evidence from the plastid and mitochondrial genomes. Am J Bot 98:306–325
9. Soltis DE et al (2011) Angiosperm phylogeny: 17 genes, 640 taxa. Am J Bot 98:704–730
10. Cook CDK, Rutishauser R (2007) Podostemaceae. In: Kubitzki K (ed) The families and genera of vascular plants, IX. Springer, Berlin
11. Esau K (1977) Anatomy of seed plants, 2nd edn. Wiley, New York
12. Willis JC (1902) Studies in the morphology and ecology of the Podostemaceae of Ceylon and India. Ann Roy Bot Gar Peradeniya 1:267–465
13. Mohan Ram HY, Sehgal A (1997) In vitro studies on developmental morphology of Indian Podostemaceae. Aquat Bot 57:97–132
14. Sehgal A, Sethi M, Mohan Ram HY (2002) Origin, structure, and interpretation of the thallus in *Hydrobryopsis sessilis* (Podostemaceae). Int J Plant Sci 163:891–905
15. Suzuki K, Kita Y, Kato M (2002) Comparative developmental anatomy of seedling in nine species of Podostemaceae (subfamily Podostemoideae). Ann Bot 89:755–765
16. Imaichi R, Maeda R, Suzuki K, Kato M (2004) Developmental morphology of foliose shoots and seedlings of *Dalzellia zeylanica* (Podostemaceae) with special reference to their meristems. Bot J Linn Soc 144:289–302
17. Koi S, Kato M (2010) Developmental morphology of seedling and shoot and phylogenetic relationship of *Diplobryum koyamae* (Podostemaceae). Am J Bot 97:373–387
18. Kita Y, Kato M (2005) Seedling developmental anatomy of an undescribed *Malaccotristicha* species (Podostemaceae, subfamily Tristichoideae) with implications for body plan evolution. Plant Syst Evol 254:221–232
19. Ota M, Imaichi R, Kato M (2001) Developmental morphology of the thalloid *Hydrobryum japonicum* (Podostemaceae). Am J Bot 88:382–390
20. Koi S, Kato M (2003) Comparative developmental anatomy of the root in three species of *Cladopus* (Podostemaceae). Ann Bot 91:927–937
21. Hammond BL (1936) Regeneration of *Podostemon ceratophyllum*. Bot Gaz 97:834–845
22. Rutishauser R (1995) Developmental patterns of leaves in Podostemaceae compared with more typical flowering plants: saltational evolution and fuzzy morphology. Can J Bot 73:1305–1317
23. Rutishauser R, Grubert M (1999) The architecture of *Mourera fluviatilis* (Podostemaceae): developmental morphology of inflorescences, flowers, and seedlings. Am J Bot 86:907–922
24. Rutishauser R, Grubert M (2000) Developmental morphology of *Apinagia multibranchiata* (Podostemaceae) from the Venezuelan Guyanas. Bot J Linn Soc 132:299–323
25. Jäger-Zürn I (1999) Morphology of Podostemaceae. V. Developmental morphology of the shoot system of *Podostemum subulatum* (Podostemaceae: Podostemoideae). Beitr Biol Pflanzen 71:281–334
26. Jäger-Zürn I (2000) Morphology of Podostemaceae. VI. Crustose root and root-born shoots of *Zeylanidium olivaceum* (Podostemaceae: Podostemoideae). Flora 195:61–82

27. Jäger-Zürn I (2002) Comparative studies in the morphology of *Crenias weddelliana* and *Maferria indica* with reference to *Sphaerothylax abyssinica* (Podostemaceae: Podostemoideae). Bot J Linn Soc 138:63–84
28. Jäger-Zürn I (2005) Shoot apex and spathella: two problematical structures of Podostemaceae–Podostemoideae. Plant Syst Evol 253:209–218
29. Imaichi R, Hiyama Y, Kato M (2005) Leaf development in absence of shoot apical meirstem in *Zeylanidium subulatum* (Podostemaceae). Ann Bot 96:51–58
30. Koi S, Imaichi R, Kato M (2005) Endogenous leaf initiation in the apical-meristemless shoot of *Cladopus queenslandicus* (Podostemaceae) and implications for evolution of shoot morphology. Int J Plant Sci 166:199–206
31. Katayama N, Koi S, Kato M (2008) Developmental anatomy of the reproductive shoot in *Hydrobryum japonicum* (Podostemaceae). J Plant Res 121:417–427
32. Kita Y, Kato M (2001) Infrafamilial phylogeny of the aquatic angiosperm Podostemaceae inferred from the nucleotide sequences of the *matK* gene. Plant Biol 3:156–163
33. Jäger-Zürn I (1970) Morphologie der Podostemaceae I. *Tristicha trifaria (*Bory ex Willd.) Spreng. Beitr Biol Pflanzen 47:11–52
34. Jäger-Zürn I (1992) Morphologie der Podostemaceae II. *Indotristicha ramosissima* (Wight) van Royen (Tristichoideae). Beitr Biol Pflanzen 80:1–48
35. Jäger-Zürn I (1997) Comparative morphology of the vegetative structures of *Tristicha trifaria*, *Indotristicha ramosissima* and *Dalzellia ceylanica* (Podostemaceae, Tristichoideae): a review. Aquat Bot 57:71–96
36. Sharma BD, Karthikeyan S, Shetty BV (1974) *Indotristicha tirunelveliana* Sharma, Karthik. & Shetty—a new species of Podostemaceae from South India. Bull Bot Surv India 16:157–161
37. Cusset G, Cusset C (1988) Etude sur les Podostemales. 10. Structures florales et végétatives des Tristichaceae. Bull Mus Natl Hist Nat Sect B Adansonia Bot Phytochim 10:179–218
38. Rutishause R, Huber KA (1991) The developmental morphology of *Indotristicha ramosissima* (Podostemaceae, Tristichoideae). Plant Syst Evol 178:195–223
39. Imaichi R, Ichiba M, Kato M (1999) Developmental morphology and anatomy of the vegetative organs in *Malaccotristicha malayana* (Podostemaceae). Int J Plant Sci 160:253–259
40. Koi S, Kato M (2007) Developmental morphology of shoot in *Weddellina squamulosa* (Podostemaceae) and implications for shoot evolution in the Podostemaceae. Ann Bot 99:1121–1130
41. Fujinami R, Imaichi R (2009) Developmental anatomy of *Terniopsis malayana* (Podostemaceae, subfamily Tristichoideae), with implications for body plan evolution. J Plant Res 122:551–558
42. Clark SE, Running MP, Meyerowitz EM (1995) *CLAVATA3* is a specific regulator of shoot and floral meristem development affecting the same processes as *CLAVATA1*. Development 121:2057–2067
43. Brand U et al (2000) Dependence of stem cell fate in *Arabidopsis* on a feedback loop regulated by *CLV3* activity. Science 289: 617–619
44. Schoof H et al (2000) The stem cell population of Arabidopsis shoot meristems is maintained by a regulatory loop between the *CLAVATA* and *WUSCHEL* genes. Cell 100:35–644
45. Kieffer M et al (2006) Analysis of the transcription factor WUSCHEL and its functional homologue in *Antirrhinum* reveals a potential mechanism for their roles in meristem maintenance. Plant Cell 18:560–573
46. Chen S-K et al (2009) The association of homeobox gene expression with stem cell formation and morphogenesis in cultured *Medicago truncatula*. Planta 230:827–840
47. Liang D et al (2009) Molecular dissection of the pea shoot apical meristem. J Exp Bot 60:4201–4213
48. Stuurman J, Jäggi F, Kuhlemeier C (2002) Shoot meristem maintenance is controlled by a GRAS-gene mediated signal from differentiating cells. Genes Dev 16:2213–2218
49. Barton MK, Poethig RS (1993) Formation of the shoot apical meristem in *Arabidopsis thaliana*: an analysis of development in the wild type and in the shoot meristemless mutant. Development 119:823–831
50. Long JA et al (1996) A member of the KNOTTED class of homeodomain proteins encoded by the *STM* gene of *Arabidopsis*. Nature 379:66–69
51. Long JA, Barton MK (2000) Initiation of axillary and floral meristems in *Arabidopsis*. Dev Biol 218:341–353
52. Waites R et al (1998) The *PHANTASTICA* gene encodes a MYB transcription factor involved in growth and dorsoventrality of lateral organs in *Antirrhinum*. Cell 93: 779–789
53. Timmermans MCP et al (1999) ROUGH SHEATH2: a Myb protein that represses *knox*

homeobox genes in maize lateral organ primordia. Science 284:151–153

54. Tsiantis M et al (1999) The maize *rough sheath2* gene and leaf development programs in monocot and dicot plants. Science 284:154–156
55. Byrne ME et al (2000) *Asymmetric leaves1* mediates leaf patterning and stem cell function in *Arabidopsis*. Nature 408:967–971
56. Ori N et al (2000) Mechanisms that control *knox* gene expression in the *Arabidopsis* shoot. Development 127:5523–5532
57. Guo M et al (2008) Direct repression of *KNOX* loci by the ASYMMETRIC LEAVES1 complex of *Arabidopsis*. Plant Cell 20:48–58
58. Landolt E (1998) Anatomy of the Lemnaceae (duckweeds). In: Landolt E, Jäger-Zürn I, Schnell RAA (eds) Extreme adaptations in angiospermous hydrophytes. Gebrüder Borntraeger, Berlin
59. Lemon GD, Posluszny U (2000) Comparative shoot development and evolution in the Lemnaceae. Int J Plant Sci 161:733–748
60. Jäger-Zürn I (1995) Morphologie der Podostemaceae. III. *Dalzellia ceylanica* (Gard.) Wight (Tristichoideae). Trop Subtrop Pflanzen 92:7–77
61. Jäger-Zürn I (2002) Morphology and morphogenesis of ensiform leaves in *Apinagia multibranchiata* and *Mourera fluviatilis* (Podostemaceae–Podostemoideae). Flora 197:394–407
62. Jäger-Zürn I (2005) Structural analysis of the dissected ensiform leaves and shoot morphology of *Marathrum foeniculaceum* (Podostemaceae). Flora 200:229–244
63. Hiyama Y, Tsukamoto I, Imaichi R et al (2002) Developmental anatomy and branching of roots of four *Zeylanidium* species (Podostemaceae), with implications for evolution of foliose roots. Ann Bot 90:735–744

Chapter 6

Brown Algae as a Model for Plant Organogenesis

Kenny A. Bogaert, Alok Arun, Susana M. Coelho, and Olivier De Clerck

Abstract

Brown algae are an extremely interesting, but surprisingly poorly explored, group of organisms. They are one of only five eukaryotic lineages to have independently evolved complex multicellularity, which they express through a wide variety of morphologies ranging from uniseriate branched filaments to complex parenchymatous thalli with multiple cell types. Despite their very distinct evolutionary history, brown algae and land plants share a striking amount of developmental features. This has led to an interest in several aspects of brown algal development, including embryogenesis, polarity, cell cycle, asymmetric cell division and a putative role for plant hormone signalling. This review describes how investigations using brown algal models have helped to increase our understanding of the processes controlling early embryo development, in particular polarization, axis formation and asymmetric cell division. Additionally, the diversity of life cycles in the brown lineage and the emergence of *Ectocarpus* as a powerful model organism, are affording interesting insights on the molecular mechanisms underlying haploid-diploid life cycles. The use of these and other emerging brown algal models will undoubtedly add to our knowledge on the mechanisms that regulate development in multicellular photosynthetic organisms.

Key words: Brown algae, Development, Polarization, Asymmetric cell division, Auxin, Hormone, Life cycle, Cell cycle, *Fucus*, *Ectocarpus*

1. Introducing Brown Algae

The brown algae are a group of photosynthetic organisms or "plant systems" belonging to the Heterokonts (Fig. 1) (1). Heterokonts are an extremely diverse kingdom that includes photosynthetic as well as non-photosynthetic protists classified in 17 classes, including diatoms, chrysophytes (golden algae), and xanthophytes (yellow-green algae), but also oomycetes or water molds, a group of pseudo-fungi many of which are notorious plant parasites such as *Phytophthora* and *Pythium* causing late blight of potato and seed rot, respectively (2, 3). Photosynthetic heterokonts have characteristic

Ive De Smet (ed.), *Plant Organogenesis: Methods and Protocols*, Methods in Molecular Biology, vol. 959,
DOI 10.1007/978-1-62703-221-6_6, © Springer Science+Business Media New York 2013

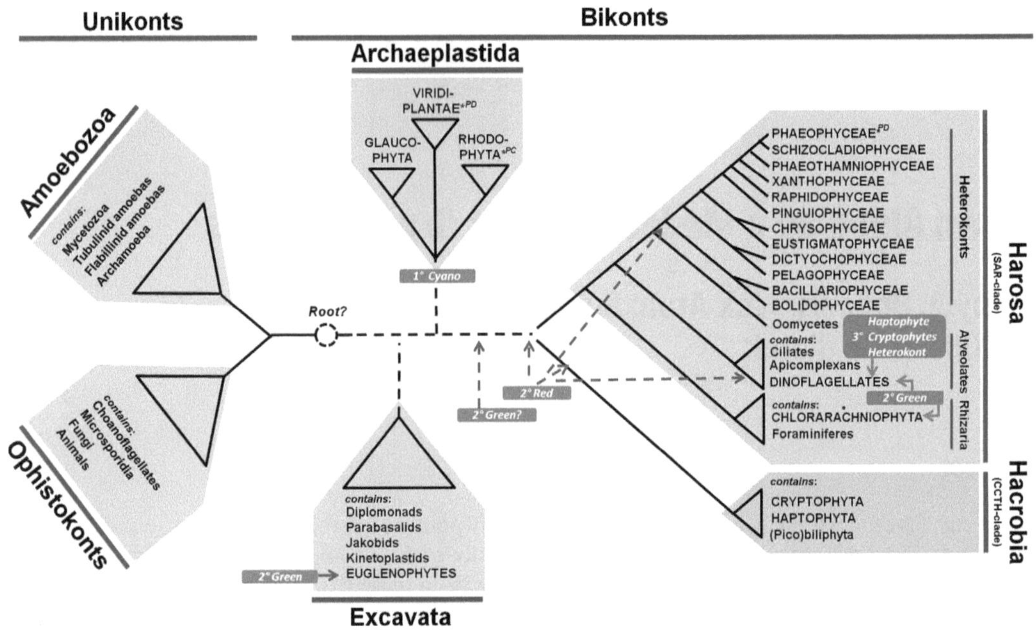

Fig. 1. Relationships between major eukaryotic lineages (after (3, 15, 19, 20, 212–217)) indicating photosynthetic lineages and complex multicellularity. Endosymbiotic gene transfer is indicated with grey rectangles specifying the nature of the organelle donor. *Dashed lines*, and *question marks* refer to controversial phylogenetic relationships among the hypothesized "supergroups." Unresolved relationships are indicated as *polytomies*. Clades containing photosynthetic eukaryotes are marked with capitals. Parenchymatous photosynthetic lineages are marked by an asterisk; 1°, Primary endosymbiosis; 2°, secondary endosymbiosis; 3°, tertiary endosymbiosis; PD, multicellularity by means of plasmodesmata (Phaeophyceae and Viridiplantae); PC, multicellularity by means of pit connections (Rhodophyta); endosymbiotic relationships are indicated by *grey rectangles*. Branch lengths are not proportional to time.

chlorophyll *c*-containing plastids surrounded by four membranes and thylakoids grouped in stacks of three (2). Chloroplast multigene trees established that heterokonts and other chlorophyll *c*-containing algae have acquired their plastids by a secondary endosymbiotic event involving a heterotrophic eukaryote and a red alga (4–8). The relationships among chlorophyll *c*-containing organisms and whether these acquired their secondary plastids independently or from a common ancestor remain uncertain (8–13).

Interestingly, brown algae are one of only five groups of eukaryotes that acquired complex multicellularity, the others being red algae (Rhodophyta), green algae (Chloroplastida, including the land plants), animals (Metazoa), and fungi (Fungi). Being Heterokonts brown algae share a common ancestor with land plants well over 1,500 mya (14). A genomic repertoire distinct from land plants is the result of this long independent evolution, only occasionally disturbed by horizontal and endosymbiotic gene transfer (9, 15). Remarkably, and despite of their evolutionary distance, brown algae and land plants share many common developmental features: (a) following mitosis, daughter cells remain attached to each other, producing a multicellular body plan often

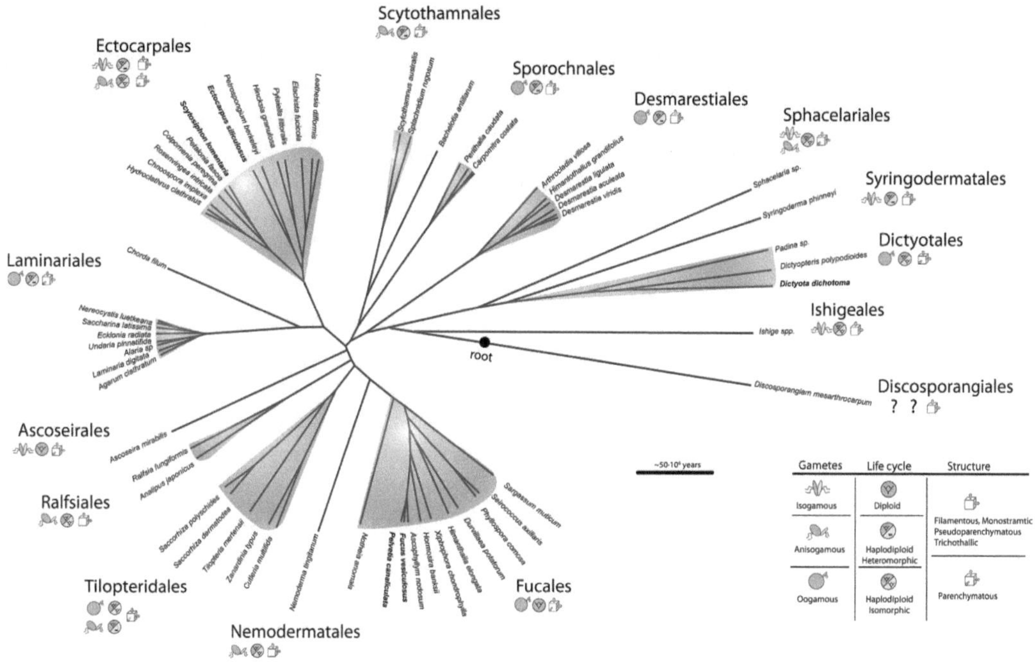

Fig. 2. Schematic representation of brown algal diversity based on phylogenies by Kawai et al. (43) and Silberfeld et al. (18). Branch lengths are roughly proportional to time based on a time-calibrated phylogeny by Silberfeld et al. (18). Gamete dimorphism, type of life history and thallus structure are indicated for the most important brown algal orders. Definitions of thallus structure follow Niklas (28, 200) other character states are gleaned from Silberfeld et al. (18).

showing a considerable amount of cell differentiation. (b) Cells remain embedded in a rigid extracellular matrix which contains cellulose fibrils. (c) Individual cells maintain cytoplasmic continuity through the formation of plasmodesmata. (d) In general, both land plants as well as brown algae use an open-growth strategy in which a potential infinite sequence of new organs is produced by meristematic cells. (e) Some lineages have evolved meristems that are able to divide in three dimensions with daughter cells establishing also secondary cytoplasmic continuity with their direct neighbors (parenchymatous tissues). (f) Multicellular development relies on positional control and clear cell lineages are absent. (g) Unlike animals, totipotency or the ability to dedifferentiate upon isolation is frequent. In contrast, other cell biological features such as the absence of a preprophase band, the presence of centrioles (16), and production of eicosanoid oxylipins (17) set them clearly apart from land plants and are more reminiscent of fungal and animal cells.

Brown algae contain about 1,800 species belonging to 285 genera, all of which are multicellular (18) (Fig. 2). Phaeophyceae is a sister group of the Schizocladiophyceae (19). The latter is a monotypic lineage that shares many characteristics with brown algae, but differs by the absence of plasmodesmata connecting neighboring cells (19). The latter is considered significant with regard to the evolution of complex multicellularity. Intercellular connections, enabling transport of photosynthates and signalling molecules, are crucial in the development of a multicellular, morphologically differentiated thallus. The most recent calibrated phylogenies indicate that the Schizocladiophyceae and Phaeophyceae diverged in the lower Jurassic, ca. 190 mya (20). Subsequent diversification of the Phaeophyceae in the Cretaceous resulted in a bewildering diversity of thallus morphologies, life cycles, and sexual strategies (21–23).

The size of the thallus varies dramatically among species. The sporophytes of kelp species, large brown algae belonging to the order Laminariales, may attain sizes of up to 60 m and grow as fast as half a meter per day (1, 23). They are able to translocate photosynthates, nitrogen, and phosphorous across all parts of the plant using differentiated cells, trumpet hyphae, resembling sieve tubes in land plants (24, 25). Other brown algae are small epi- or endophytes that are rarely to be noticed with the naked eye (26, 27). Thallus complexity varies from uniseriate, branched filaments with diffuse growth over filamentous thalli with localized meristematic zones or filaments uniting to form a pseudoparenchyme, to complex thalli whereby a single or a group of meristematic cells divides in three dimensions and the produced daughter cells establish also secondary symplastic continuity with their direct neighbors to produce parenchymatous tissue (28).

In this review chapter we focus on how brown algal models can be attractive and powerful systems that will offer a wider phylogenetic perspective of the underlying processes controlling development. The similarities in development between brown algae and land plants and the advantages of brown algae in terms of manipulation for a number of experimental protocols have led researchers to use fucoids, such as *Fucus* and *Silvetia* (formerly called *Pelvetia*), as general models for a number of mechanisms such as embryogenesis, polarity, and asymmetric cell division (29–33). Additionally, the diversity of life cycles among brown alga, and the particularly complex haploid–diploid life cycle of *Ectocarpus* has drawn the interest of evolutionary ecologists as well as functional biologists (21, 34, 35).

In recent years, following the selection of *Ectocarpus* as a model for the brown algae (31), a considerable effort was invested in the development of genomic and genetic tools for this organism. The most important of these was the assembly and analysis of the complete 214 Mbp genome sequence (36), together with the development of a range of molecular tools including whole genome tiling arrays, deep sequencing of small RNAs, mutant screens (31), microarray analysis of gene expression (37), a sequence-tagged genetic map (38), stramenopile-adapted bioinformatic tools (39), proteomics (40), and metabolomic methodologies. Finally, we also address the question of whether plant hormones, such as auxin, are involved in the growth and development of brown algae. The latter question has been raised and addressed several times in the past (41), but despite numerous reports indicating the presence of plant hormones in brown algae the effects on development remain controversial.

2. Life Cycle and Development

Sexual life cycles in the eukaryotes involve a cyclic alternation between diploid and haploid generations with meiosis mediating the transition from the diploid to the haploid state and cell fusion (syngamy) reconstituting a diploid genome. Brown algae are highly interesting models for studies on the life cycles not only because of their evolutionary distance from other eukaryotic lineages but also because they exhibit a broad variety of life cycles, ranging from isomorphic haploid–diploid life cycles, in which both gametophyte and sporophyte generations exhibit multicellular development, to diploid life cycles, where only the diploid generation of the life cycle is multicellular (reviewed in ref. 34).

The ancestral brown algal sexual life cycle most likely involved alternation between two isomorphic generations: the diploid sporophyte and the haploid gametophyte (42, 43). However, many deviations from this situation are found in extant brown algae, which have independently evolved a remarkably fluid range of life cycles, usually involving morphological differentiation and reduction of one of the two generations (Fig. 3). For instance, in the kelps the gametophyte generation is reduced but still develops independently of the sporophyte. In the Fucoids, the gametophyte generation has been completely lost, producing a diploid life cycle. Variations in life cycle structure can even be seen within a single order, as demonstrated by the Ectocarpales in which some families have life cycles with isomorphic generations (the Acinetosporaceae) and other families have strongly heteromorphic generations, with either the gametophyte (Chordariaceae, Adenocystaceae) or the sporophyte (Scytosiphonaceae) generation being microscopic (44).

Gametes of brown algae may be of equal size (isogamous), unequal size (anisogamous), or differentiated into female eggs and male sperm cells. Isogamy most likely presents the ancestral condition (42) and several lineages have evolved independently towards anisogamous and oogamous sexual strategies. Interestingly, the life history of the model brown alga *Ectocarpus* involves alternation between two morphologically distinct, sporophyte and gametophyte generations, both free living and macroscopic (Fig. 2). In addition to this sexual cycle, an asexual cycle presents evidence of the extraordinary reproductive and developmental plasticity of *Ectocarpus.* Gametes that do not meet a partner of the opposite sex can still germinate parthenogenetically to produce haploid parthenosporophytes, a proportion of which will endoreduplicate to produce homozygous diploid parthe-

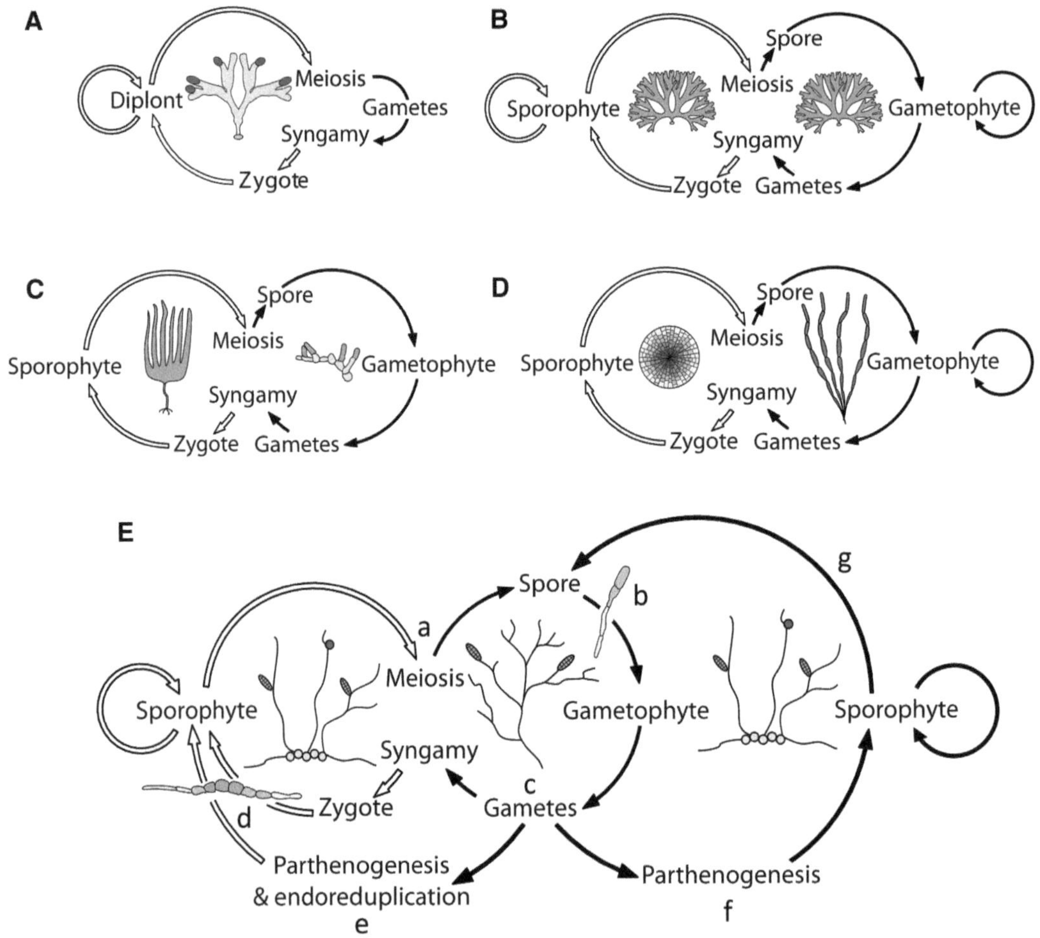

Fig. 3. Schematic representation of brown algal life cycles. Diploid and haploid stages are marked by *white* and *black arrows*, respectively. (**A**) Fucales are characterized by a diplontic life cycle. Meiosis in the reproductive tissue is immediately followed by gametogenesis and syngamy producing a diploid zygote (218). (**B–D**) Most brown algae exhibit a diplohaplontic life cycle where a haploid gametophyte alternates with a diploid sporophyte. Here following meiosis the resulting spore develops into a multicellular organism. Both haploid and diploid phases may be of identical morphology (isomorphic) (**B**, *Dictyota dichotoma*) (219), or one of both phases may develop differently (heteromorphic) with either the gametophyte (**C**, *Laminaria digitata*) (220) or the sporophyte (**D**, *Scytosiphon lomentaria*) (221) being microscopic. (**E**) Simplified diplohaplontic life cycle of *Ectocarpus siliculosus* (31, 45). Meiosis (a) takes place in the sporophyte (diploid) to produce haploid spores. First cell division in germinating spores is asymmetric (b) and they grow into multicellular gametophytes. Gametophytes produce morphologically identical but physiologically differentiated male and female gametes (c), which fuse to form a zygote. After a symmetric first cell division (d) the zygote grows into a diploid sporophyte. Alternatively, gametes that do not meet a partner of opposite sex grow parthenogenetically into diploid parthenosporophytes by means of parthenogenesis combined with endoreduplication (e) or in a haploid parthenosporophyte (f). The latter produce meiospores via a nonreductive apomeiotic event (g). Loops connecting a generation with itself denote asexual reproduction mediated by vegetative reproduction (e.g., fragmentation, propagule formation) (**A**, **B**) (219, 222), the formation of asexual spores (mitospores) (**B**, **E**) (45, 223), or parthenogenetic development of unfertilized gametes (**D**) (221).

nosporophytes (45). Of particular significance is the fact that haploid or endoreduplicated parthenosporophytes and diploid heterozygous sporophytes (produced from a zygote) are morphologically indistinguishable, and no obvious correlation exists between ploidy level and cell volume. This is in contrast to what is typically described for most land plants where the level of ploidy is closely associated with the size of cells (46).

Gametophyte and sporophyte generations in *Ectocarpus* develop independently of the parent organism from a single progenitor cell that is released into the surrounding sea water. Remarkably, the morphological dissimilarity between generations is already discernible at the first cell division, which is asymmetrical in the gametophyte (just as in fucoids, producing a rhizoid cell and a "thallus" cell) as opposed to symmetrical in the sporophyte (31). It would be tempting to speculate that the pattern of this first cell division is linked to the developmental fate as a sporophyte or as a gametophyte. Mutant studies suggest that this may not be the case: the sporophyte generation of the *Ectocarpus* life cycle mutant *immediate upright* exhibits an asymmetrical first cell division (typical of a gametophyte) yet still develops into a fully functional sporophyte (31).

What are the biological functions of the two heteromorphic generations in haploid–diploid life cycles? Possible advantages of possessing distinct gametophyte and sporophyte developmental patterns include allowing the two generations to exploit alternative ecological niches or to avoid biotic aggressors due to differential susceptibility. Interestingly, the degree of morphological dissimilarity between the generations is highly variable between ecotypes, although a relationship between the degree of dissimilarity and the homogeneity of the environment, in terms of niche or pathogen prevalence, remains to be demonstrated.

How are the sporophyte and gametophyte developmental programs coordinated with the alternation between meiosis and syngamy? Molecular approaches using established model systems with haploid–diploid life cycles such as *Arabidopsis* would be expected to shed light on this matter. From a practical point of view, however, identification of genes that synchronize sporophyte/gametophyte development with life cycle progression can be hampered by the fact that mutations in such genes would most probably induce gametophyte or embryo lethality. In theory, though, a mutation in a gene that regulates the transition between the two generations could also lead to the development of the "wrong" generation (for example, production of a gametophyte where a sporophyte would be expected). Redirection from a gametophyte to a sporophyte developmental program occurs during androgenesis and gynogenesis in land plants but the molecular mechanisms that regulate these processes have not yet been accessed genetically. *Ectocarpus*, therefore, represents a promising system to search for mutations that affect the switch between the sporophyte and gametophyte generations.

In the brown algae, the relationship between ploidy and the alternation of generations during the life cycle is not absolute, and the "classic" idea that gametophytes are haploid and sporophytes are diploid is not strictly applicable to these organisms. In *Ectocarpus*, both gametophytes and sporophytes can be haploid or diploid, showing that nuclear ploidy and life cycle generation are uncoupled (47). Therefore, the "choice" to deploy the gametophyte or the sporophyte developmental program is not determined by the ploidy of the initial cell but rather is under genetic control. Two single-locus mutations supporting this interpretation have recently been isolated in *Ectocarpus*. These mutations cause partial and complete conversion, respectively, of the sporophyte into a gametophyte developmental program (31, 35). The identification of these two master regulatory genes responsible for the switch between life cycle generations will surely add to our knowledge on how development is controlled at the level of the whole organism.

3. Cell Polarization and Asymmetric Cell Division

Asymmetric cell division (ACD) is a nearly universal key mechanism that generates cellular diversity in complex multicellular systems. Because of its importance, the mechanisms involved in ACD have been investigated in a wide range of model systems. Cell polarity, the asymmetrical distribution of organelles, is a fundamental prerequisite for ACD, and it enables tip growth (48), lobe formation (49), polar diffuse growth (50), localized secretion (51), cell motility (52), and polar transport of morphogens such as auxins (53).

Zygotes of brown algae have served as practical model systems for cell polarization and ACD, a cell division which produces two or more daughter cells with different cell fates, since the late nineteenth century (e.g., refs. 54, 55). In particular, zygotes of *Fucus* and *Silvetia* have proved to be tractable cell biological models with important advantages over angiosperm systems. Studies on the establishment of polarity in angiosperms are technically challenging because asymmetrically dividing cells are often embedded in the mother tissue or surrounding cells (e.g., the zygote and lateral root initiation respectively) and polarity is often predetermined (e.g., pollen tubes). Fucoid eggs, conversely, are released in the surrounding seawater as radial symmetric spheres. Polarity is established after fertilization and can be manipulated by external cues. Development of embryos is highly synchronous, facilitating pharmacological treatments, cell cycle and biochemical studies (56). These characteristics have enabled researchers to manipulate the polarization axis according to the needs of the experimental setup. In addition, zygotes of oogamous brown algae are relatively large (50–100 μm) making it relatively easy to visualize the internal cytological organization and making them amenable to micromanipulation (microinjection, patch-clamping). The released eggs provide a "natural" system to study synchronous cell wall differentiation, in contrast to protoplasts in land plants which are produced using potentially deleterious agents such as lytic enzymes. Finally, in vitro cultured zygotes maintain their polarity, while most of the cells from multicellular plants dedifferentiate and lose their original polarization axis upon isolation and in vitro culture (57).

The historic use of fucoid reproduction and early development as a model system has provided a number of crucial findings. In fucoid zygotes it has been clear for decades that ACD is dependent on both internal and external factors (58). Fucoid zygotes, being able to sense a whole range of environmental gradients and polarize accordingly, have put the concept of multiple polarization signals at the forefront (59). The widespread and fundamental role of free cytoplasmic calcium (Ca^{2+}cyt) gradients (60, 61), the cytoskeleton (32, 62), and their interplay in the establishment of cell polarization, tip growth and fate determination in plant systems resulted to a considerable extent from insights gained on fucoid zygotes. The same applies to the importance of the plasma membrane (63) and targeted secretion to the cell wall (64, 65). Additionally the fucoid zygotes were the first plant systems to show the importance of increases in free Ca^{2+}cyt and membrane depolarization in egg activation (reviewed in refs. 66, 67).

ACDs are traditionally divided in two categories each using a different mechanism (68, 69). The first mechanism involves the asymmetrical distribution of fate determinants according to a polarization axis established prior or during cell division. Subsequently, orientation of the mitotic spindle in parallel with the polarization axis ensures that one daughter cell will inherit most of the determinants. Because the cell fate is controlled by cell fate determinants within the cytoplasm, this mechanism is called "intrinsic" or "cell-autonomous." The second mechanism, in contrast, involves a cell producing two initially identical daughter cells, which differentiate via positional signalling. Because cell fate in the latter scenario is controlled by external sources, the mechanism is termed "extrinsic" or "non-cell-autonomous."

The polarization mechanism and ACD in the fucoid zygote has shown that both intrinsic and extrinsic cues can operate simultaneously (58) (Fig. 4). After fertilization, if there are no detectable environmental cues, the sperm entry site will determine the axis of polarity (70, 71). However, the zygote remains responsive to an impressive range of environmental cues (blue light, gravity, ionic, electrical and osmotic gradients, presence of neighboring zygotes, etc.) until the polarization axis is fixed about 15–18 h after fertilization (72–77). The cell will preferentially polarize according to these environmental cues and consequently its mRNA (78, 79), endomembrane system (80), sulfated fucans in the cell wall (81), dihydropyridine receptors (presumably calcium channels) (81, 82), and vitronectine (83) will become asymmetrically distributed.

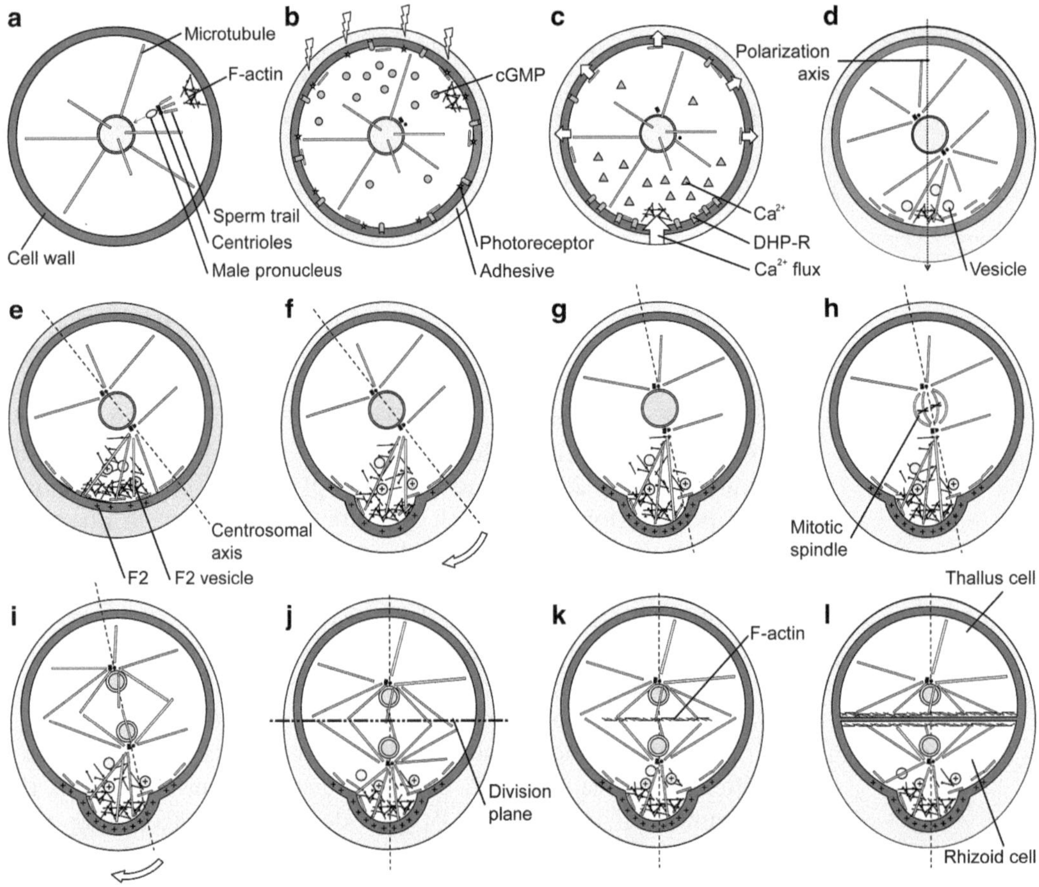

Fig. 4. Schematic overview of a working model of photopolarizaton and asymmetric cell division in fucoid zygotes. (**a**) Following fertilization the sperm pronucleus migrates towards the egg pronucleus. At the site of sperm entry, an F-actin/Arp2/3 patch is established and marks the default rhizoid pole. The microtubular cytoskeleton is mainly nucleated at the level of the nuclear envelope. (**b**) A couple of hours after fertilization the zygote secretes a uniform adhesive and attaches to the substrate. The cell becomes sensitive to the direction of the light. Rhodopsin-like receptors perceive the direction of the incoming light and establish a cGMP gradient decreasing from the thallus towards the rhizoid pole. In the meantime karyogamy has taken place and the centriole pair is deposited near the nuclear envelope. (**c**) The centrioles migrate towards two opposite poles independently of the future polarization axis (*dotted arrow*). A new F-actin patch is established at the shaded side of the zygote. Putative calcium channels (DHP-R, dihidropyridine receptors) accumulate progressively at the rhizoid pole. A calcium gradient is established decreasing from the future rhizoid pole towards the thallus pole. (**d**) Concurrently, the endomembrane system becomes asymmetrically distributed at the rhizoid pole and a polar secondary adhesive is deposited in addition to the uniform primary one. Also the cortical and cytoplasmic microtubular cytoskeleton become polarly distributed at the rhizoid pole early in the polarization process. The centrioles have arrived at opposite poles and the two centrosomes have become the most prominent nucleating centers of the cell. (**e**) During the polarization process the F-actin patch expands to form a cone extending from the cortex at the rhizoid pole towards the nucleus. Targetted secretion towards the rhizoid pole establishes the ASC ("axis stabilizing complex"). Among others a highly sulfated fucan (F2) is secreted at the rhizoid pole. (**f**) Following F2 localization tip growth starts. Microtubules that get stabilized at the level of the ASC result in an asymmetry in the microtubular distribution which is thought to generate the force for the nuclear rotation which aligns the centrosomal axis with the polarization axis. (**g**) Prior to the metaphase the centrosomal axis is partially aligned. (**h**) The mitotic spindle is established in a partially aligned orientation. (**i**) After metaphase the alignment is completed. (**j**) Out of the two centrosomes microtubules radiate and delineate the plane of cell division at the zone where they interdigitate. (**k**) F-actin is deposited at this plane and the cell plate is deposited in a centrifugal way. (**l**) Similar to land plants, the asymmetric cell division of the zygote establishes the apical–basal pattern with a large thallus cell and a smaller rhizoid cell.

These asymmetries are established before cell division and therefore represent intrinsic control over ACD which produces two daughter cells with a different fate (one rhizoid cell and one thallus cell).

Nowadays it is clear that every ACD depends on both intrinsic and extrinsic factors (84). In spite of the fact that this categorical framework has been fruitful in the research of both internal and external factors, it stretches the truth in the sense that the framework concentrates on the state of the daughter cells as its starting point rather than approaching ACD as a developmental process (85). Therefore it neglects the fact that the asymmetrical state of the *Fucus* thallus and rhizoid cell is dependent on external information in the first place. Similarly it has been shown that the orientation of the theoretically extrinsic ACD of *Drosophila* male germline stem cells depends on the intrinsic asymmetrical retention of the mother centrosome at the stem cell niche and migration of the daughter centrosome towards the opposite cell pole (86).

The dependence of the ACD on cues as light, gravity, presence of neighboring tissue and the sperm entry site shows the fucoid zygotes are receptive to multiple polarization cues rather than only one. ACDs in plant systems are also receptive to multiple cues, but being in a multicellular context these signals are rather ligand–receptor mediated signals (87) or cell-to-cell movement transcription factors and miRNAs instead of physical factors as light and gravity like in fucoids (*see* ref. 59 for a review).

The mechanisms underlying the perception of the environmental signals in fucoids remain obscure, although a rhodopsin-like protein (88), cGMP (89) and redox activity at the plasma membrane (90) have been proposed to be involved in perception of the light direction (*see* refs. 91, 92 for a review). A speculative model involving calcium channel untethering at the lit side and tethering at the dark side in response to local cGMP concentration has also been proposed (92).

3.1. The Role of Ca^{2+}

Calcium appears to be an important player in cell polarity signalling during for example pollen growth, polarized cell expansion of pavement cells and trichome development in *A. thaliana* (57). The involvement of Ca^{2+}cyt in asymmetric cell division of angiosperms, however, has not yet been unequivocally established (84), and this is mostly due to difficulties in cell manipulation. The pioneering work of Jaffe and coworkers (93–95) showed that fucoid zygotes establish endogenous ionic currents consisting of mainly Ca^{2+}, K^{+}, and Cl^{-} soon after fertilization. This research drew interest towards the important role for Ca^{2+}cyt in polarization and early development (61, 96).

The first evidence for a Ca^{2+} influx at the future rhizoid pole and efflux at the thallus pole resulting in a high Ca^{2+}cyt at the differentiating rhizoid pole came from the use of $^{45}Ca^{2+}$ isotopes as early as 1975 (97). Later, it was shown that rhizoids preferentially germinate towards the side exposed to higher concentrations of Ca^{2+} ionophores (98). The use of calcium sensitive microelectrodes, calcium reporter Quin-2 (60) and Fura-2 (99) established directly a high Ca^{2+}cyt at the rhizoid tip. Just as in other tip growing systems, the high Ca^{2+}cyt in polarizing and tip growing zygotes is regulated by reactive oxygen production (ROS), illustrating the commonality of the ROS-Ca^{2+}cyt motif in development despite phylogenetic distances (29, 30).

The role of calcium influx during polarization of *Fucus* zygotes has been studied using different calcium buffers (100), calcium reporters (101), patch and current clamp techniques (102, 103) and ion substitutions or additions to the medium (104, 105). By quantitative microinjection of Calcium Crimson dextran, high concentrations of Ca^{2+}cyt were detected already within 1 h after exposure of the cells to unilateral light (101). These elevations are F-actin dependent (106). Pharmacological data and antibody microinjections suggest a role for the conserved calmodulin protein downstream of Ca^{2+}cyt in the polarization process of fucoids (101, 105). Further information on the role of calcium and calmodulin in fucoid polarization can be found in several excellent reviews (92, 107, 108).

Upon selection of the polarization axis based on external factors the axis is still labile and can be changed for example if the current axis is not parallel to the current light cue (109), indicating that polarity is acquired in distinct phases (59). Later, the axis is amplified by a positive feedback loop involving local increases of Ca^{2+}cyt and secretion at the rhizoid pole (see ref. 110 for a review). It is still unclear how this positive feedback loop is achieved. It has been suggested that Ca^{2+}cyt induces local secretion of Ca^{2+} channels at the rhizoid pole (91). This hypothesis, however, contradicts the observation that inhibiting secretion does not affect redistribution of dihydropiridine receptors (putative Ca^{2+}-channels) (111). An alternative hypothesis involves Ca^{2+} elevations which induces cell wall softening, secretion and consequently allows turgor pressure to locally expand the cell wall and the plasma membrane while activating stretch-sensitive calcium channels (see ref. 96 for a review). Today, fucoid zygotes remain a valuable system for Ca^{2+} research (96) and serve in the development of new technical advances such as biolistic delivery of Ca^{2+} dyes into plant cells (112).

3.2. The Role of the Cytoskeleton

A lot of effort has been invested in determining the role of the cytoskeleton in polarity acquisition and ACD in *Fucus* zygotes (see ref. 113 for a review). F-actin, nucleated by the Arp 2/3 complex (109), localizes at the future rhizoid pole within minutes after axis selection (110). The actin patch then increases and expands to form a cone extending from the cortex at the rhizoid pole towards the nucleus (114). A similar importance of F-actin for the acquisition of polarity is corroborated by studies on protoplast regeneration in *Macrocystis pyrifera* (115) and on the ACD during branch initiation of *Sphacelaria rigidula* (116). Interestingly, asymmetric cell division of stomatal subsidiary mother cell in *Zea mays* also involves an F-actin patch extending towards the nucleus to localize to the pole close to the guard mother cell (117). F-actin plays an important role during polar growth of for example the *A. thaliana* root hair (118), *Nicotiana tabacum* pollen tubes (48), *A. thaliana* pavement cells (119), and *A. thaliana* trichomes (120). In these systems, F-actin is highly dynamic, oscillating between a polymerized and depolymerized state. These dynamics are regulated by effectors of ROP1, a Rho-like GTPase (119, 121, 122). These small GTPases act as molecular switches in many processes, both in animals, plants and yeast (123, 124). In *Fucus distichus* two small GTPases, FdRac1 and FdRab8, have been detected using degenerated PCR. Sequence analysis indicates that these proteins are members of respectively the Rho-family and the Rab-family (125). FdRac1 has been shown to accumulate at the rhizoid pole of polarized zygotes (125) and the Rac1 inhibitor NSC23766 has been shown to inhibit endomembrane polarization and adhesive polarization (126). The finding that Rac1 has a function in ACD in a heterokont alga supports the hypothesis that the role of Rho GTPases in cell polarity is a common feature in eukaryotes (125).

In land plants the orientation of cellulose microfibril deposition in the cell wall is controlled by the cortical microtubular cytoskeleton (127). The orientation of these microfibrils determines the direction of polar cell expansion (127, 128). Brown algae also contain cellulose and hexameric cellulose synthase complexes (129–131); however, cortical F-actin determines the orientation of the cellulose microfibrils (132–134). Cortical microtubules in fucoid zygotes have been detected using immunofluorescence (135), and microinjection of fluorescent tubulin monomers (136). It has been proposed that they have a role in tip growth (136). Additionally, the cytoplasmic microtubular skeleton has been shown to polarize and co-localize with the endoplasmatic reticulum and to direct exocytosis towards the rhizoid pole (137). The finding that microtubules are important for exocytosis and tip growth, just as in land plants (138), is corroborated by studies on the tip growing apical meristem cells of *Sphacelaria*. The polarization of the microtubular skeleton in fucoid zygotes is regulated by phospholipase D, which demonstrates the conservation of phospholipase D signalling in microtubule regulation in eukaryotes (139) and illustrates the similarities with polarity establishment during tip growth in land plants (see ref. 33 for a recent review).

3.3. Division Plane Determination and Cytokinesis

In contrast to animals cells and brown algae, land plants do not have centrioles in their microtubule organizing complex (140). In animal cells and budding yeast it is the position of the centrosomal axis that determines the orientation of the division plane. The spindle is reoriented by a "search and capture" mechanism that aligns the centrosomal axis with the polarization axis and off-center placement of the nucleus can be achieved by stronger pulling forces at the pole of the smaller cell (141, 142). It is still unclear how the plane of cytokinesis is determined in angiosperm cells. In some land plant cells the future cell plate is positioned by the pre-prophase band (PPB), a dynamic microtubular band beneath the plasma membrane that predicts the position of the plasma membrane (see ref. 143 for a review). The land plant PPB and the animal spindle have opposite relations and therefore it has been argued that the "plant polarity proteins" that are asymmetrically distributed towards one pole of the polarization axis are not able to coordinate the division plane orientation during asymmetric cell division like they do in animals and yeast. Consequently plant division mechanisms are regarded as intrinsically different (85).

In brown algae, correct positioning of the plane of cytokinesis is of fundamental importance, as misaligned asymmetrical division in fucoid zygotes disrupts normal development (81). In contrast to land plants, the position of the cell plate is determined in a way similar to furrow plane determination in animals and budding yeast. Similar to animals, in brown algae centrioles serve as the most important nucleation sites for microtubuli. One important difference, however, is that the centrioles duplicate and migrate to opposite poles soon after the last cytokinesis and not during S-phase of the cell cycle. Therefore, almost all vegetative cells have two pairs of centrioles instead of only one. Using polyspermic zygotes of *Scytosiphon* and zygotes of *Fucus* it has been shown that the position of the cytokinetic plane is determined by the two centrosomes out of which microtubules radiate and delineate the plane of cell division at the zone where they interdigitate (144, 145). During the first cell cycle of fucoid zygotes, the centrosomal axis is rotated until it is parallel with the polarization axis using a mechanism similar to the "search and capture" mechanism of microtubuli in animals and budding yeasts (see ref. 32 for a review). A similar reorientation of the centrosomal axis parallel to the new polarization axis has also been observed during branch formation in *Ectocarpus siliculosus*, *Sphacelaria rigidula*, and *Macrocystis pyrifera* (116, 146). In highly vacuolated *Macrocystis pyrifera* protoplasts and vegetative cells, however, the cytokinetic mechanism has been modified (147) in a way more reminiscent to the one in land plants (148).

The finding that fucoid zygotes use a division mechanism similar to ophistokonts suggests that the direct specification of the plane perpendicular to the polarization axis is not inherently coupled to plant developmental biology. In land plant cells with a PPB it remains completely unknown how the PPB is positioned (149). In other land plants cells like pollen (150) and *Physcomitrella patens* protonemata (151) no PPBs are present, still the cell plate is positioned according to the polarization axis of the cell. This indicates that an additional mechanism should be involved in division plane determination in land plants (59). Interestingly it has been recently suggested that the division mechanism in land plants and animals does not differ that substantially as previously thought. During cell division of *A. thaliana* GFP-tagged plus-end binding EB1a and EB1b were reported to form polar caps during prophase and to penetrate inside the nuclear envelope (152–154). The polar caps were suggested to take up a role similar to centrioles in ophistokont and heterokonts. During tobacco pollen development, which lacks PPBs, also a microtubular cap structure was reported during the ACD and the EB1 gene product has been suggested to interact with TOBACCO MICROTUBULE BUNDLING POLYPEPTIDE OF 200 kDa (TMBP200), a MAP215/DIS1-family member which was shown to be essential for asymmetric cell division (150).

Brown algal cytokinesis usually occurs by centrifugal outgrowth of a partition membrane which gets progressively filled with cell wall material delivered by the Golgi bodies and transported by microtubules, as in land plants. However, representatives of the Dictyotales–Sphacelariales–Syringodermatales clade such as *Dictyota* and *Halopteris* (no clear pattern) and *Sphacelaria* (centripetal) (155) present an exception to this pattern. There are also additional differences compared to land plants like the involvement of ER derived flat cisternae (156) that are supported by a peculiar actin structure at the future division plane (133, 157) and probably serve as the main membrane source for the partition membrane. Golgi derived vesicles in contrast rather serve as the main source of cell wall material (recently reviewed in refs. 155, 158).

3.4. The Axis Stabilizing Complex

Fucoid zygotes have been instrumental in appreciating the importance of the plasma membrane, cortical sites, cell wall and targeted secretion for patterning and morphogenesis in plant systems (reviewed in refs. 63, 65, 159). In earlier work it has been established that the period during which the zygotes can reorient their polarization axis is prolonged as long as the cell wall is repeatedly removed (160). Similarly when the cell wall of a rhizoid cell is removed, the cell is able to regenerate but loses its polarity (161). Sulfated F2 fucans and at least one binding protein are preferentially secreted at the rhizoid pole with the same kinetics as the final alignment (63, 111). Also tip growing meristem cells of *Sphacelaria* lose their polarity following cell wall removal and divide symmetrically (162). Together with the idea that F-actin localizes to the rhizoid pole, this led to the hypothesis that transmembrane interactions between the cytoskeleton and the cell wall at the rhizoid pole (collectively termed the "axis stabilizing complex" or "ASC") are involved in polar growth. This positional information at the

ASC may also play a role in the rotation of the centrosomal axis that aligns it with the polarization axis, because inhibiting polarized secretion to the cell wall with brefeldin A inhibits not only tip growth (as expected) but also alignment of the centrosomal axis with the polarization axis (81). Inhibiting polarized secretion, however, does not inhibit polar localization of dihydropyridine receptors and F-actin. Next to a role in cell plate determination, polar growth and fixation of the polarization axis, the ASC also plays a role in cell fate specification. Laser-ablation of one cell of two-celled zygotes established that the remnants of cell walls are able to induce a thallus or a rhizoid cell fate in the daughter cell of the regenerating remaining cell that comes into contact with this cell wall (64). The finding that intrinsic factors are not necessarily transcription factors, mRNA's, membrane proteins and kinases but might be secreted cell wall components is intriguing and calls for additional research (163).

Also for land plants the role of targeted secretion of fate determinants towards cortical sites is relevant (reviewed in ref. 65). In addition the recently discovered BREAKING OF ASYMMETRY IN THE STOMATAL LINEAGE (BASL) in *Arabidopsis* meristemoid mother cell ACD (164) and the LRR-RLK PANGLOSS1 (PAN1) in maize subsidiary mother cell ACD (165), whose localization overlaps with the F-actin patch, also have polarized distribution patterns at a cortical domain and are reported to control ACD. Polar localization of PIN-FORMED (PIN) proteins is dependent on the actin cytoskeleton and on localized secretion towards cortical sites (53, 166). The apical–basal pattern of zygotes of tobacco that develop in vitro is dependent on the presence of the original cell wall, and even cell walls that hang loosely around the zygote are enough to signal the apical–basal axis (167). Interestingly arabinogalactan proteins were reported to be asymmetrically distributed in the tobacco zygote cell wall before ACD and pharmacological interference with these arabinogalactan proteins increases the frequency of symmetrical cell divisions (168).

3.5. Cell Cycle and Development

Zygotes of brown algae serve as excellent model systems for cell cycle research as they are naturally synchronized by fertilization (56). For example the inhibition of CDK proteins through a tyrosine phosphorylation at the S/M DNA replication checkpoint was first demonstrated in fucoid zygotes (56) and later confirmed in *A. thaliana* (169). Because fucoid zygotes polarize and divide asymmetrically free from the surrounding tissue, they are particularly interesting models for the study of the interactions between cell cycle progression and the early patterning in a unicellular context because the first cell cycle of the zygote is concomitant with cell polarization. Cell cycle alterations during late embryogenesis and organogenesis of animals and adult development of land plants seem to have little effect on patterning. In contrast, cell cycle alternation during early embryogenesis in animals and budding in the unicellular yeast seems to result in severe effects on patterning because of a tight coordination between cell cycle and polarization. Indeed, the G1/S checkpoint in fucoid zygotes simultaneously controls S-phase entry and polarization, confirming that this trend also extends to plant systems. Coordination of cell cycle and polarization in *Fucus* resembles the cell cycle of the unicellular budding yeast in its presence of a DNA replication checkpoint, a G1/S checkpoint which coordinates cell cycle and polarization (170), but it differs in the lack of a morphogenesis checkpoint (171).

Fucoid zygotes are excellent systems to test whether there are differences in cell cycle regulation between early embryos and somatic cells in plant systems. The cell cycle in early embryos of animals differs substantially from cell cycle regulation in somatic cells. S-phases and M-phases alternate in quick succession without G1 and G2 phases which is regulated by periodic synthesis and degradation of cyclin B which activates the CDK CDC2 (172, 173). In somatic cells during organogenesis or upon terminal differentiation, however, the cell cycle appears to be more tightly controlled by CDKs then in the early embryos (reviewed in ref. 174). In *Fucus* zygotes, however, the cell cycle is also tightly regulated and parallels a somatic cell cycle (170). This confirms the general idea that embryogenesis in plant systems in contrast to animals is not a distinct and independent stage in a plants life history, illustrating how brown algae can serve as an phylogenetic independent source of information for understanding plant systems.

4. Hormonal and Positional Control

4.1. A Long-Standing Controversy

In animal systems cell fate decisions during embryogenesis are based on interactions between the cells or parts of the embryo ("regulative development") or by predetermined segregation of the intrinsic determinants specifying cell lineages ("mosaic development"). Plant systems typically develop in a more indeterminate way in which the plant lays out only the basic body plan during embryogenesis and adapts its adult shape and physiology to the environment by responding to external stimuli. Consequently positional information rather than cell lineages play the most important role in cell fate decisions of plant systems (175). Positional information is often controlled by gradients of diffusible factors which are interpreted by the target tissue (176). Some phytohormones such as auxin take up the role of such a diffusible factor comparable to animal morphogens or play an important role in growth regulation (177).

From regeneration experiments after ablation or wounding using *Sphacelaria* (178) or *Dictyota* meristems (179) and *Fucus* embryos (180) it is clear that brown algal development is under positional control. Whether the classical phytohormones of land plants play a role in organogenesis of brown algae is a long-standing but intriguing question. Many old and new reports provide arguments for a function for auxins, cytokinins, abscisic acid, and gibberellins in a wide range of brown algal developmental processes such as tropisms, polarity acquisition, gametogenesis, apical dominance, and branching (reviewed in refs. 181–187). Most attention has been drawn towards IAA (indole-3-acetic acid), an auxin, and consequently we focus on this phytohormone.

Despite the many reports of a role for auxins in brown algal development their conclusiveness remains highly debated. A lot of the older reports involve bioassays in which purified algal extracts are applied to *Avena* coleoptiles and the degree of curvature is tested. Despite being indications for the presence of phytohormones in these extracts, positive reactions may result from aspecific activity rather than presence of auxins (188). More recent evidence involves the identification of the phytohormones in the extracts using GC-MS or HPLC. In fruiting tips of *F. distichus* concentrations up to 10 ng/g fresh weight have been recently reported, which are in the same order of magnitude as in land plants (189, 190). However, the mere detection of phytohormones does not provide a conclusive argument for its regulative potential on brown algal organogenesis. For example auxin has been detected (using GC-MS) also in unikonts, including human cerebrospinal fluid, at concentrations up to 15 ng/ml (191). Therefore these compounds may just be ordinary intermediates or breakdown products of the tryptophan metabolism. In addition, most if not all of the studies used field collected material or non-axenic cultures, which raises the question whether the detected phytohormones are actually produced by brown algae and not by epiphytic bacteria (188). Axenic cultures are difficult to obtain and often seem to be nonviable. Axenic cultures have been obtained from *Fucus spiralis* and *Ascophyllum nodosum*, but the thalli had an aberrant morphology and limited growth relative to the non-axenic controls (192, 193). Interestingly in *Ascophyllum* the aberrant phenotype could be (at least partially) rescued by respectively applying phenylacetic acid (PAA) and *p*-hydroxy-phenylacetic acid (*p*-OHPAA) and combinations of indole-3-acetic acid (IAA), 6-(γ,γ-dimethylallylamino)-purine (2iP), and zeatin at concentrations ranging around 10^{-7} M (193). PAA and IAA are next to indole-3-butyric acid (IBA) and 4-chloroindole-3-acetic acid (4-Cl-IAA) one of four endogenous auxins of higher plants (194). Zeatin and 2iP are cytokinins. These observations suggest that part of the detected auxins might be of bacterial origin. Regardless of their origin, however, these findings do illustrate the regulatory effect of phytohormones on brown algal organogenesis.

Many studies examined the effect of adding exogenous phytohormones and auxin transport inhibitors on the phenotype of brown algae. Care needs to be taken interpreting data obtained from application of transport inhibitors, since the effects may be totally unrelated to a putative auxin function in brown algae. For example the inhibitory effects of triiodobenzoic acid (TIBA) and genistein on polarization of fucoid zygotes can be understood in the context of respectively a breakdown product of TIBA affecting sulfhydryl-groups of the sulfated fucans that fix the polarization axis (see above) (195) or dependence of the polarization process on a protein tyrosine kinase in *Fucus* zygotes (196). Second, the overall evidence is often conflicting and equivocal (188). Sometimes positive reports of effects are reported to be not reproducible (see e.g., ref. 197).

Despite these objections, a number of recent studies are providing additional evidence for the role of auxins. For example radioactive (^{3}H)IAA pulse–chase experiments provided a strong argument for the effect of NPA and TIBA on auxin transport in *F. distichus* zygotes (189). Moreover, the auxin transport and F-actin cytoskeleton were shown to interact during photopolarization of *F. distichus* embryos (77). Immunolocalization of IAA showed IAA also to be present inside *E. siliculosus* cells and putative homologs for enzymes of the tryptamine (TAM) and indole-3-acetaldoxime (IAOx) Trp-dependent IAA biosynthesis pathways were identified in the *E. siliculosus* genome. An auxin responsive gene, *EsGRP1*, was identified using microarray analysis, which showed increased expression levels in mutants affected in their apical dominance (198).

Tracing the auxin action in brown algae may provide valuable answers to evolutionary developmental questions in plant systems. For example, it has been argued that tracing the evolutionary patterns of auxin action can more easily uncover how regulation of IAA action relates to organogenesis as more basal plants show a simpler regulation than land plants (199, 200). The brown algal clade might provide an independent test case for the hypothesis that simple body plans like the filamentous *Ectocarpus* or the parenchymatous *Dictyota* or *Fucus* are characterized by homeostasis established by regulation of IAA biosynthesis and degradation rather than by conjugation (190, 198, 199).

4.2. Auxin Function as a Homoplasious or Homologous Character?

Various explanations may be applied to account for the putative function of auxin as a phytohormone in brown algae. In silico analyses of whole genome sequences and EST libraries demonstrated that the canonical auxin response system and auxin transport mechanism of land plants evolved gradually in the streptophyte lineage (201). Homologues of key enzymes (e.g., ARF's, TIR1, PIN-like proteins) were not found in the genomes of unicellular green algae. An AUX–IAA-ARF-mediated signalling system is absent from other unicellular photosynthetic algal lineages and from *Ectocarpus* (41, 198). These results, therefore, do not support a common origin of an AUX–IAA-ARF-mediated signalling mechanism.

Alternatively, several physicochemical characteristics render IAA an ideal signalling molecule and therefore make convergent evolution a plausible scenario. Its size leads to rapid diffusion in aqueous solutions. When protonated (p*K*a 4.75) IAA becomes lipophilic, it readily diffuses through cell walls and most eukaryotes have the machinery to synthesize IAA (41). It is therefore possible that Phaeophyceae evolved convergently alternative auxin-dependent signalling and transport mechanisms. The absence of a common origin of an AUX–IAA-ARF-mediated signalling mechanism in *E. siliculosus* and unicellular Viridiplantae, however, does necessarily mean that auxin function in brown algae has no homologous origin (by either vertical or horizontal gene transfer) and that the auxin pathway had to evolve from scratch. First, Cooke et al. (199) proposed that IAA initially served as a growth-regulating pheromone in photosynthetic unicellular aquatic organisms. The possibility of alternative pathways for auxin signalling in unicellular green algae suggested by putative INDOLE-3-BUTYRIC ACID RESPONSE5 (IBR5) or AUXIN BINDING PROTEIN1 (ABP1) homologues (201) and reports on IAA function in polarization (e.g., of *Fucus* zygotes), or cell division of unicellular organisms or rhizoid growth suggest that auxin function predates the advent of multicellularity (41, 201). Second, the only whole genome sequences available from respectively the red and the brown lineage are the ones of *Cyanidioschyzon merolae* (unicellular) and *E. siliculosus* (filamentous) which have both a relatively simple body plan and a relatively small genome (36, 202, 203). As complexity in green plants is thought to be linked to the number of auxin genes (204), this developmental simplification and genome reduction might have resulted in the secondary loss of auxin related genes, leaving only scarce traces of the common origin and subsequent independent divergence. Indeed some putative homologs of the ABCB efflux IAA transporter family, in *E. siliculosus* (198) and single celled members of Chlorophyta (201) have been found. Also a putative homolog for BIG, a callosin-like protein needed for auxin mediated inhibition of PIN protein endocytosis (205), was found in *E. siliculosus.*

4.3. Future Perspectives

Over one century of research on algal development, mainly focussing on embryogenesis and patterning of fucoid zygotes, has strongly positioned brown algae as widely used model organisms to study polarization and asymmetrical cell division. In addition, the rich variation in life cycles and associated developmental control of the gametophyte and sporophyte generations make the Phaeophyceae an extremely interesting group of organisms. Studies on the relations between life cycle and developmental patterns is spearheaded by *E. siliculosus*, which has the benefit of having its whole genome sequenced as well as being easily cultured. Classical genetics have been proven to be a valuable tool for the research in the mechanisms underlying *E. siliculosus* development and the control of its life cycle (31, 35, 198).

Importantly, in spite of the extensive knowledge of brown algal embryogenesis and development, tools enabling functional characterization (e.g., transformation, RNAi) have yet to be developed for brown algal model systems. Development of transformation protocols may highlight the need for additional species which in contrary to the fucoids can be cultured throughout their life cycle. Completion of the fucoid life cycle in laboratory conditions is extremely complicated (206), posing obvious problems for performing mutant analysis and stable transformations. From a comparison of several brown algae as candidates for genomic research (207), *Dictyota dichotoma* emerged as a valuable option which may be complementary to *Fucus*. Just like the latter, *Dictyota* has evolved parenchymatous growth (Fig. 2), though the tissue of *Dictyota* remains thin enough to allow lab cultures and consequently has been cultured in the lab for decades (208, 209). The ability to culture *Dictyota* throughout its life cycle enables crossing experiments, and potentially maintenance of isolated mutants or transformations for longer time periods. In addition, *Dictyota* is one of few oogamous brown algae where large eggs are suspended in the medium after egg release. Preliminary findings suggest that different but complementary insights on polarization and asymmetric cell division can be gained by using *Dictyota* (K.A. Bogaert and O. De Clerck, unpublished data). In addition to its tractability, the distinct phylogenetic position as a heterokont in the eukaryotic tree of life (Fig. 1) and as a member of the early diverging SSDO group in the brown algal phylogeny (210) makes *Dictyota* an interesting physiological and phylogenomic data point for unravelling a plethora of evolutionary developmental questions, such as for example the evolutionary origin of auxin action in the regulation of the plant body plan.

The limited rather "patchy" taxonomic sampling of plants systems concentrating almost exclusively on the green plant lineage (Viridiplantae) is hampering comparative genomics and phylogenomics (15, 211). Sequencing projects of multicellular "brown" and "red" plant systems as well as the further development of tools enabling functional characterization of genes will undoubtedly provide important insights on the development of complex multicellular organisms outside of the well-trodden domain of "higher plants" and offer a more inclusive view on plant morphogenesis.

Acknowledgments

The authors are indebted to the Research Foundation Flanders (FWO) (PhD fellowship to KB), the Erasmus Mundus External Cooperation Window (EMECW) (PhD fellowship to AA). The Algal Genetics lab is supported by the Centre National de Recherche Scientifique, the University Pierre and Marie Curie, the Groupement d'Interet Scientifique Génomique Marine, the Agence Nationale de la Recherche (ANR), and the Interreg Programme (Marinexus).

References

1. van den Hoek C, Mann DG, Jahns HM (1995) Algae: an introduction to phycology. Cambridge University Press, Cambridge, pp. 623
2. Andersen RA (2004) Biology and systematics of heterokont and haptophyte algae. Am J Bot 91:1508–1522
3. Riisberg I et al (2009) Seven gene phylogeny of heterokonts. Protist 160:191–204
4. Khan H et al (2007) Plastid genome sequence of the cryptophyte alga *Rhodomonas salina* CCMP1319: lateral transfer of putative DNA replication machinery and a test of chromist plastid phylogeny. Mol Biol Evol 24:1832–1842
5. Sanchez-Puerta M, Delwiche CF (2007) A hypothesis for plastid evolution in chromalveolates. J Phycol 44:1097–1107
6. Wang Y, Joly S, Morse D (2008) Phylogeny of dinoflagellate plastid genes recently transferred to the nucleus supports a common ancestry with red algal plastid genes. J Mol Evol 66:175–184
7. Le Corguillé G et al (2009) Plastid genomes of two brown algae, *Ectocarpus siliculosus* and *Fucus vesiculosus*: further insights on the evolution of red-algal derived plastids. BMC Evol Biol 9:253
8. Janouskovec J et al (2010) A common red algal origin of the apicomplexan, dinoflagellate, and heterokont plastids. Proc Natl Acad Sci USA 107:10949–10954
9. Green BR (2011) After the primary endosymbiosis: an update on the chromalveolate hypothesis and the origins of algae with Chl c. Photosynth Res 107:103–115
10. Baurain D et al (2010) Phylogenomic evidence for separate acquisition of plastids in cryptophytes, haptophytes and stramenopiles. Mol Biol Evol 27:1698–1709
11. Cavalier-smith T (1999) Principles of protein and lipid targeting in secondary symbiogenesis: euglenoid, dinoflagellate, and sporozoan plastid origins and the eukaryote family tree. J Eukaryot Microbiol 46:347–366
12. Delwiche CF (2007) Algae in the warp and weave of life: bound by plastids. In: Brodie J, Lewis J (eds) Unravelling the algae: the past, present, and future of algal systematics, pp. 7–20. Taylor and Francis: Boca Raton, FL
13. Yoon HS et al (2002) The single, ancient origin of chromist plastids. Proc Natl Acad Sci USA 99:15507–15512
14. Yoon HS et al (2004) A molecular timeline for the origin of photosynthethic eukaryotes. Mol Biol Evol 21:809–818
15. Elias M, Archibald JM (2009) Sizing up the genomic footprint of endosymbiosis. BioEssays 31:1273–1279
16. Nagasato C, Motomura T, Ichimura T (1999) Influence of centriole behavior on the first spindle formation in zygotes of the brown alga *Fucus distichus* (Fucales, Phaeophyceae). Dev Biol 208:200–209
17. Ritter A et al (2008) Copper stress induces biosynthesis of octadecanoid and eicosanoid oxygenated derivatives in the brown algal kelp *Laminaria digitata*. New Phytol 180:809–821
18. Silberfeld T et al (2010) A multi-locus time-calibrated phylogeny of the brown algae (Heterokonta, Ochrophyta, Phaeophyceae): investigating the evolutionary nature of the "brown algal crown radiation". Mol Phylogenet Evol 56:659–674
19. Kawai H et al (2003) *Schizocladia ischiensis*: a new filamentous marine chromophyte belonging to a new class, Schizocladiophyceae. Protist 154:211–228
20. Brown JW, Sorhannus U (2010) A molecular genetic timescale for the diversification of

autotrophic stramenopiles (Ochrophyta): substantive underestimation of putative fossil ages. PLoS One 5:11
21. Bell G (1997) The evolution of the life cycle of brown seaweeds. Biol J Linn Soc 60:21–38
22. de Reviers B, Rousseau F, Draisma SGA (2007) Classification of the Phaeophyceae from past to present and current challenges, in Unravelling the algae: the past, present, and future of algal systematics. In: Brodie J, Lewis J (eds) The systematics association special volume series, pp. 267–284. Taylor and Francis: Boca Raton, FL
23. Graham M, Vásquez J (2007) Global ecology of the giant kelp Macrocystis: from ecotypes to ecosystems. Oceanogr Mar Biol 45:39–88
24. Parker BC (1965) Translocation in the giant kelp *Macrocystis* I. rates, direction, quantity of C14-labeled products and fluorescein. J Phycol 1:41–46
25. Parker BC, Huber J (1965) Translocation in *Macrocystis*. II. Fine structure of the sieve tubes. J Phycol 1:172–179
26. Peters AF, Burkhardt E (1998) Systematic position of the kelp endophyte *Laminarionema elsbetiae* (Ectocarpales sensu lato, Phaeophyceae) inferred from nuclear ribosomal DNA sequences. Phycologia 37: 114–120
27. Burkhardt E, Peters AF (1998) Molecular evidence from nrDNA its sequences that *Laminariocolax* (Phaeophyceae, Ectocarpales sensu lato) is a worldwide clade of closely related kelp endophytes. J Phycol 34:682–691
28. Niklas K (2000) The evolution of plant body plans—a biomechanical perspective. Ann Bot London 85:411–438
29. Coelho SM et al (2002) Spatiotemporal patterning of reactive oxygen production and Ca2+ wave propagation in *Fucus* rhizoid cells. Plant Cell 14:2369–2381
30. Coelho SMB, Brownlee C, Bothwell JHF (2008) A tip-high, Ca(2+)-interdependent, reactive oxygen species gradient is associated with polarized growth in *Fucus serratus* zygotes. Planta 227:1037–1046
31. Peters AF et al (2008) Life-cycle-generation-specific developmental processes are modified in the immediate upright mutant of the brown alga *Ectocarpus siliculosus*. Development 135:1503–1512
32. Bisgrove SR (2007) Cytoskeleton and early development in fucoid algae. J Integr Plant Biol 49:1192–1198
33. Hable WE, Hart PE (2010) Signaling mechanisms in the establishment of plant and fucoid algal polarity. Mol Reprod Dev 77:751–758
34. Coelho SM et al (2007) Complex life cycles of multicellular eukaryotes: new approaches based on the use of model organisms. Gene 406:152–170
35. Coelho SM et al (2011) OUROBOROS is a master regulator of the gametophyte to sporophyte life cycle transition in the brown alga *Ectocarpus*. Proc Natl Acad Sci USA 108:11518–11523
36. Cock JM et al (2010) The *Ectocarpus* genome and the independent evolution of multicellularity in brown algae. Nature 465:617–621
37. Dittami SM et al (2009) Global expression analysis of the brown alga *Ectocarpus siliculosus* (Phaeophyceae) reveals large-scale reprogramming of the transcriptome in response to abiotic stress. Genome Biol 10:R66
38. Heesch S et al (2010) A sequence-tagged genetic map for the brown alga *Ectocarpus siliculosus* provides large-scale assembly of the genome sequence. New Phytol 188:42–51
39. Gschloessl B, Guermeur Y, Cock JM (2008) HECTAR: a method to predict subcellular targeting in heterokonts. BMC Bioinformatics 9:393
40. Ritter A et al (2010) Copper stress proteomics highlights local adaptation of two strains of the model brown alga *Ectocarpus siliculosus*. Proteomics 10:2074–2088
41. Lau S et al (2009) Auxin signaling in algal lineages: fact or myth? Trends Plant Sci 14:182–188
42. Cho GY, Lee SH, Boo SM (2004) A new brown algal order, Ishigeales (Phaeophyceae), established on the basis of plastid protein-coding rbcL, psaA, and psbA region comparisons. J Phycol 40:921–936
43. Kawai H et al (2007) Molecular phylogeny of *Discosporangium mesarthrocarpum* (Phaeophyceae) with a reinstatement of the Order Discosporangiales. J Phycol 43:186–194
44. Peters AF, Ramirez ME (2001) Molecular phylogeny of small brown algae, with special reference to the systematic position of *Caepidium antarcticum* (Adenocystaceae, Ectocarpales). Cryptogamie Algol 22:187–200
45. Bothwell JH et al (2010) Role of endoreduplication and apomeiosis during parthenogenetic reproduction in the model brown alga *Ectocarpus*. New Phytol. 188:111–21 doi: 10.1111/j.1469-8137.2010.03357.x
46. John PCL, Qi R (2008) Cell division and endoreduplication: doubtful engines of vegetative growth. Trends Plant Sci 13:121–127
47. Müller DG (1967) Generationswechsel, Kernphasenwechsel und Sexualität der

Braunalge Ectocarpus siliculosus im Kulturversuch. Planta 75:39–54
48. Cheung AY, Wu H-M (2008) Structural and signaling networks for the polar cell growth machinery in pollen tubes. Annu Rev Plant Biol 59:547–572
49. Xu T et al (2010) Cell surface- and Rho GTPase-based auxin signaling controls cellular interdigitation in *Arabidopsis*. Cell 143: 99–110
50. Szymanski DB (2009) Plant cells taking shape: new insights into cytoplasmic control. Curr Opin Plant Biol 12:735–744
51. Young RE et al (2008) Analysis of the Golgi apparatus in *Arabidopsis* seed coat cells during polarized secretion of pectin-rich mucilage. Plant Cell 20:1623–1638
52. Li L, Saga N, Mikami K (2008) Phosphatidylinositol 3-kinase activity and asymmetrical accumulation of F-actin are necessary for establishment of cell polarity in the early development of monospores from the marine red alga *Porphyra yezoensis*. J Exp Bot 59:3575–3586
53. Dhonukshe P et al (2008) Generation of cell polarity in plants links endocytosis, auxin distribution and cell fate decisions. Nature 456:962–966
54. Rosenvinge MLK (1889) Influence des agents extérieurs sur l'organisation polaire dorsiventrale des plantes. Rev Gen Bot 1:135
55. Farmer JB, Williams LJ, Scott DH (1898) Contributions to our knowledge of the Fucaceae: their life history and cytology. Philos Trans R Soc B 190:623–645
56. Corellou F et al (2000) A S/M DNA replication checkpoint prevents nuclear and cytoplasmic events of cell division including centrosomal axis alignment and inhibits activation of cyclin-dependent kinase-like proteins in fucoid zygotes. Development 127:1651–1660
57. Yang Z (2008) Cell polarity signaling in *Arabidopsis*. Annu Rev Cell Dev Bi 24: 551–575
58. Kropf DL (1992) Establishment and expression of cellular polarity in fucoid zygotes. Microbiol Rev 56:316–339
59. Paciorek T, Bergmann DC (2010) The secret to life is being different: asymmetric divisions in plant development. Curr Opin Plant Biol 13:1–9
60. Brownlee C, Wood J (1986) A gradient of cytoplasmic free calcium a in growing rhizoid cells of *Fucus serratus*. Nature 320:624–626
61. Hepler PK (2005) Calcium: a central regulator of plant growth and development. Plant Cell 17:2142–2155
62. Kropf D (1994) Cytoskeletal control of cell polarity in a plant zygote. Dev Biol 165:361–371
63. Fowler JE, Quatrano RS (1997) Plant cell morphogenesis: plasma membrane interactions with the cytoskeleton and cell wall. Annu Rev Cell Dev Biol 13:697–743
64. Berger F, Taylor A, Brownlee C (1994) Cell fate determination by the cell wall in early *Fucus* development. Science 263:1421–1423
65. Belanger KD, Quatrano RS (2000) Polarity: the role of localized secretion. Curr Opin Plant Biol 3:67–72
66. Brownlee C (1994) Tansley review no. 70 signal transduction during fertilization in algae and vascular plants. New Phytol 127: 399–423
67. Dumas C, Gaude T (2006) Fertilization in plants: is calcium a key player? Semin Cell Dev Biol 17:244–253
68. Knoblich JA (2001) Asymmetric cell division during animal development. Nat Rev Mol Cell Bio 2:11–20
69. Knoblich JA (2008) Mechanisms of asymmetric stem cell division. Cell 132:583–597
70. Knapp E (1932) Entwicklungsphysiologische Untersuchungen an Fucaceen-Eiern I. Zur Kenntnis der Polarität der Eier von Cystosira barbata. Planta 14:731–751
71. Hable WE, Kropf DL (2000) Sperm entry induces polarity in fucoid zygotes. Development 127:493–501
72. Hurd AM (1920) Effect of unilateral monochromatic light and group orientation on the polarity of germinating *Fucus* spores. Bot Gaz 70:25–50
73. Bentrup F, Sandan T, Jaffe L (1967) Induction of polarity in *Fucus* eggs by potassium ion gradients. Protoplasma 64:254–266
74. Bentrup FW, Jaffe LF (1968) Analyzing the "Group effect": rheotropic responses of developing *Fucus* eggs. Protoplasma 65:25–35
75. Novak B, Bentrup FW (1973) Orientation of *Fucus* egg polarity by electric a. c. and d. c. fields. Biophysik 9:253–260
76. Love J, Brownlee C, Trewavas AJ (1997) Ca2+ and calmodulin dynamics during photopolarization in *Fucus serratus* zygotes. Plant Physiol 115:249–261
77. Sun H et al (2004) Interactions between auxin transport and the actin cytoskeleton in developmental polarity of *Fucus distichus* embryos in response to light and gravity. Plant Physiol 135:266–278
78. Bouget F-Y, Gertula S, Quatrano RS (1995) Spatial distribution of poly(A) RNA during

polarization of *Fucus* zygotes is dependent upon microfilaments. Dev Biol 171:258–261
79. Bouget FY et al (1996) Localization of actin mRNA during the establishment of cell polarity and early cell divisions in *Fucus* embryos. Plant Cell 8:189–201
80. Hadley R, Hable WE, Kropf DL (2006) Polarization of the endomembrane system is an early event in fucoid zygote development. BMC Plant Biol 6:5
81. Shaw SL, Quatrano RS (1996) Polar localization of a dihydropyridine receptor on living *Fucus* zygotes. J Cell Sci 109:335–342
82. Berger F, Brownlee C (1993) Ratio confocal imaging of free cytoplasmic calcium gradients in polarising and polarised *Fucus* zygotes. Zygote 1:9–15
83. Wagner VT, Brian L, Quatrano RS (1992) Role of a vitronectin-like molecule in embryo adhesion of the brown alga *Fucus*. Proc Natl Acad Sci USA 89:3644–3648
84. De Smet I, Beeckman T (2011) Asymmetric cell division in land plants and algae: the driving force for differentiation. Nat Rev Mol Cell Biol 12:177–188
85. Abrash EB, Bergmann DC (2009) Asymmetric cell divisions: a view from plant development. Dev Cell 16:783–796
86. Yamashita YM et al (2007) Asymmetric inheritance of mother versus daughter centrosome in stem cell division. Science 315:518–521
87. De Smet I et al (2009) Receptor-like kinases shape the plant. Nat Cell Biol 11:1166–1173
88. Gualtieri P, Robinson KR (2002) A rhodopsin-like protein in the plasma membrane of *Silvetia* compressa eggs. Photochem Photobiol 75:76–78
89. Robinson KR, Miller BJ (1997) The coupling of cyclic GMP and photopolarization of Pelvetia zygotes. Dev Biol 187:125–130
90. Berger F, Brownlee C (1994) Photopolarization of the *Fucus* sp. zygote by blue light involves a plasma membrane redox chain. Plant Physiol 105:519–527
91. Kropf D, Bisgrove S, Hable W (1999) Establishing a growth axis in fucoid algae. Trends Plant Sci 4:490–494
92. Robinson K et al (1999) Symmetry breaking in the zygotes of the fucoid algae: controversies and recent progress. Curr Top Dev Biol 44:101–125
93. Jaffe LF, Robinson KR, Nuccitelli R (1974) Local cation entry and self-electrophoresis as an intracellular localization mechanism. Ann NY Acad Sci 238:372–389
94. Nuccitelli R, Jaffe LF (1974) Spontaneous current pulses through developing fucoid eggs. Proc Natl Acad Sci USA 71: 4855–4859
95. Nuccitelli R, Jaffe LF (1976) The ionic components of the current pulses generated by developing fucoid eggs. Dev Biol 49:518–531
96. Jaffe LF (1999) Organization of early development by calcium patterns. BioEssays 21:657–667
97. Robinson K, Jaffe L (1975) Polarizing fucoid eggs drive a calcium current through themselves. Science 187:70–72
98. Robinson KR, Cone R (1980) Polarization of fucoid eggs by a calcium ionophore gradient. Science 207:77–78
99. Brownlee C, Pulsford AL (1988) Visualization of the cytoplasmic Ca2+ gradient in *Fucus serratus* rhizoids: correlation with cell ultrastructure and polarity. J Cell Sci 91:249–256
100. Speksnijder JE et al (1989) Calcium buffer injections block fucoid egg development by facilitating calcium diffusion. Proc Natl Acad Sci USA 86:6607–6611
101. Pu R, Robinson KR (1998) Cytoplasmic calcium gradients and calmodulin in the early development of the fucoid alga *Pelvetia compressa*. J Cell Sci 111:3197–3207
102. Taylor AR et al (1996) Spatial organization of calcium signaling involved in cell volume control in the *Fucus* rhizoid. Plant Cell 8: 2015–2031
103. Taylor A, Brownlee C (1993) Calcium and potassium currents in the *Fucus* egg. Planta 189:109–119
104. Nuccitelli R, Jaffe LF (1976) The ionic components of the current pulses generated by developing fucoid eggs. Dev Biol 49:518–531
105. Robinson K (1996) Calcium and the photopolarization of *Pelvetia* zygotes. Planta 198:378–384
106. Pu R, Wozniak M, Robinson KR (2000) Cortical actin filaments form rapidly during photopolarization and are required for the development of calcium gradients in *Pelvetia compressa* zygotes. Dev Biol 222:440–449
107. Taylor A, Roberts S, Brownlee C (1992) Calcium and related channels in fertilization and early development of *Fucus*. Philos Trans R Soc B 338:97–104
108. Brownlee C, Bouget FY, Corellou F (2001) Choosing sides: establishment of polarity in zygotes of fucoid algae. Semin Cell Dev Biol 12:345–351
109. Hable W, Kropf D (1998) Roles of secretion and the cytoskeleton in cell adhesion and polarity establishment in zygotes. Dev Biol 198:45–56

110. Alessa L, Kropf DL (1999) F-actin marks the rhizoid pole in living *Pelvetia compressa* zygotes. Development 126:201–209
111. Shaw SL, Quatrano RS (1996) The role of targeted secretion in the establishment of cell polarity and the orientation of the division plane in *Fucus* zygotes. Development 122:2623–2630
112. Bothwell JHF et al (2006) Biolistic delivery of Ca^{2+} dyes into plant and algal cells. Plant J 46:327–335
113. Bisgrove S (2007) Asymmetric cell divisions: zygotes of fucoid algae as a model system. Plant Cell Monogr 9:323–341
114. Hable WE, Kropf DL (2005) The Arp2/3 complex nucleates actin arrays during zygote polarity establishment and growth. Cell Motil Cytoskel 61:9–20
115. Varvarigos V, Galatis B, Katsaros C (2007) Radial endoplasmic reticulum arrays co-localize with radial F-actin in polarizing cells of brown algae. Eur J Phycol 42:253–262
116. Katsaros C, Karyophyllis D, Galatis B (2006) Cytoskeleton and morphogenesis in brown algae. Ann Bot London 97:679–693
117. Panteris E, Apostolakos P, Galatis B (2006) Cytoskeletal asymmetry in *Zea mays* subsidiary cell mother cells: a monopolar prophase microtubule half-spindle anchors the nucleus to its polar position. Cell Motil Cytoskel 63:696–709
118. Baluska F et al (2000) Root hair formation: F-actin-dependent tip growth is initiated by local assembly of profilin-supported F-actin meshworks accumulated within expansin-enriched bulges. Dev Biol 227:618–632
119. Fu Y et al (2005) *Arabidopsis* interdigitating cell growth requires two antagonistic pathways with opposing action on cell morphogenesis. Cell 120:687–700
120. Mathur J et al (2003) Mutations in actin-related proteins 2 and 3 affect cell shape development in Arabidopsis. Plant Cell 15:1632–1645
121. Fu Y, Wu G, Yang Z (2001) Rop GTPase-dependent dynamics of tip-localized F-actin controls tip growth in pollen tubes. J Cell Biol 152:1019–1032
122. Gu Y et al (2005) A Rho family GTPase controls actin dynamics and tip growth via two counteracting downstream pathways in pollen tubes. J Cell Biol 169:127–138
123. Brembu T et al (2006) A RHOse by any other name: a comparative analysis of animal and plant Rho GTPases. Cell Res 16:435–445
124. Perez P, Rincón SA (2010) Rho GTPases: regulation of cell polarity and growth in yeasts. Biochemical J 426:243–253
125. Fowler JE et al (2004) Localization to the rhizoid tip implicates a *Fucus distichus* Rho family GTPase in a conserved cell polarity pathway. Planta 219:856–866
126. Hable WE, Reddy S, Julien L (2008) The Rac1 inhibitor, NSC23766, depolarizes adhesive secretion, endomembrane cycling, and tip growth in the fucoid alga, *Silvetia compressa*. Planta 227:991–1000
127. Hamant O, Traas J (2010) The mechanics behind plant development. New Phytol 185:369–385
128. Kropf DL, Bisgrove SR, Hable WE (1998) Cytoskeletal control of polar growth in plant cells. Curr Opin Cell Biol 10:117–122
129. Katsaros C, Reiss HD, Schnepf E (1996) Freeze-fracture studies in brown algae: putative cellulose-synthesizing complexes on the plasma membrane. Eur J Phycol 31:41–48
130. Schubetaler A, Hirn S, Katsaros C (2003) Cellulose synthesizing terminal complexes and morphogenesis in tip-growing cells of *Syringoderma phinneyi* (Phaeophyceae). Phycol Res 51:35–44
131. Michel G et al (2010) The cell wall polysaccharide metabolism of the brown alga *Ectocarpus siliculosus*. Insights into the evolution of extracellular matrix polysaccharides in Eukaryotes. New Phytol 188:82–97
132. Katsaros C (2003) F-actin cytoskeleton and cell wall morphogenesis in brown algae. Cell Biol Int 27:209–210
133. Karyophyllis D, Katsaros C, Dimitriadis I (2000) F-actin organization during the cell cycle of *Sphacelaria rigidula* (Phaeophyceae). Eur J Phycol 35:25–33
134. Bisgrove SR, Kropf DL (2001) Cell wall deposition during morphogenesis in fucoid algae. Planta 212:648–658
135. Motomura T (1994) Electron and immunofluorescence microscopy on the fertilization of *Fucus distichus* (Fucales, Phaeophyceae). Protoplasma 178:97–110
136. Corellou F et al (2005) Spatial re-organisation of cortical microtubules in vivo during polarisation and asymmetric division of *Fucus* zygotes. J Cell Sci 118:2723–2734
137. Peters NT, Kropf DL (2010) Asymmetric microtubule arrays organize the endoplasmic reticulum during polarity establishment in the brown alga *Silvetia compressa*. Cytoskeleton 67:102–111
138. Sieberer BJ et al (2005) Microtubules guide root hair tip growth. New Phytol 167:711–719
139. Peters NT et al (2007) Phospholipase D signaling regulates microtubule organization in

the fucoid alga *Silvetia compressa*. Plant Cell Physiol 48:1764–1774

140. Shimamura M et al (2004) Gamma-tubulin in basal land plants: characterization, localization, and implication in the evolution of acentriolar microtubule organizing centers. Plant Cell 16:45–59
141. McCarthy EK, Goldstein B (2006) Asymmetric spindle positioning. Curr Opin Cell Biol 18:79–85
142. Yamashita YM, Fuller MT (2008) Asymmetric centrosome behavior and the mechanisms of stem cell division. J Cell Biol 180:261–266
143. Van Damme D (2009) Division plane determination during plant somatic cytokinesis. Curr Opin Plant Biol 12:745–751
144. Nagasato C, Motomura T (2002) Influence of the centrosome in cytokinesis of brown algae: polyspermic zygotes of *Scytosiphon lomentaria* (Scytosiphonales, Phaeophyceae). J Cell Sci 115:2541–2548
145. Bisgrove S, Henderson D, Kropf D (2003) Asymmetric division in fucoid zygotes is positioned by telophase nuclei. Plant Cell 15:854–862
146. Katsaros C (1992) Immunofluorescence study of microtubule organization in some polarized cell types of selected brown algae. Bot Acta 105:400–406
147. Varvarigos V, Galatis B, Katsaros C (2005) A unique pattern of F-actin organization supports cytokinesis in vacuolated cells of *Macrocystis pyrifera* (Phaeophyceae) gametophytes. Protoplasma 226:241–245
148. Panteris E et al (2004) A cortical cytoplasmic ring predicts the division plane in vacuolated cells of *Coleus*: the role of actomyosin and microtubules in the establishment and function of the division site. New Phytol 163:271–286
149. Müller S, Wright AJ, Smith LG (2009) Division plane control in plants: new players in the band. Trends Cell Biol 19:180–188
150. Oh SA et al (2010) The tobacco MAP215/Dis1-family protein TMBP200 is required for the functional organization of microtubule arrays during male germline establishment. J Exp Bot 61:969–981
151. Doonan JH, Cove DJ, Lloyd CW (1985) Immunofluorescence microscopy of microtubules in intact cell lineages of the moss, Physcomitrella patens. I. Normal and CIPC-treated tip cells. J Cell Sci 75:131–147
152. Lloyd C, Chan J (2006) Not so divided: the common basis of plant and animal cell division. Nat Rev Mol Cell Biol 7:147–152
153. Dhonukshe P, Vischer N, Gadella TWJ (2006) Contribution of microtubule growth polarity and flux to spindle assembly and functioning in plant cells. J Cell Sci 119:3193–3205
154. Chan J et al (2005) Localization of the microtubule end binding protein EB1 reveals alternative pathways of spindle development in *Arabidopsis* suspension cells. Development 17:1737–1748
155. Katsaros C et al (2009) Diaphragm development in cytokinetic vegetative cells of brown algae. Bot Mar 52:150–161
156. Nagasato C et al (2010) Membrane fusion process and assembly of cell wall during cytokinesis in the brown alga, *Silvetia babingtonii* (Fucales, Phaeophyceae). Planta 232:287–298
157. Bisgrove SR, Kropf DL (2004) Cytokinesis in brown algae: studies of asymmetric division in fucoid zygotes. Protoplasma 223:163–173
158. Motomura T, Nagasato C, Kimura K (2010) Cytoplasmic inheritance of organelles in brown algae. J Plant Res 123:185–192
159. Fowler J, Quatrano R (1995) Cell polarity, asymmetric division, and cell fate determination in brown algal zygotes. Semin Dev Biol 6:347–358
160. Kropf D, Kloareg B, Quatrano R (1988) Cell wall is required for fixation of the embryonic axis in *Fucus* zygotes. Science 239:187–190
161. Kropf DL, Coffman HR, Kloareg B, Glenn P, Allen VW (1993) Cell wall and rhizoid polarity in *Pelvetia* embryos. Dev Biol 160: 303–314
162. Rusig AM, Guyader H, Ducreux G (1993) Microtubule organization in the apical cell of *Sphacelaria* (Phaeophyceae) and its related protoplast. Hydrobiologia 260–261: 167–172
163. Scheres B, Benfey PN (1999) Asymmetric cell division in plants. Annu Rev Plant Phys 50:505–537
164. Dong J, MacAlister CA, Bergmann DC (2009) BASL controls asymmetric cell division in *Arabidopsis*. Cell 137:1320–1330
165. Cartwright HN, Humphries JA, Smith LG (2009) PAN1: a receptor-like protein that promotes polarization of an asymmetric cell division in maize. Science 323:649–651
166. Muday GK (2001) Maintenance of asymmetric cellular localization of an auxin transport protein through actin cytoskeleton. J Plant Growth Regul 19:385–396
167. He Y-C et al (2007) Tobacco zygotic embryogenesis in vitro: the original cell wall of the zygote is essential for maintenance of cell polarity, the apical-basal axis and typical suspensor formation. Plant J 49:515–527
168. Qin Y, Zhao J (2006) Localization of arabinogalactan proteins in egg cells, zygotes, and

two-celled proembryos and effects of beta-D-glucosyl Yariv reagent on egg cell fertilization and zygote division in *Nicotiana tabacum* L. J Exp Bot 57:2061–2074
169. De Schutter K et al (2007) *Arabidopsis* WEE1 kinase controls cell cycle arrest in response to activation of the DNA integrity checkpoint. Plant Cell 19:211–225
170. Corellou F et al (2001) Cell cycle in the *Fucus* zygote parallels a somatic cell cycle but displays a unique translational regulation of cyclin-dependent kinases. Plant Cell 13:585–598
171. Bothwell JHF et al (2008) Ca2+ signals coordinate zygotic polarization and cell cycle progression in the brown alga *Fucus serratus*. Development 135:2173–2181
172. Kishimoto T (1988) Metaphase by a maturation-promoting factor. Dev Growth Differ 30:105–115
173. Arion D, Meijer L (1989) M-phase-specific protein kinase from mitotic sea urchin eggs: cyclic activation depends on protein synthesis and phosphorylation but does not require DNA or RNA synthesis. Exp Cell Res 183:361–375
174. Budirahardja Y, Gönczy P (2009) Coupling the cell cycle to development. Development 136:2861–2872
175. van den Berg C et al (1995) Cell fate in the *Arabidopsis* root meristem determined by directional signalling. Nature 378:62–65
176. Jaeger J, Irons D, Monk N (2008) Regulative feedback in pattern formation: towards a general relativistic theory of positional information. Development 135:3175–3183
177. Bhalerao RP, Bennett MJ (2003) The case for morphogens in plants. Nat Cell Biol 5:939–943
178. Katsaros CI (1995) Apical cells of brown algae with particular reference to Sphacelariales, Dictyotales and Fucales. Phycol Res 43:43–59
179. Gaillard J, L'Hardy-Halos MT (1984) Morphogenesis of *Dictyota dichotoma* (Huds.) Lamouroux (Phaeophyceae, Dictyotales). Ann Sci Nat Bot Biol 6:111–133
180. Bouget FY, Berger F, Brownlee C (1998) Position dependent control of cell fate in the *Fucus* embryo: role of intercellular communication. Development 125:1999–2008
181. Augier H (1976) Les hormones des algues. Etat actuel des connaissances. I-Recherche et tentatives d'identification des auxines. Bot Mar 19:127–144
182. Augier H (1976) Les hormones des algues. Etat actuel des connaissances. II—Recherche et tentatives d'identification des gibbérellines, des cytokinines et de diverses autres substances de nature hormonale. Bot Mar 19:245–254
183. Augier H (1976) Les hormones des algues. Etat actuel des connaissances. III-Rôle des hormones dans les modalités de croissance et de développement des thalles. Bot Mar 19:351–378
184. Augier H (1977) Les hormones des algues. Etat actuel des connaissances. IV—Rôle des hormones dans les divers métabolismes cellulaires et dans les mécanismes de reproduction sexuée et asexuée; rôle écologique. Bot Mar 20:1–12
185. Augier H (1977) Les hormones des algues. Etat actuel des connaissances. V-Index alphabétique par espèces des travaux de caractérisation des hormones endogènes. Bot Mar 20:187–204
186. Augier H (1977) Les hormones des algues. Etat actuel des connaissances. VI-Index alphabétique par espéces des travaux sur le rôle des hormones dans la vie des algues. Bot Mar 20:363–380
187. Augier H (1978) Les hormones des algues. Etat actuel des connaissances. VIl-Applications, conclusion, bibliographie. Bot Mar 21:175–197
188. Evans LV, Trewavas AJ (1991) Is algal development controlled by plant growth substances? J Phycol 21:322–326
189. Basu S et al (2002) Early embryo development in *Fucus distichus* is auxin sensitive. Plant Physiol 130:292
190. Stirk W, Novák O, Hradecká V (2009) Auxins and abscisic acid in *Ulva fasciata* (Chlorophyta) and *Dictyota humifusa* (Phaeophyta): towards understanding their biosynthesis and homoeostasis. Eur J Phycol 44:231–240
191. Bertilsson L, Palmer L (1972) Indole-3-acetic acid in-human cerebrospinal fluid identification and quantification by mass fragmentography. Science 177:74–76
192. Fries L (1977) Growth regulating effects of phenylacetic acid and p-hydroxy phenylacetic acid on *Fucus spiralis* L. (Pheophyta, Fucales) in axenic culture. Phycologia 16:451–455
193. Fries L (1988) Ascophyllum nodosum (Phaeophyta) in axenic culture and its response to the endophytic fungus *Mycosphaerella ascophylli* and epiphytic bacteria. J Phycol 24:333–337
194. Simon S, Petrášek J (2011) Why plants need more than one type of auxin. Plant Sci 180:454–460
195. Novotny AM, Forman M (1974) The relationship between changes in cell wall composition and the establishment of polarity in *Fucus* embryos. Dev Biol 40:162–173

196. Corellou F et al (2000) Inhibition of the establishment of zygotic polarity by protein tyrosine kinase inhibitors leads to an alteration of embryo pattern in *Fucus*. Dev Biol 219:165–182
197. Bouget F-Y, Corellou F, Kropf DL (2001) Fucoid algae as model organisms for investigating early embryogenesis. Cah Biol Mar 42:101–107
198. Le Bail A et al (2010) Auxin metabolism and function in the multicellular brown alga *Ectocarpus siliculosus*. Plant Physiol 153: 128–144
199. Cooke TJ et al (2002) Evolutionary patterns in auxin action. Plant Mol Biol 49:319–338
200. Niklas KJ, Kutschera U (2009) The evolutionary development of plant body plans. Funct Plant Biol 36:682–695
201. De Smet I et al (2010) Unraveling the evolution of auxin signaling. Plant Physiol 155: 209–221
202. Matsuzaki M et al (2004) Genome sequence of the ultrasmall unicellular red alga *Cyanidioschyzon merolae* 10D. Nature 428:653–657
203. Nozaki et al (2007) A 100%-complete sequence reveals unusually simple genomic features in the hot-spring red alga *Cyanidioschyzon merolae*. BMC Biol 5:28
204. Rensing SA et al (2008) The *Physcomitrella* genome reveals evolutionary insights into the conquest of land by plants. Science 319:64–69
205. Paciorek T et al (2005) Auxin inhibits endocytosis and promotes its own efflux from cells. Nature 435:1251–1256
206. McLachlan J, Chen LC, Edelstein T (1971) The culture of four species of *Fucus* under laboratory conditions. Can J Bot 49:1463–1469
207. Waaland JR, Stiller JW, Cheney DP (2004) Macroalgal candidates for genomics. J Phycol 40:26–33
208. Schreiber E (1935) Über Kultur und Geschlechtsbestimmung von *Dictyota dichotoma*. Planta 24:266–275
209. Müller D (1962) Über jahres- und lunarperiodische Ersheinungen bei einigen Braunalgen. Bot Mar 4:140–155
210. Bittner L et al (2008) Molecular phylogeny of the Dictyotales and their position within the Phaeophyceae, based on nuclear, plastid and mitochondrial DNA sequence data. Mol Phylogenet Evol 49:211–226
211. Dorrell RG, Smith AG (2011) Do red and green make brown? Perspectives on plastid acquisitions within chromalveolates. Eukaryot Cell 10:856–868
212. Hampl V et al (2009) Phylogenomic analyses support the monophyly of Excavata and resolve relationships among eukaryotic "supergroups". Proc Natl Acad Sci USA 106:3859–3864
213. Burki F et al (2007) Phylogenomics reshuffles the eukaryotic supergroups. PLoS One 2:e790
214. Burki F, Shalchian-Tabrizi K, Pawlowski J (2008) Phylogenomics reveals a new "megagroup" including most photosynthetic eukaryotes. Biol Lett 4:366–369
215. Burki F et al (2009) Large-scale phylogenomic analyses reveal that two enigmatic protist lineages, telonemia and centroheliozoa, are related to photosynthetic chromalveolates. Genome Biol Evol 1:231–238
216. Koonin EV (2010) The origin and early evolution of eukaryotes in the light of phylogenomics. Genome Biol 11:209
217. Patron NJ, Inagaki Y, Keeling PJ (2007) Multiple gene phylogenies support the monophyly of cryptomonad and haptophyte host lineages. Curr Biol 17:887–891
218. Evans LV, Callow JA, Callow M (1982) The biology and biochemistry of reproduction and early development in *Fucus*. In: Round FE, Chapman DJ (eds) Progress in phycological research, 1st edn, pp. 68–110. Elsevier, Amsterdam
219. De Clerck O (2003) The genus Dictyota in the Indian Ocean. Opera Botanica Belgica 13: 1–205
220. Bartsch I et al (2008) The genus *Laminaria* sensu lato: recent insights and developments. Eur J Phycol 43:1–86
221. Parente M, Neto A (2003) Morphology and life history of *Scytosiphon lomentaria* (Scytosiphonaceae, Phaeophyceae) from the Azores. J Phycol 39:353–359
222. Tatarenkov A et al (2005) Intriguing asexual life in marginal populations of the brown seaweed *Fucus vesiculosus*. Mol Ecol 14:647–651
223. Hwang I-K, Kim H-S, Lee WJ (2005) Polymorphism in the brown alga *Dictyota dichotoma* (Dictyotales, Phaeophyceae) from Korea. Mar Biol 147:999–1015

Chapter 7

Microscopic Analysis of Ovule Development in *Arabidopsis thaliana*

Balaji Enugutti, Maxi Oelschner, and Kay Schneitz

Abstract

Ovules are the major female reproductive organs in higher plants. In addition, ovules of *Arabidopsis thaliana* are successfully used as model system to study plant organogenesis. Here we describe two microscopic techniques to analyze Arabidopsis ovule development from the organ to the cellular level in a rapid and reproducible fashion. Both methods are of great value when comparing the morphology of wild-type and mutant ovule development.

Key words: Arabidopsis, Light microscopy, Methods, Microscopy, Confocal laser scanning microscopy, Ovule, Ovule prep, Ovule mutants, Scanning electron microscopy

1. Introduction

The ovule is the major female plant organ involved in sexual reproduction. It is the organ where the egg cell originates, where fertilization takes place, and where the embryo develops during seed development. Therefore, it comes as no surprise that the genetic, molecular, and evolutionary basis of ovule development has been the focus of intense study (1–4). The remarkable success in this endeavor has also been made possible by a number of technical advantages that render the Arabidopsis ovule an excellent model to study plant organogenesis. An individual plant can produce several hundreds to thousands of ovules. As their development is independent a researcher can quickly analyze a large number of ovules at all developmental stages from a single mutant plant. Importantly, Arabidopsis ovules undergo a stereotypic mode of development (5). This feature allows easy detection of developmental aberrations in mutant analysis. Moreover, ovules within a gynoecium develop in a largely synchronous fashion, except during

Ive De Smet (ed.), *Plant Organogenesis: Methods and Protocols*, Methods in Molecular Biology, vol. 959,
DOI 10.1007/978-1-62703-221-6_7, © Springer Science+Business Media New York 2013

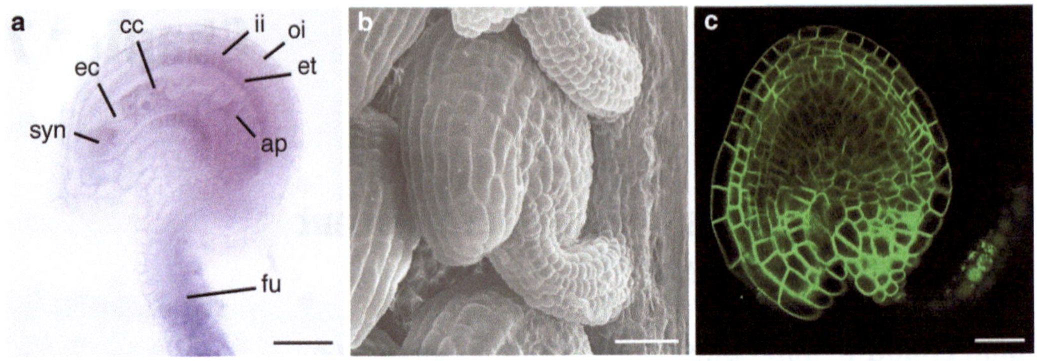

Fig. 1. Micrographs of mature Arabidopsis ovules. (**a**) Mid-optical section of a wild-type stage 3-VI ovule stained with Mayer's hemalum (Col). (**b**) Scanning electron micrograph of a wild-type stage 3-VI/4-I ovule (Col). (**c**) Confocal laser scanning micrograph of a GFP-labeled stage 3-VI ovule from a L*er* plant carrying a transgene encoding a plasma-membrane-anchored EGFP (9) under the control of a ubiquitin promoter (*UBQ10::EGFP:LTI6b*). The ovule was dissected as described (*see Note 1*), mounted in PBS, and viewed immediately. *ap* antipodal cells, *cc* central cell, *ec* egg cell, *et* endothelium, *fu* funiculus, *ii* inner integument, *oi* outer integument, *syn* synergid. Size bars: 20 μm.

the short phase of meiosis. Finally, ovules are devoid of chlorophyll and consist of at the most six cell layers, including the embryo sac, before fertilization. Thus, morphological inspection is rapidly achieved using either standard light microscopy on fixed and cleared specimens or confocal microscopy. Except for ultrastructural studies there is no need to embed and section ovules to analyze their morphology in Arabidopsis.

Here we describe two simple microscopic techniques commonly used in our laboratory to study the morphology of ovule development in Arabidopsis. The first method is quite straightforward and involves cleared whole-mount specimens and conventional light microscopy (5) (Fig. 1a). The second technique uses scanning electron microscopy (SEM) to provide a more three-dimensional view of ovule development (6) (Fig. 1b). Furthermore, using the general notes and remarks outlined in this chapter one can study fixed ovules of all stages by confocal laser scanning microscopy (CLSM) using published protocols developed for floral tissue without modifications (7). Similarly, live images of ovules expressing a protein coupled to a fluorescent tag, such as GFP, can be obtained easily (Fig. 1c).

2. Materials

2.1. Mayer's Hemalum Staining Components

1. 3.7% FAA fixative: 3.7% formaldehyde (FA) from a 37% stock, 5% glacial acetic acid in 1× PBS.
2. Phosphate-buffered saline (PBS), pH 7.2: 3 mM NaH_2PO_4, 7 mM Na_2HPO_4, 130 mM NaCl.

3. Mayer's hemalum (MH), modified: Dissolve 20 g aluminum potassium sulfate in 1 L water. Add 1 g hematoxylin, 0.2 g sodium iodate. Wrap in aluminum foil. The stain is ready to use at once. Filter through paper filter if any metallic scum is floating on top of the solution. Prepare new batch every 2–3 months. Store at room temperature.
4. Methyl benzoate (MB).
5. Spurr's medium: Use Low Viscosity Epoxy Embedding Kit (Electron Microscopy Sciences, USA). Mix 10 g ERL4221, 6 g DER 736, 26 g NSA, 0.4 g DMAE. Store in 1 mL aliquots at −80°C.
6. Canada Balsam.
7. U-100 insulin syringes, 0.5 ml/27 G (Becton, Dickinson and Company).
8. Forceps, class 5 Dumont INOX.
9. Double-sticky tape.
10. Tungsten needle: Cut a 5 cm piece of tungsten wire (0.25 mm in diameter, Luma Metals) and introduce it into the tip of a 25 G needle (orange). Push it up towards the rear end of the needle so that 1–2 cm stick out of the rear end of the needle. Cut the tip of the needle including a small piece of tungsten wire with a pair of scissors (this will prevent the needle from closing up at the tip due to deformation). Introduce a small kink on the rear end side of the tungsten wire and push the wire back through the tip. Pull the wire in this direction until it is tightly fixed in the needle. Connect the needle to a power supply (using a clamp and wire). Dip the tungsten needle into a 10% $NaNO_2$ (sodium nitrite) solution in H_2O, which is connected to the other pole of the power supply. Pass enough current through the needle to let it glow. Slowly retract the needle, repeat until the desired shape of the tip has been created. To remove the oxidation debris created in the process, reduce the current.
11. Razor blade.
12. KimWipe.
13. Microscope slides 76 × 26 × 1 mm.
14. Microscope cover glasses 22 × 22 mm Nr 1.

2.2. SEM Components

1. Acetone.
2. Glutaraldehyde.
3. Sodium cacodylate trihydrate.
4. Osmium tetroxide (EM grade) (Electron Microscopy Sciences, USA).
5. Aluminum specimen stubs, 0.5″ (Agar Scientific, UK).

6. Leit adhesive carbon discs, 12 mm (Agar Scientific, UK).
7. SEM pin stub specimen mount gripper tweezers (Agar Scientific, UK).
8. SEM pin stub specimen storage box (Agar Scientific, UK).
9. Cotton.
10. Parafilm.
11. Desiccator.
12. Glass tubes No. 2775/1 (Hecht-Assistent, Germany).
13. Glass scribe (Agar Scientific, UK).
14. U-100 insulin syringes, 0.5 ml/27 G (Becton, Dickinson and Company).
15. Tungsten needle.
16. Forceps, class 5 Dumont INOX.
17. Double-sticky tape.

3. Methods

Detailed staging information of Arabidopsis ovule development, also relative to flower development, is given in (5). Ovules within a pistil develop in a quite synchronous fashion. In wild-type *Ler* plants each pistil carries about 50 ovules. The first opened flower (floral stage 13 (8)) contains ovules that have just been fertilized and show zygotes and early endosperm development. Starting with the last or oldest closed flower bud (floral stage 12) and counting "inwards" the next 4–6 consecutively younger closed flowers include ovules at various stages of development. The 2–3 oldest closed buds carry stage 3 ovules (integument and embryo sac development). The next two buds carry stage 2 ovules (megaspore mother cell formation, meiosis, integument initiation, and early outgrowth) while the youngest buds carry stage 1 ovules (fingerlike ovule protrusions). For stage 13 flowers and the next 2–3 younger flowers it is feasible to dissect individual pistils. For younger flowers, just slightly press open the flower bud; if possible poke small holes in the pistil with a fine needle, and process the flower as a whole. Before observation dissect out pistil and ovules on the slide (for Mayer's hemalum staining) or on aluminum specimen stubs covered with adhesive carbon discs (for SEM).

3.1. Mayer's Hemalum Staining of Fixed Ovules (Ovule Preps)

Use 1.5 mL Eppendorf tubes filled with 1 mL of each solution.

1. Dissect pistils and remove ovary walls under a dissecting microscope using insulin syringes with fine needles or tungsten needles (work carefully without hurting the ovules but also reasonably

fast as ovules rapidly dry out in the strong light of the dissecting scope) (see Note 1).

2. Fix in freshly prepared 3.7% FAA fixative for at least 2 h at room temperature or overnight at 4°C. Tissue can be stored for extended periods of time at 4°C (weeks/months) (see Note 2).
3. Wash 2× in 1× PBS for 10 min (see Note 3).
4. Place tissue into Mayer's hemalum staining solution (MH, filtered) for 30 min to 2 h (see Note 4).
5. Wash tissue 2× in H_2O for 5 min.
6. Destain in 1.5% acetic acid for 45 min.
7. Wash in H_2O for 5 min.
8. Dehydrate with an ethanol series (p.a. 5%, 25%, 50%, 75%, 100%) for 15–30 min/each (see Note 5).
9. Put tissue into ethanol/methyl benzoate (1:1) for 30 min under the hood.
10. Put tissue into MB for 30 min.
11. Put tissue into MB/Spurr's medium (8:1) overnight at room temperature; you can store the tissue at 4°C for weeks and months (see Note 6).
12. Prepare slide for observation: Fix one layer of sticky tape onto the slide and cut out a rectangular section using a razor blade to create a small (about 1 to 1.5 cm on the bottom side) "empty" cavity on the slide (to prevent squashing the ovules) (see Note 7).
13. Place a small drop of MB/Spurr's medium into the empty cavity and disperse it with a pipette tip or a needle.
14. Add the tissue to a drop of MB/Spurr's medium on the slide (within the cavity), dissect the ovules, carefully cover with a coverslip, and avoid bubbles by carefully placing the left edge of the coverslip on the tape, fixing its position with forceps and slowly lowering the right side of the coverslip using a second pair of tweezers or a needle (see Note 8).
15. If the MB/Spurr's medium does not reach the edge of the coverslip add some by carefully pipetting a small amount of MB/Spurr's medium beneath one edge of the coverslip. If there is too much MB/Spurr's medium, i.e., the coverslip floats too much, try to remove some using a piece of kimwipe.
16. For long-lasting storage seal with Canada Balsam and let dry overnight (see Note 9).
17. Observe with 20× (air) or 40× (air or oil) Differential Interference Contrast (DIC or Nomarski) objectives on a good research light microscope, such as the Olympus BX61/63, Leica DM6000 B, or Nikon Eclipse 80i (see Note 10) (Fig. 1a).

3.2. SEM of Ovules

Use 1.5 mL Eppendorf tubes filled with 1 mL of each solution. Unless mentioned, all steps are carried out at room temperature under a fume hood.

Fixation I:

1. Fix dissected carpels or whole flower buds in freshly prepared glutaraldehyde (2%), acetone (69%) fixative in H_2O for 6 h to overnight (see Notes 1, 3, and 11).
2. Wash 4× in 70% acetone for 15 min/each.
3. Wash 6× in 70% acetone for 30 min/each. If necessary, tissue can be stored in refrigerator at this point.

Fixation II:

1. Wash for 15 min with 50% acetone in 50 mM sodium cacodylate buffer (pH 7.0) (see Note 12).
2. Wash for 10 min in 25% acetone/50 mM cacodylate buffer (pH 7.0).
3. Wash for 10 min in 10% acetone/50 mM cacodylate buffer (pH 7.0).
4. Wash for 5 min in 50 mM cacodylate buffer (pH 7.0).
5. Fix for 2 h with 2% osmium tetroxide in 50 mM cacodylate buffer (pH 7.0) (see Notes 12 and 13).
6. Wash 2× in 50 mM cacodylate buffer (pH 7.0) for 15 min/each.
7. Repeat step 3.
8. Wash for 10 min in 10% acetone.
9. Dehydrate with an acetone series (p.a. 20%, 40%, 60%, 70%) for 30 min/each. Tissue can be stored for extended periods of time at 4°C (weeks/months).
10. Wash in 100% acetone for 3–4 times with 2-h interval between each exchange before critical point drying.

Critical Point Drying (C.P.D.):

1. Label the glass tubes by engraving using a glass scribe (see Note 14).
2. Transfer the tissue to the labeled glass tubes in 100% acetone.
3. Without introducing any air bubbles plug the neck of the vial with very loose cotton to avoid the sample mix-up if you are drying different specimens.
4. Submerge all the sample vials in acetone tank.
5. To completely dry the sample while maintaining the morphological features, the C.P.D. procedure (using Critical Point Dryer, such as from Polaron, Quorum Technologies, UK) is carried out for 2–3 h using liquid CO_2.

6. Seal the vials with parafilm and place in the desiccator until mounting of samples (weeks/months).

Specimen preparation and scanning:

1. Stick the Leit adhesive carbon disc on the aluminum specimen stub by holding it with SEM specimen mount gripper tweezers on an Eppendorf rack.
2. Pistils of all stages are mounted in desired orientation on the carbon discs and dissected with fine tungsten needles or fine 0.5 ml/27 G needles under a dissecting microscope (see Notes 1 and 15).
3. Do the gold sputtering of specimens for 60 s, according to the instructions on the sputter instrument (such as Leica EM SCD 005, Germany).
4. Scan the specimens with 5–15 voltage levels at the desired magnification on a good SEM, such as the JEOL JSM-5900L scanning electron microscope (Fig. 1b).

4. Notes

1. Prepare a slide covered with a double-stick tape. Remove the pistil from the flower and place it on the sticky tape. For further dissection use a conventional dissecting scope with 5–40× magnification lenses. If possible use a ring-light source as this extremely facilitates the observation of the specimen. Orient the pistil with the replum pointing upwards (i.e., the valves directed sideways). Cut off the pedicel and the bottom first millimeter of the pistil and with a fine needle poke a small hole next to the replum and at the top of each valve. Then use the needle to slice along the edges of the valves (next to the replum) and remove the valves. Use forceps to pick up the pistil at the stigma and quickly place the pistil with the attached ovules in the fixative.
2. The addition of acetic acid results in a better morphology of, for example, the egg cell.
3. Make sure that the tissue is always submerged, also for consecutive steps.
4. Duration of MH staining may vary with the age of ovules or type of mutation. During later integument development (mid-to-late stage 3) Arabidopsis ovule integuments become less permeable to MH. Typically, mature normal ovules need about 2 h of staining. Ovules younger than early stage 3 (i.e., before the mononuclear embryo sac stage) do not need to be counterstained/destained at all. Staining of early stage 3 ovules

takes about 30 min. Mature ovules of mutants with a defect in integument development may only need 15 min of incubation in MH. Mature ovules of mutants with a defect restricted to sporogenesis need about 2 h of staining. The appropriate staining period needs to be determined individually.

5. If pressed for time one can abbreviate the treatment by either using shorter incubation periods (5–10 min) or skipping intermediate steps of the series.
6. In our experience the use of MB/Spurr's results in the best contrast. Furthermore, the addition of Spurr's medium makes the MB solution slightly less fluid (but can also be omitted). Finally, using MB (rather than for example chloral hydrate) leaves the specimens intact and they can be stored and reanalyzed for years.
7. Make sure that the sticky tape is tightly pressed on the upper surface of the microscope slide (particularly in the alignment next to the cavity; no air bubbles, otherwise poke in the air bubble with a fine needle).
8. For younger ovules, dissect free the pistil and remove much of the other floral parts. Then either dissect free the ovules or just cut the pistil into many pieces and look for appropriately oriented ovules later in the microscope. Collect pistils from several flowers. For older pistils, make sure that you cut off and remove the stigmas as well. Stigmas often are too hard and bulky and can create problems with the coverslip. Be patient, the dissecting steps require experience and may not work the first time.
9. Use a yellow pipet tip that has its bottom third cleaved off. The Canada Balsam will dry overnight and seal the edges of the coverslip (not necessary if slide will be discarded after preparation and immediate analysis). Try not to contaminate the pipet with Canada Balsam. If so, clean immediately.
10. Take care of the working distance of the objective. Be careful not to squeeze the ovules when using a 40× objective.
11. It is important to use glutaraldehyde, which was bottled under N_2. Upon opening the bottle dispense quickly into 1 mL aliquots and immediately store at −20 ° C.
12. Wear gloves and take care when handling cacodylic acid and osmium tetroxide. Avoid inhalation and contact with skin. Both chemicals are very poisonous!
13. Osmium is a heavy metal and has to be disposed after neutralizing with twice the volume of vegetable oils containing high percentage of unsaturated bonds, such as corn oil, to the one volume of 2% osmium for several hours. Neutralization is monitored with a piece of filter paper by dipping into the

osmium tetroxide/corn oil mix. Upon complete neutralization the filter paper keeps its color; otherwise it turns black.

14. Critical point drying can be done in many sorts of glass vials/tubes, even in Eppendorf tubes.
15. Take care not to damage the samples. Pistils of stage 8–9 flowers (carrying stage 1 ovules) are very difficult to dissect, requiring quite some practice. Be patient. Carefully try to break open small pieces of the carpel wall generating small holes through which the ovules can be observed. Despite careful handling one often observes depressions on the ovule primordia generated by the contact with the needles.

Acknowledgments

Our work is currently supported by grants SCHN 723/1-3, SCHN 723/3-2, and SCHN 723/6-1 from the German Research Council (DFG) to KS.

References

1. Chevalier D, Sieber P, Schneitz K (2002) The genetic and molecular control of ovule development. In: O' Neill SD, Roberts JA (eds) Plant reproduction. Sheffield Academic Press, Sheffield
2. Colombo L, Battaglia R, Kater MM (2008) Arabidopsis ovule development and its evolutionary conservation. Trends Plant Sci 13:444–450
3. Kelley DR, Gasser CS (2009) Ovule development: genetic trends and evolutionary considerations. Sex Plant Reprod 22:229–234
4. Shi D, Yang W (2011) Ovule development in Arabidopsis: progress and challenge. Curr Opin Plant Biol 14:74–80
5. Schneitz K, Hülskamp M, Pruitt RE (1995) Wild-type ovule development in *Arabidopsis thaliana*: a light microscope study of cleared whole-mount tissue. Plant J 7:731–749
6. Schneitz K et al (1997) Dissection of sexual organ ontogenesis: a genetic analysis of ovule development in *Arabidopsis thaliana*. Development 124:1367–1376
7. Running MP, Clark SE, Meyerowitz EM (1995) Confocal microscopy of the shoot apex. Methods Cell Biol 49:217–229
8. Smyth DR, Bowman JL, Meyerowitz EM (1990) Early flower development in *Arabidopsis*. Plant Cell 2:755–767
9. Kurup S et al (2005) Marking cell lineages in living tissues. Plant J 42:444–453

Chapter 8

Imaging of Phenotypes, Gene Expression, and Protein Localization During Embryonic Root Formation in Arabidopsis

Cristina Llavata-Peris, Annemarie Lokerse, Barbara Möller, Bert De Rybel, and Dolf Weijers

Abstract

Plants grow elaborate architectures by repeatedly initiating new organs post-embryonically. The competence to do so depends on the activity of meristems, stem cell niches located at the tips of shoot and root. These meristems are first specified early during embryogenesis. Therefore, important insight into the activity of factors that are central to the establishment of stem cell niches in plants can be gained from studying early embryogenesis. However, embryos are not directly accessible to microscopic observation since they are embedded within the seed, which is itself enveloped by the fruit. Here we describe a suite of methods for the analysis of mutant phenotypes, fluorescent reporter gene expression and protein localization in Arabidopsis embryos, and show how these methods can be used to visualize key factors in embryonic root formation.

Key words: Arabidopsis embryo, DIC, Fluorescence microscopy, Immunofluorescence, Expression analysis, Phenotype analysis, Confocal microscopy

1. Introduction

During embryogenesis in plants one cell called the zygote undergoes a series of controlled divisions that will generate the basic body plan of the seedling, containing all tissue types. Therefore, a strong control of this process is of vital importance. One of the major developmental processes leading up to a fully formed embryo is the specification of an apical-basal axis and the formation of a radial pattern (1). To define these different regions, specific genes are activated to execute different genetic programs.

Ive De Smet (ed.), *Plant Organogenesis: Methods and Protocols*, Methods in Molecular Biology, vol. 959,
DOI 10.1007/978-1-62703-221-6_8, © Springer Science+Business Media New York 2013

A critical step in the characterization of the function of selected genes during embryogenesis is the determination of their expression pattern at cellular resolution. A second important step is to understand the function of these genes. A straightforward method is the phenotypic analysis of knockout or overexpression lines.

Combining these two relatively simple approaches, elucidation of expression pattern, and assessment of mutant or misexpression phenotypes can give rich source of information, such as non-cell autonomous gene function.

In this chapter we discuss the use of two microscopic techniques in order to elucidate spatial and functional information of proteins expressed during *Arabidopsis* embryogenesis, with emphasis on the process of root initiation. In the first section we describe the analysis of knockout phenotypes in whole mount preparations of developing seeds using a microscope equipped with differential interference contrast (DIC) optics. The second section is devoted to studying the pattern of expression of genes using transcriptional or translational fusions to a fluorescent protein using confocal microscopy. The last section explains the use of immunofluorescence techniques for the visualization of low-abundant proteins using specific antibodies.

During embryogenesis, the root meristem is formed at the boundary between the proembryo and the suspensor. The suspensor is a transient filamentous structure of which only the uppermost cell, the hypophysis, contributes to the mature embryo by giving rise to the quiescent center and the columella stem cells (2). The correct specification of the hypophysis strictly requires the action of the MONOPTEROS (MP) protein, as *mp* mutants show plants with no root and abnormal hypophysis (3). MP/ARF5 is a member of a family of transcription factors, the auxin response factors (ARF), whose activity is controlled by the plant hormone auxin. In fact the accumulation pattern of this protein shows that MP is not expressed in the hypophysis itself, but in the adjacent proembryo (4), which suggests a non-cell autonomous function in hypophysis specification through (a) secondary signal(s). Recently, several direct targets of MP were identified (5). Two of these, *TARGET OF MONOPTEROS5* (*TMO5*) and *TMO7*, encode transcription factors of the bHLH family. While both genes are expressed in the proembryo, the TMO7 protein accumulates in the hypophysis, consistent with being a mobile protein (5). At the same time during this process auxin is transported from the proembryo to the uppermost suspensor cell, which can be visualized by the activation of an ARF-dependent gene expression reporter, DR5 (6).

2. Materials

2.1. Phenotypic Analysis Using DIC Microscopy

1. Siliques from plants with a mutation in the gene of interest (usually heterozygous if mutation is embryo- or seedling-lethal).
2. Microscope equipped with DIC optics. In this case we use a Leica DMRB, but any alternative is possible.
3. CCD camera and computer with image-processing software.
4. Stereomicroscope.
5. Tweezers (e.g., Rubis Switzerland 5-SA).
6. Needle (e.g., Microlance; 3 0.6 × 25 mm).
7. Double-sided adhesive tape.
8. Microscope slides.
9. Microscope coverslips 20 × 20 mm.
10. Clearing/mounting solution: Chloral hydrate:water:glycerol (w/v/v) 8:3:1.

2.2. Analysis of Expression Patterns by Confocal Microscopy

1. Transgenic plants carrying a reporter construct (protein or promoter fused to the fluorescent protein of choice such as GFP).
2. Stereomicroscope.
3. Tweezers (e.g., Rubis Switzerland 5-SA).
4. Needle (e.g., Microlance; 3 0.6 × 25 mm).
5. Double-sided adhesive tape.
6. Microscope slides.
7. Microscope coverslips 24 × 50 mm.
8. Mounting solution: PBS with 4% paraformaldehyde (PFA), 5% glycerol, and FM4-64 (stock solution: 15.5 mM; working dilution: 4:1,000). The FM 4-64 is used as a counterstaining since it fluoresces intensely upon binding the plasma membrane.
9. Zeiss Axiovert 100 M (LSM510/Confocor 2) confocal microscope or any alternative that allows imaging of GFP and FM4-64 fluorescence.

2.3. Immunostaining of Embryos

1. 10× concentrated PBS buffer: 80 g NaCl, 2 g KCl, 14.4 g Na_2HPO_4, 2.4 g KH_2PO_4 in 1 l dH_2O. Dilute 10 × PBS 10 times to get 1× PBS and adjust pH to 6.9 with HCl.
2. Fixative solution: 4% PFA in 1 × PBS, 0.1% Triton-X-100.
   - Fixative stock (10% PFA): Add 1 g of PFA in 10 ml dH_2O. Heat in a water bath until PFA dissolves. Adjust pH to 6.9.

3. 3% driselase (Sigma-Aldrich, Cat. No. D9515) solution in 1× PBS. Vortex thoroughly, centrifuge for 2 min at maximum speed. Only use the supernatant.
4. Permeabilization solution: 20% DMSO, 3% NP40 in 1× PBS.
5. Blocking solution: 3% bovine serum albumin (BSA) in 1× PBS, sterile-filter before use.
6. Primary antibody dilution: α-GFP 1:600 in blocking solution. Dilution has to be chosen empirically; at the beginning try a range of different dilutions.
7. Different possibilities for primary antibody (examples):
   - Anti-GFP from mouse, monoclonal (Roche, Cat. No. 11814460001).
   - Anti-GFP from rabbit, polyclonal (Invitrogen, Cat. No. A6455).
   - GFP booster ATTO 488 (ChromoTek, Cat. No. gba488 (this antibody is already coupled to a fluorescent dye; consequently no secondary antibody is necessary).
8. Secondary antibody solution: For example, Goat α-Rabbit-Alexa488, 1:600 (Invitrogen, Cat. No. A11008) in blocking solution. The choice of the secondary antibody depends on the first antibody used and the characteristics of the microscope that you are going to use for the analysis. Alternatives are Alexa546-coupled antibodies (red fluorescence) or anti-Mouse antibodies.
9. FM4-64 (stock solution: 15.5 mM; working dilution: 4:1,000).
10. Mounting solution: 90% glycerol, 10% 1 × PBS pH 8.5, 25 mg/ml 1,4-Diazabicyclo(2.2.2)octane (DABCO).
11. SuperFrost object slides (Thermo Scientific, Cat. No. J1800AHNZ).
12. PAP-pen (Sigma-Aldrich, Cat. No. Z377821).
13. Humid chamber (e.g., StainTray; Sigma-Aldrich, Cat. No. Z670146).
14. Oven set at 37°C.
15. Parafilm.
16. 20 × 20 mm coverslips.
17. 24 × 50 mm coverslips.
18. Stereomicroscope.
19. Tweezers (e.g., Rubis Switzerland 5-SA).
20. Needle (e.g., Microlance; 3 0.6 × 25 mm).
21. Double-sided adhesive tape.
22. Razor blade.

23. Eppendorf LoBind tubes (0.5 ml).
24. Zeiss Axiovert 100 M (LSM510/Confocor 2) confocal microscope or any alternative that allows imaging of GFP and FM4-64 fluorescence.

3. Methods

3.1. Differential Interference Contrast Microscopy for Analysis of Mutant Phenotypes

Direct visualization of plant embryos is not an easy task. Embryos are embedded in maternal tissue (the seed), which is enclosed in a protective structure, the silique. A better alternative for observing embryos therefore is the use of DIC microscopy also known as Nomarski microscopy. Using DIC it is possible to use whole mount preparation of isolated immature seeds without a need for staining. This means that developing seeds can be collected from the siliques, directly deposited in a drop of the mounting solution and with no further manipulation these can be observed by microscopy. DIC uses polarized light that strongly interacts with the sample, generating a high contrast with the background. The only prerequisite for the use of DIC microscopy is that the samples have to be transparent and not too thick. In *Arabidopsis thaliana*, the small size of developing seeds precludes the need to make histological sections. To render the opaque seeds translucent, a clearing solution is used that contains chloral hydrate and glycerol. This not only generates transparent samples but also creates an excellent refraction index for the use of DIC. Here we demonstrate this technique using a mutant of the auxin response factor MP/ARF5. This mutation produces an aberrant phenotype in the hypophysis (Fig. 1a, b).

1. Collect siliques at the right stage of embryo development (see Note 1).
2. Stick the siliques to a piece of double-sided adhesive tape attached to a petri dish or a microscope slide. Open the silique along the septum using a hypodermic needle, separate the 2 valves of the silique, and make them stick to the tape. At this point the developing seeds are exposed. Be careful not to use a stereomicroscope with bottom-lighting, as this will heat up the sample quickly.
3. Carefully collect the developing seeds with the shaft of the needle (the opening should face away from the seeds) and deposit the seeds into a drop (~50 μl) of clearing/mounting solution on a slide (see Note 2). It is possible to collect seeds from 1 to 3 siliques per drop, but when embryos from the same stage need to be compared it is preferred to collect seeds from a single silique in individual drops. Carefully put a coverslip on top of the seeds. Be careful not to squash the seeds and

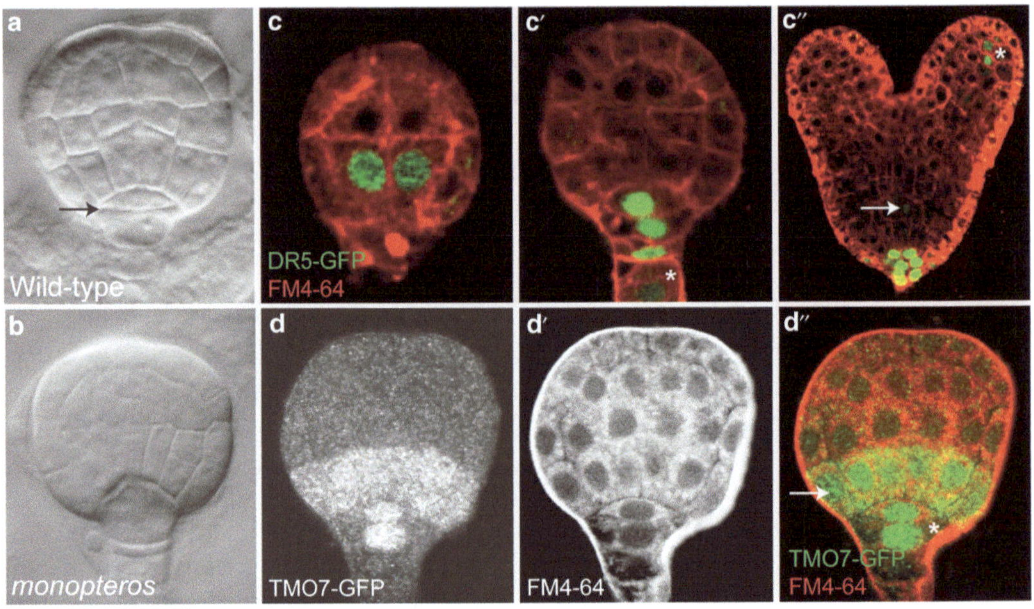

Fig. 1. Imaging phenotypes, gene expression, and protein localization in *Arabidopsis* embryos. (**a**, **b**) Nomarski imaging of wild-type (**a**) and *monopteros* mutant (**b**) embryos at globular stage. Note that due to the high contrast of this imaging technique, cell walls and cell organelles such as vacuoles and nuclei are clearly visible. The arrow indicates the asymmetric division of the hypophysis cell in wild type. This division does not occur in the *monopteros* mutant embryo shown here. (**c**–**c″**) Expression of nuclear-localized triple *GFP* (*n3GFP*), driven from the auxin-responsive *DR5rev* promoter. Green nuclear signals highlight those cells that display active auxin response. Membranes are stained with the FM4-64 dye (*red*). Note that at the 16-cell stage (**c**) *DR5rev* is active in the inner cells in the lower half of the embryo. This activity is displaced to the hypophysis and subtending cells (also note weak signal indicated with an *asterisk*) at globular stage (**c′**). This maximum is maintained through the heart stage (**c″**), when additional expression peaks are established in the vascular cylinder (*arrow*), and in the cotyledon primordia (*asterisk*). (**d**–**d″**) Detection of TMO7-GFP fusion protein by indirect immunofluorescence in a globular stage embryo. Green TMO7-GFP signal (**d**) and *red* FM4-64 membrane staining (**d′**) are separately shown, as well as combined in a merged image (**d″**). TMO7-GFP is detected in cytosol and nuclei of the lower tier of the embryo (*arrow* in (**d″**)), and in the nuclei of the cells derived from hypophysis division (*asterisk* in (**d″**)).

add more clearing/mounting solution if necessary. Drain excess solution with a piece of filter paper if necessary. Leave at least 30 min at room temperature before observation. Alternatively, slides can be kept at 4°C for several days.

4. Observe the sample using a DIC microscope. An LCD camera is combined with the microscope and the software Mightex Capture is used for obtaining images. Make sure that the microscope settings are optimal.

3.2. Analysis of Expression Patterns Using Confocal Microscopy

Fluorescently tagged proteins are routinely used to monitor the presence, localization, abundance, transport, or degradation of a protein of interest. This technique makes use of the capacity of fluorescent proteins to absorb light at a particular wavelength and emit at a longer wavelength. Emitted photons can cover different regions of the spectra, representing different colors. The first fluorescent protein isolated was the green fluorescent protein GFP from the jellyfish *Aequorea victoria*. This protein has an excitation peak at 395 nm and a weaker excitation peak at 475 nm, and emits at 508 nm in the green range. Subsequently different mutants of the GFP protein have been created, with different chromophores and consequently with different spectra for absorption and emission.

The use of GFP includes the construction of transcriptional fusions that report the transcription pattern of a gene, and translational fusions that can be used to demonstrate protein localization. Here we show the use of transcriptional fusions as a reporter for the pattern of expression of the ARF-dependent gene expression reporter, *pDR5::GFP* (Fig. 1c–c″; ref. (4)).

1. Collect siliques at the right developmental stage of embryo (see Note 1).
2. Stick siliques to a piece of double-sided adhesive tape (on microscope slide or petri dish). Using a hypodermic needle open the silique along the septum, separate the 2 valves of the silique, and stick these to the tape. At this point the developing seeds are exposed. Be careful not to use a stereomicroscope with bottom-lighting, as this will heat up the sample quickly.
3. Carefully collect the developing seeds with the needle and deposit them in a drop (~50 μl) of mounting solution on a slide (see Note 3).
4. Place a 24 × 50 mm coverslip on top of the seeds.
5. Gently squash the developing seeds using the tip of a pencil in order to break the seed coat and release the embryo from the seed. This step requires practise. The efficacy of embryo extrusion can be checked under a microscope.
6. Samples can be observed using a confocal laser-scanning microscope; we use a Carl Zeiss LSM510. If you are using GFP (see Note 4) the settings that can be used are Argon laser line at 488 nm, band-pass filter ranging from 505 to 530 nm for the GFP signals, while for FM4-64 signals a long-pass filter of 650 nm can be used.

3.3. Immunostaining

In some cases the protein of interest may have a very low level of expression, and might not be visible with the confocal microscope when fused to a fluorescent protein reporter. A way to overcome this is the use of indirect immunofluorescence. Using a monoclonal or a polyclonal antiserum directed against the tag (in this case GFP) and a fluorescently labeled secondary antibody will increase the signal since multiple antibodies are now bound to a single tagged protein. A range of first and secondary antibodies is commercially available.

The essential steps in immunofluorescence are first of all tissue fixation and immobilization to slides. To facilitate antibody penetration, the cell wall is first degraded by a mix of fungal enzymes. Next, the cell membrane needs to be permeabilized using the detergent NP-40 and the solvent DMSO. After blocking the excess of non-target proteins with BSA, antibody incubation steps are performed. It is of vital importance to perform extensive washes with PBS buffer between different incubation steps to get rid of the different solutions. Also take care that the samples are always covered by liquid, as drying of the sample may affect the efficiency of the immunostaining process.

Here, we show the use of immunostaining to detect the TMO7 protein (5). Expression of this protein is too weak to be detected by direct observation. Immunofluorescence, however, allows detection of this protein (Fig. 1d–d″). The immunofluorescence protocol is adapted from the one described by ref. (8).

1. Prepare 0.5 ml Eppendorf LoBind tubes with 300–400 μl fixative solution (see Note 5). The use of LoBind tubes limits the loss of seeds by adhesion to the surface of the tube.
2. Isolate developing seeds from siliques (as in Subheadings 3.1 and 3.2) and transfer these to the Eppendorf tubes with the fixative solution (see Note 1). Keep tubes on ice while harvesting seeds. Note that during the procedure, there will be loss of embryos at several steps. Make sure to collect as many as possible, at least 5–10 siliques per developmental stage. Seeds of different stages can be harvested and processed separately to facilitate analysis of embryos of similar stage.
3. Incubate under vacuum for 1 h on ice.
4. Wash four times for 10 min with dH_2O.
5. Use a glass Pasteur pipette to carefully transfer the seeds to SuperFrost object slides (see Note 6). Avoid air bubbles as this leads to excessive sticking of seeds to the glass wall. Use a needle to move seeds so that they are contained within a small area (<18 × 18 mm) and cover with a clean 18 × 18 mm coverslip.
6. Firmly squash the seeds with the tip of a pencil in order to extrude the embryos.
7. Freeze the slide in liquid nitrogen until bubbling stops. Lift the coverslip with a razor blade. If the coverslip is clean, the buffer containing seeds should (mostly) stick to the slide, not the coverslip.
8. Let the slides dry overnight at room temperature (RT).
9. The following morning place the slides in a humid chamber (see Note 7).
10. Use the PAP-pen to draw a circle around the plant material on the glass slides (see Note 8). Let the solvent evaporate.
11. Rehydrate the tissue with 1× PBS pH 6.9 for at least 5 min. In this and all following steps remove the solution by tilting the

humid chamber. If necessary, drain excess liquid by holding a piece of filter paper next to the embryos.

12. Add ~200 µl 3% Driselase solution per slide. Incubate for 30–45 min at 37°C in a humid chamber.
13. Wash 5–6× for 5 min with 1× PBS pH 6.9.
14. Add ~200 µl Permeabilization solution per slide. Incubate for 1 h at RT.
15. Wash 5–6× for 5 min with 1× PBS pH 6.9.
16. Add ~200 µl Blocking solution per slide and incubate for 2–3 h at 37°C in a humid chamber.
17. Add ~100 µl 1st antibody solution per slide. Carefully cover the slides with a piece of Parafilm no bigger than the slide. Transfer to a humid chamber and incubate for 2 h at 37°C, and then overnight at 4°C (see Note 9).
18. Carefully remove Parafilm strips and wash 5–6 times for 5 min with 1× PBS pH 6.9.
19. Add ~100 µl 2nd antibody solution per object slide. Cover the slides with Parafilm, transfer to the humid chamber, and incubate for 2 h at 37°C (see Note 10).
20. Wash 3× for 5 min with dH_2O.
21. Add ~200 µl 1% FM4-64 for 4 min.
22. Wash 3× for 10 min with dH_2O.
23. Remove liquid and add a few drops of mounting solution.
24. Add a 24×50 mm coverslip and analyze directly with confocal microscope or store at 4°C in the dark (see Note 11).

4. Notes

1. For some studies it is important to choose the right developmental stage of the embryo. A way to identify the stage is to count the number of siliques from the first open flower down. As guidance, 3–4 siliques down from the first open flower usually contain globular stage embryos. You should also take into account the age of the plant. In older plants the developing seeds can be in a more advanced stage of development than expected.
2. The time needed for the clearing of the developing seeds is around half an hour, but it is also possible to prepare the slides and store them at 4°C overnight, observing them under the microscope the day after. Mild heating (30–40°C) will speed up clearing, but can also lead to enhanced tissue disruption.

3. At this point it is possible to look directly at the slides under the microscope, but it is also possible to store them at 4°C for a couple of days. If the slides are stored it is important to make sure that the seeds are covered in sufficient mounting solution.
4. A factor desirable in a fluorescent protein is a high brightness that can be easily detected by eye or using photodetectors when using a fluorescence microscope. The brightness depends on two factors: the efficiency in the absorption of the excitation light, called the extinction coefficient, and the ratio of photons emitted to photons absorbed, named the quantum yield. If you are doing long measurements where photobleaching (the photochemical destruction of the fluorophore) can pose a problem, the most advantageous situation is a molecule with high extinction coefficient because then it is possible to use lower excitation light levels, reducing the photobleaching. High quantum yield also means lower excitation light levels, resulting in higher brightness using the same intensity of excitation light. The enhanced versions of the GFP and YFP (eGFP and eYFP) have a similar quantum yield, but eYFP has a higher extinction coefficient, making it less susceptible to photobleaching. eCFP, the enhanced cyan form, is less susceptible to photobleaching but is less bright than eGFP and eYFP. For the choice of which fusion to use you can take into account these characteristics; it is also possible to try several different ones and choose for the one that gives better results.
5. The fixative solution should be freshly prepared or small aliquots may be stored at −20°C. PFA is a rapid fixative and a suspected carcinogen. It is recommended to wear protective gloves and work in a fume hood when handling the powder.
6. These slides possess a permanent positive charge that electrostatically binds the fresh frozen tissue to the slides.
7. The use of a humid chamber is recommended in order to keep the right humidity conditions and avoid drying out of samples during long incubation steps. There are commercial humid chambers that can be purchased, basically consisting of a container with holders where the slides can be temporarily fixed and a deposit on the bottom part that can be filled with water. The slides can be fixed to these supports, facilitating the exchange of liquids in washing steps by just tilting the box. In some cases this is not recommended because it can cause loss of sample, so we advice using a glass Pasteur pipette to carefully add or remove liquids.
8. This creates a hydrophobic barrier around the developing seeds, so the amount of antibody reagent per slide can be reduced. This also helps to minimize drying out of the slides.

9. Most commercially available antibodies directed against the GFP are also able to recognize YFP and CFP. If the antibodies are polyclonal, different epitopes of the same protein can be recognized. Hence, more than one antibody can bind to the GFP, leading to an increase of the signal. We have noticed that there can be large differences in results depending on the manufacturer and batch of the primary antibody. It is advisable to include negative and positive controls, and if possible test multiple antibodies.
10. There is a large range of commercially available secondary antibodies. For the immunofluorescence experiments we use an Alexa488-coupled antibody, which emits fluorescence in the green range. Other forms are available where the excitation and emission spectra cover the visible spectrum and extend into the infrared. Note that the same procedure can be used to detect multiple proteins (e.g., GFP and PIN1) (7). In this case, primary antibodies need to be from different host species, and secondary antibodies should have different fluorophores.
11. A problem that often occurs in immunostaining is the presence of fluorescence background. A good negative control is to incubate with only the secondary antibody.

Acknowledgments

This work was supported by funding from the European Commission 7th Framework Program (Initial Training Network "SIREN"; Contract no. 214788) and by the Netherlands Organization for Scientific Research (NWO; ALW-816.02.014 and ALW-VIDI-864.06.012) to D.W and by a long-term FEBS fellowship and a Marie Curie long-term FP7 Intra-European Fellowship (IEF-2009-252503) to B.D.R.

References

1. Jürgens G, Mayer U (1994) Arabidopsis. In: Bard JBL (ed) EMBRYOS, color atlas of development. Mosby–Year Book Limited, London, pp 7–22
2. Scheres B, Wolkenfelt H, Willemsen V, Terlouw M, Lawson E, Dean C, Weisbeek P (1994) Embryonic origin of the *Arabidopsis* primary root and root meristem initials. Development 120:2475–2487
3. Berleth T, Jürgens G (1993) The role of the *monopteros* gene in organising the basic body region of the *Arabidopsis* embryo. Development 118:575–587
4. Weijers D, Schlereth A, Ehrismann JS, Schwank G, Kientz M, Jürgens G (2006) Auxin triggers transient local signaling for cell specification in *Arabidopsis* embryogenesis. Dev Cell 10:265–270
5. Schlereth A, Möller B, Liu W, Kientz M, Flipse J, Rademacher EH, Schmid M, Jürgens G, Weijers D (2010) MONOPTEROS controls embryonic root initiation by regulating a mobile transcription factor. Nature 464:913–916
6. Ulmasov T, Murfett J, Hagen G, Guilfoyle TJ (1997) Aux/IAA proteins repress expression of reporter genes containing natural and highly

active synthetic auxin response elements. Plant Cell 9:1963–1971
7. Steinmann T et al (1999) Coordinated polar localization of auxin efflux carrier PIN1 by GNOM ARF-GEF. Science 286:316–318
8. Lauber MH, Waizenegger I, Steinmann T, Schwarz H, Mayer U, Lukowitz W, Jürgens G (1997) The *Arabidopsis* KNOLLE protein is a cytokinesis-specific syntaxin. J Cell Biol 139: 1485–1493

Chapter 9

Inducible System for Lateral Roots in *Arabidopsis thaliana* and Maize

Leen Jansen, Boris Parizot, and Tom Beeckman

Abstract

The study of biological processes contributing to plant growth can be complicated by the small number of cells involved and by the brief and sudden appearance of some crucial developmental steps. Given such troublesome circumstances, methods to monitor the timing or to increase the number of concerned cells can be of great advantage to researchers. Lateral root initiation is a process taking place endogenously in a discrete number of cells of the parent root. It represents the onset of the formation of a new meristem and provides the below ground part of the plant the ability to react on environmental conditions such as nutrient and water availability. Insights into the underlying mechanisms of this developmental event can be of major importance to provide means of improving tolerance to nutrient and water deficient conditions. The exact timing and site of lateral root initiation are, however, impossible to predict, hampering exhaustive studies of the sequence of events directing this process. Here, we present a method to synchronize the induction of lateral roots in *Arabidopsis thaliana* and maize. By initially preventing the formation of laterals in young seedlings and subsequently inducing lateral root initiation, this method not only allows controlling the process in time but also enlarges significantly the population of cells involved, opening up the way to systems biology approaches.

Key words: Lateral roots, *Arabidopsis thaliana*, Maize, Auxin, Paper rolls

1. Introduction

In spite of being underestimated for a long time, plant roots gain nowadays more and more interest from researchers because their importance for plant growth has become increasingly clear (1, 2). Some crop species, like maize and rice, have an extensive root system consisting of embryonic and post-embryonic roots. A common feature of all these root types is the formation of lateral roots, which significantly contributes to the proliferation of the total root system. Lateral roots represent a means for plants to screen the surrounding soil more efficiently for water and nutrients, thereby increasing the absorptive capacity of the root (3, 4).

Ive De Smet (ed.), *Plant Organogenesis: Methods and Protocols*, Methods in Molecular Biology, vol. 959,
DOI 10.1007/978-1-62703-221-6_9, © Springer Science+Business Media New York 2013

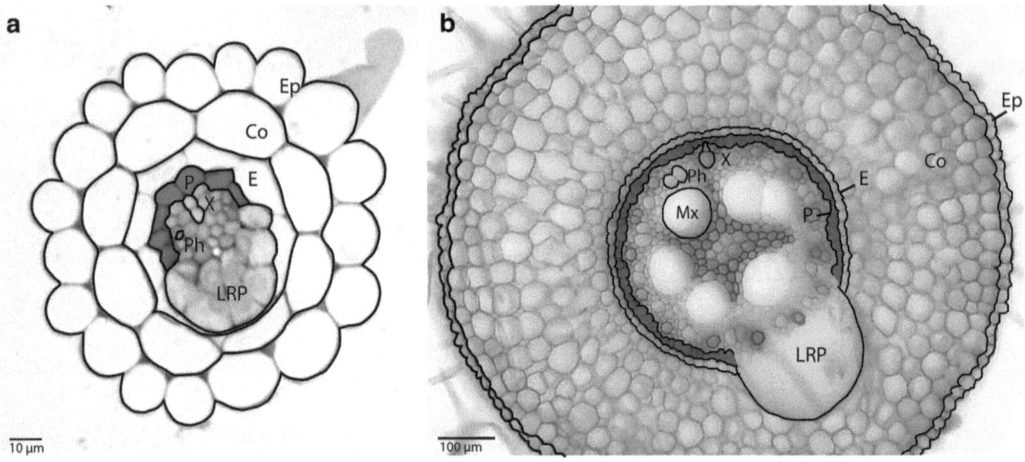

Fig. 1. Transverse sections through an Arabidopsis (**a**) and maize (**b**) root. Lateral root primordia (LRP) are initiated in pericycle (P) cells. The pericycle encloses the vascular bundle with xylem (X), phloem (Ph), and meta-xylem (Mx) vessels. Endodermis (E), cortex (Co), and epidermis (Ep) surround the pericycle.

Plant roots are composed of different cell types organized in concentric layers surrounding the vascular tissue. In the model plant *Arabidopsis thaliana* this vasculature is diarch, with two xylem and phloem poles (Fig. 1a) (5). In other plants, like maize and rice, the vasculature has a poly-arch structure with multiple alternating xylem and phloem elements (Fig. 1b) (6). The pericycle encloses the vascular tissues, which in turn is surrounded by the endodermal, the cortical, and the epidermal cell layers. Lateral root formation in angiosperms is initiated by a discrete number of pericycle cells, called "founder cells." In Arabidopsis, but also in onion, radish, and sunflower, these founder cells are located at the xylem poles, while in maize, rice, and carrot, pericycle cells at the phloem poles are particularly prone to lateral root initiation (5, 7–9). The first step in lateral root formation is an asymmetric anticlinal division that occurs simultaneously in 3 or more neighboring pericycle founder cells (10–12). Further periclinal and anticlinal divisions lead to the formation of a dome-shaped primordium which eventually emerges from the parent root (8). The timing and site of occurrence of the first divisions are however hard to predict (13). Together with the small number of cells involved, it significantly complicates the study of lateral root initiation.

The plant hormone auxin plays a crucial role in different aspects of lateral root formation. The first asymmetric divisions are preceded by locally increased auxin levels (14). During primordium formation the establishment of an auxin gradient is indispensable to obtain a functional lateral root (15). Finally, the emergence of the lateral from its parent root is facilitated by the expression of auxin-induced genes (16). Overproduction or exogenous application of auxin leads to increased lateral root formation, while inhibition or defects in auxin transport and response prevent lateral root initiation or lead to aberrantly shaped primordia. The strong lateral-root-inductive feature of auxin has led to the establishment of a lateral root-inducible system (LRIS) for Arabidopsis (17). Germination of the seeds on the auxin transport inhibitor N-1-naphthylphthalamic acid (NPA) prevents the initiation of lateral roots. Transfer of NPA-incubated seedlings to the synthetic auxin naphthalene acetic acid (NAA) rapidly activates pericycle cells over a large part of the root that will subsequently divide and give rise to lateral roots. Such an inducible system provides the opportunity to steer the lateral root initiation process allowing to disentangle the course of events in time, and significantly increases the number of cells involved.

Here, we provide a detailed description of the LRIS for use in Arabidopsis, and we explain how the system can be extrapolated to a cereal crop species such as maize. As the role of auxin in the initiation and further development of lateral roots seems to be highly conserved among plants, this system will be applicable to many other plant species, thereby providing a versatile method to study root architecture.

2. Materials

2.1. *Arabidopsis*

1. 70% Ethanol.
2. Prepare a fresh sterilization solution: 3.85 ml NaOCl stock (12% Cl), 5 μl Polysorbate 20 (Tween® 20), up to 10 ml with water.
3. Sterile distilled water.
4. Square Petri dishes, 120 × 120 × 17 mm (e.g., Greiner Bio-One cat. no. 688102).
5. Breathable tape to seal the Petri dishes (e.g., 3 M Micropore™ cat. no. 1530-1).
6. Growth medium (1 L): 2.154 g Murashige and Skoog (MS) salt mixture, 10 g sucrose, 0.1 g myo-inositol, 0.5 g 2-(N-morpholino)ethanesulfonic acid (MES) in distilled water. Adjust the pH to 5.7 with 1 M KOH, add 8 g agar. Autoclave.
7. Prepare a 25 mM stock solution NPA and a 50 mM stock solution NAA in dimethylsulfoxide (DMSO) (see Note 1).
8. Sterile forceps, curved (e.g., Dumont style 5/45).
9. 200 μl micropipette with sterile tips (see Note 2).
10. Growth cabinet 21°C, continuous light.

2.2. Maize

1. Sterilization solution: 6% NaOCl.
2. Sterile distilled water.
3. Glass tubes, approx. 6 cm diameter, 14 cm height, e.g., centrifuge tubes 250 ml (Duran® catalog no. 21 601 36) (see Note 3).
4. Absorbing paper or germination paper (e.g., Scott Rollwipe 6681 or Anchor Seed Germination Paper) (see Note 4).
5. Forceps with blunt ends (e.g., VWR cat. no.232-1982).
6. Prepare a 25 mM stock solution NPA and a 50 mM stock solution NAA in DMSO (see Note 1).
7. Growth cabinet 27°C, continuous light, 70% rel. humidity (see Note 5).

3. Methods

3.1. Sterilization of Arabidopsis Seeds

1. Distribute the seeds in micro-centrifuge tubes (see Note 6).
2. Add 1 ml of 70% ethanol. Invert or gently shake for 2 min.
3. Replace the ethanol by 1 ml sterilization solution for 15 min. Invert the tubes several times to make sure that all seeds come in contact with the solution (see Note 7).
4. In a laminar flow cabinet, rinse the seeds 5 times for 5 min with 1 ml sterile distilled water. Before each new rinsing step the seeds can be collected at the bottom of the tube by a short centrifugation step.
5. Store the seeds at 4°C in the dark for at least 2 days (see Note 8).

3.2. Lateral Root Induction in Arabidopsis

1. Prepare plates for in vitro growth. Count on 50 ml of growth medium per plate, and add either 20 μl of NPA stock solution or 10 μl of NAA stock per plate, to obtain a final concentration of 10 μM (see Note 9). Homogenize upon addition.
2. Sow the seeds on NPA-plates with the use of a 200 μl pipette. Pipette up and down a couple of times to resuspend the seeds, then take up approx. 150 μl water with 10–20 seeds, and wait till the seeds sediment at the bottom of the tip. When touching the agar gently with the cut tips, a drop of water with a seed will be released from the tip. Per plate approx. 100 seeds can be sown (10 rows of 10 seeds). Seal the plates with breathable tape.
3. Move the plates to the growth cabinet for 5 days (2 days of germination and 3 days of growth) in a near-vertical position (see Note 10).

4. Transfer the seedlings to the plates supplemented with 10 μM NAA (see Note 11). The arms of the curved forceps can be hooked under the cotyledons of the seedling. Picking them up as such prevents damage to the plantlets. Make sure that plant roots make good contact with the NAA-containing agar medium. This can be achieved by skimming the root over the agar surface over a small distance. If needed, press the root gently to the agar surface with the forceps.
5. Seal the new plates and place them in the growth cabinet in near-vertical position.

3.3. Surface Sterilization of Maize Kernels

Although the growth of maize does not take place in sterile conditions, kernels are surface sterilized, and sterile water is used for growth to reduce contamination to a minimum.

1. Bring the kernels in a beaker with 100 ml sterilization solution and gently stir for 5 min manually or using a magnetic stirrer or a shaker.
2. Rinse 3 times for 5 min with sterile water.

3.4. Lateral Root Induction in Maize

In this protocol we explain a system to induce and synchronize lateral root formation in the primary root of maize. It can be adapted for the induction of laterals in adventitious roots (see Note 12).

1. Fill the glass vials halfway with a solution of 50 μM NPA in distilled water. When using the centrifuge tubes this corresponds to 125 ml water and 250 μl NPA stock solution (see Note 13).
2. Prepare paper strips of about 100 cm in length, 25 cm height, and at least 4 layers thick. For the Scott Rollwipe this corresponds to two layers of two sheets, folded double over the length.
3. Distribute the kernels in one line on the paper with blunt forceps, at about 2 cm from the top, with an interspace of 7–10 cm and the radicle of the kernel pointing down (Fig. 2a). Roll the paper over the length; make sure that the kernels stay in place (see Note 14).
4. Put the paper roles in the prepared tubes. A large part of the liquid will be immediately taken up by the paper (Fig. 2b, c). Make sure that the paper is moisturized till the top (see Note 15).
5. Transfer the tubes to the growth cabinet for 64 h (see Notes 16 and 17).
6. Fill new tubes with a same volume of sterile distilled water as before, and a working concentration of 50 μM NAA (see Note 18). For the tubes this corresponds to 125 ml water and 125 μl NAA.

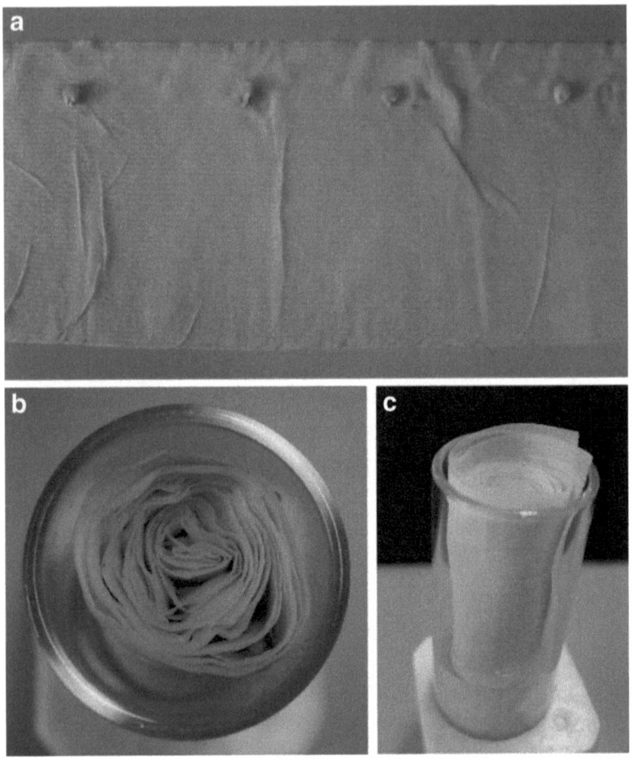

Fig. 2. Illustration of the paper roll system. (**a**) Kernels are spread on top of paper strips with their radicle pointing down. (**b, c**) The paper is rolled over the length and placed in glass tubes.

7. Transfer the paper roles with the germinated seeds to the new tubes. Gently squeeze the paper to remove excess of NPA liquid and place the tubes in the growth cabinet.

3.5. Applications

A plethora of reporter lines and mutants is available in Arabidopsis. In combination with the described inducible system they can be used to follow the expression pattern of specific genes during lateral root formation or to elucidate their function in the process (17, 18). Expression levels of specific genes can be followed by quantitative RT-PCR (qRT-PCR) (19), or transcriptome analysis can be done with micro-array or deep sequencing (18). When whole roots are sampled, it is recommended to cut off the root tip to avoid interference with processes taking place in the apical meristem. In this case 30 plants per condition yield sufficient RNA. GAL4-GFP enhancer trap lines offer the possibility to isolate specific cell types via fluorescence-activated cell sorting (FACS) (20). As such it is for instance possible to perform high-resolution transcriptome analysis on those cells competent for lateral root initiation (21). The RNA yield is influenced by the expression pattern and strength of the fluorescence signal, preparation of the samples for FACS, and efficiency of the sorting process itself. Therefore at least 1,000 plants should be sampled per condition, and an RNA amplification step might be required.

The time-points at which samples should be taken depend on the process of interest. The early events following lateral root induction in Arabidopsis have been characterized using marker lines available for different processes. Upon transfer to NAA the auxin response is rapidly induced, which can be followed with the synthetic auxin response element DR5 fused to a reporter gene (22). After about 1.5 h on NAA, auxin response can be detected in the apical part of the root, and after 3 h over the entire root. Also the pericycle cell divisions first take place in the apical part, about 6 h after the transfer, while at 12 h the complete pericycle is dividing (17).

Up to now, the choice of marker lines in maize is limited. The efficiency of the lateral root induction can be assessed by observing mitotic figures after staining the root with Feulgen DNA stain (23). In maize it is possible to separate cortex and epidermis from stele tissues, which makes the pericycle accessible for observation with an epifluorescence microscope (24). Timing of the events in maize or other species is not necessarily the same as in Arabidopsis, and can vary with the concentrations of compounds used in the experiment. The induced roots can for instance be used for gene expression analysis by qPCR, transcriptome, or proteome analysis. Separating cortex and stele tissues can help to increase the specificity of the sampled tissues. Although for maize a 2 cm root can already be sufficient for RNA extraction, it is recommended to pool material from several roots to cope with variation. Specific cells or tissues can be isolated by laser capture microdissection on frozen or paraffin sections (25).

4. Notes

1. These solutions can be stored for several weeks at 4°C or at -20°C.
2. Cut the first 1–2 mm off the tips with a scalpel or a single-edged razor blade (e.g., GEM scientific blades). The cut tips will be used for the sowing of Arabidopsis seeds, which is facilitated by a larger tip opening, better adapted to the size of the seeds.
3. The proposed centrifuge tubes are suited for the germination of up to 12 kernels. Also narrow glass beakers can be used, e.g., a 400 ml beaker with dimensions 80 × 150 mm. The volumes mentioned further in the protocol correspond to the use

of the 250 ml centrifuge tubes. Different volumes might be needed when using beakers or other vials.

4. For young seedlings of up to 3–4 days after germination, normal hand tissue paper can be used. As plants grow older, the roots tend to grow through the paper, complicating observation or sampling. In this case alternating a layer of tissue paper and germination paper is recommended, with the germination paper in contact with the roots.
5. Controlled humidity is not indispensable, but high humidity facilitates germination and growth of the seedlings and prevents evaporation of the solutions.
6. In 1.5 or 2 ml tubes, respectively, up to 20 or 40 mg of seeds can be sterilized. For large-scale experiments, sterilization of up to 1 g of seeds can be done in 15 ml Falcon tubes, using 10 ml solution in subsequent steps. The amount of seeds needed depends on the purpose of the experiment (*see* Subheading 3.5).
7. As Arabidopsis is grown in sterile in vitro conditions, all the following steps should be performed in aseptic conditions.
8. This step acts as a stratification which is needed to synchronize seed germination.
9. The compounds should be added to liquid sterilized medium that is cooled down to a temperature of about 60°C. Higher temperatures impair the stability of the compounds.
10. Racks for vertical plates can be easily made. For instance aluminum U-profiles of 2 cm width and 1.5 cm height fixed at least 3 cm from each other can be used to keep the plates in the correct position.
11. As NPA disturbs the gravitropic responses, roots might leave the surface of the agar medium. To make sure that all roots used for induction are devoid of lateral root initiation sites, only transfer those plants of which the roots were in close contact with the medium. At this step a loss of 30–40% of the plants should be taken into account.
12. For the induction of lateral roots on the first series of adventitious roots of maize, grow the plants in water till the adventitious roots start to emerge (approx. 6 days). At this point, transfer the plants to NPA for 4 days. Induction is done by transferring the plants to NAA, as described for the primary root.
13. We tested a concentration range of 10-50 μM NPA. Although we occasionally observed divisions in endodermis and pericycle, no laterals were formed after several days. However, to avoid nonrelevant cell divisions to the most, and to increase

induction, we generally use a concentration of 50 μM for the primary root and 25 μM for adventitious roots.

14. Humidifying the paper beforehand facilitates spreading of the seeds and rolling of the paper.
15. Particularly when higher paper rolls are used, it might be needed to humidify the top of the paper rolls by over-brimming them with part of the liquid, or even to increase the volume of the solution.
16. The paper should stay humid at all times. In case no controlled humidity is available in the growth cabinet, liquid might evaporate before this time. In this case it is recommended to add some extra water and NPA to avoid drying of the paper.
17. The roots of the seedlings should be about 2 cm long before proceeding to induction. Depending on the line, the quality of the kernels, and the growth conditions, seed germination and growth might be delayed. In this case the incubation time should be extended.
18. Use for NAA the same concentration as the one used for NPA.

References

1. Gewin V (2010) Food: an underground revolution. Nature 466:552–553
2. Glover JD, Cox CM, Reganold JP (2007) Future farming: a return to roots. Sci Am 297:82–89
3. Ito K, Tanakamaru K, Morita S, Abe J, Inanaga S (2006) Lateral root development, including responses to soil drying, of maize (*Zea mays*) and wheat (*Triticumaestivum*) seminal roots. Physiol Plant 127:260–267
4. Varney GT, Canny MJ (1993) Rates of water uptake into the mature root system of maize plants. New Phytol 123:775–786
5. Dolan L, Janmaat K, Willemsen V, Linstead P, Poethig S, Roberts K, Scheres B (1993) Cellular organisation of the Arabidopsis thaliana root. Development 119:71–84
6. Hochholdinger F, Park WJ, Sauer M, Woll K (2004) From weeds to crops: genetic analysis of root development in cereals. Trends Plant Sci 9:42–48
7. Casero P, Casimiro I, Lloret P (1995) Lateral root initiation by asymmetrical transverse divisions by pericycle cells in four plant species: *Raphanus sativus*, *Helianthus anuus*, *Zea mays*, and *Daucus carota*. Protoplasma 188:49–58
8. Malamy JE, Benfey PN (1997) Organization and cell differentiation in lateral roots of Arabidopsis thaliana. Development 124:33–44
9. Rebouillat J, Dievart A, Verdeil JL, Escoute J, Giese G, Breitler JC, Gantet P, Espeout S, Guiderdoni E, Périn C (2009) Molecular genetics of rice root development. Rice 2:15–34
10. Dubrovsky JG, Rost TL, Colon-Carmona A, Doerner P (2001) Early primordium morphogenesis during lateral root initiation in Arabidopsis thaliana. Planta 214:30–36
11. Laskowski MJ, Williams ME, Nusbaum HC, Sussex IM (1995) Formation of lateral root meristems is a two-stage process. Development 121:3303–3310
12. Sussex IM, Godoy JA, Kerk NM, Laskowski MJ, Nusbaum HC, Welsch JA, Williams ME (1995) Cellular and molecular events in a newly organizing lateral root meristem. Philos Trans R Soc Lond B Biol Sci 350:39–43
13. Dubrovsky JG, Gambetta GA, Hernández-Barrera A, Shishkova S, González I (2006) Lateral root initiation in *Arabidopsis*: developmental window, spatial patterning, density and predictability. Ann Bot 97:903–915
14. Dubrovsky JG, Sauer M, Napsucialy-Mendivil S, Ivanchenko MG, Friml J, Shishkova S, Celenza J, Benková E (2008) Auxin acts as a local morphogenetic trigger to specify lateral root founder cells. Proc Natl Acad Sci USA 105:8790–8794

15. Benkova E, Michniewicz M, Sauer M, Teichmann T, Seifertova D, Jurgens G, Friml J (2003) Local, efflux-dependent auxin gradients as a common module for plant organ formation. Cell 115:591–602
16. Swarup K, Benková E, Swarup R, Casimiro I, Péret B, Yang Y, Parry G, Nielsen E, De Smet I, Vanneste S, Levesque MP, Carrier D, James N, Calvo V, Ljung K, Kramer E, Roberts R, Graham N, Marillonnet S, Patel K, Jones JD, Taylor CG, Schachtman DP, May S, Sandberg G, Benfey P, Friml J, Kerr I, Beeckman T, Laplaze L, Bennett MJ (2008) The auxin influx carrier LAX3 promotes lateral root emergence. Nat Cell Biol 10:946–954
17. Himanen K, Boucheron E, Vanneste S, de Almeida Engler J, Inzé D, Beeckman T (2002) Auxin-mediated cell cycle activation during early lateral root initiation. Plant Cell 14:2339–2351
18. Vanneste S, De Rybel B, Beemster GT, Ljung K, De Smet I, Van Isterdael G, Naudts M, Iida R, Gruissem W, Tasaka M, Inzé D, Fukaki H, Beeckman T (2005) Cell cycle progression in the pericycle is not sufficient for SOLITARY ROOT/IAA14-mediated lateral root initiation in Arabidopsis thaliana. Plant Cell 17:3035–3050
19. Laplaze L, Benkova E, Casimiro I, Maes L, Vanneste S, Swarup R, Weijers D, Calvo V, Parizot B, Herrera-Rodriguez MB, Offringa R, Graham N, Doumas P, Friml J, Bogusz D, Beeckman T, Bennett M (2007) Cytokinins act directly on lateral root founder cells to inhibit root initiation. Plant Cell 19:3889–3900
20. Laplaze L, Parizot B, Baker A, Ricaud L, Martiniere A, Auguy F, Franche C, Nussaume L, Bogusz D, Haseloff J (2005) GAL4-GFP enhancer trap lines for genetic manipulation of lateral root development in Arabidopsis thaliana. J Exp Bot 56:2433–2442
21. De Smet I, Vassileva V, De Rybel B, Levesque MP, Grunewald W, Van Damme D, Van Noorden G, Naudts M, Van Isterdael G, De Clercq R, Wang JY, Meuli N, Vanneste S, Frîml J, Hilson P, Jurgens G, Ingram GC, Inzé D, Benfey PN, Beeckman T (2008) Receptor-like kinase ACR4 restricts formative cell divisions in the Arabidopsis root. Science 322:594–597
22. Ulmasov T, Murfett J, Hagen G, Guilfoyle TJ (1997) Aux/IAA proteins repress expression of reporter genes containing natural and highly active synthetic auxin response elements. Plant Cell 9:1963–1971
23. De Tomasi JA (1936) Improving the technique of the Feulgen stain. Stain Technol 11:137–144
24. Saleem M, Lamkemeyer T, Schutzenmeister A, Madlung J, Sakai H, Piepho HP, Nordheim A, Hochholdinger F (2010) Specification of cortical parenchyma and stele of maize primary roots by asymmetric levels of auxin, cytokinin, and cytokinin-regulated proteins. Plant Physiol 152:4–18
25. Liu Y, von Behrens I, Muthreich N, Schütz W, Nordheim A, Hochholdinger F (2010) Regulation of the pericycle proteome in maize (Zea mays L.) primary roots by RUM1 which is required for lateral root initiation. Eur J Cell Biol 89:236–241

Chapter 10

Adventitious Root Induction in *Arabidopsis thaliana* as a Model for In Vitro *Root* Organogenesis

Inge Verstraeten, Tom Beeckman, and Danny Geelen

Abstract

Adventitious root formation, the development of roots on non-root tissue (e.g. leaves, hypocotyls and stems) is a critical step during micropropagation. Although root induction treatments are routinely used for a large number of species micropropagated in vitro as well as for in vivo cuttings, the mechanisms controlling adventitious rooting are still poorly understood. Researchers attempt to gain better insight into the molecular aspects by studying adventitious rooting in *Arabidopsis thaliana*. The existing assay involves etiolation of seedlings and measurements of de novo formed roots on the elongated hypocotyl. The etiolated hypocotyls express a novel auxin-controlled signal transduction pathway in which auxin response factors (ARFs), microRNAs and environmental conditions that drive adventitious rooting are integrated. An alternative assay makes use of so-called thin cell layers (TCL), excised strips of cells from the inflorescence stem of *Arabidopsis thaliana*. However, both the etiolated seedling system and the TCL assay are only distantly related to industrial rooting processes in which roots are induced on adult stem tissue. Here, we describe an adventitious root induction system that uses segments of the inflorescence stems of *Arabidopsis thaliana*, which have a histological structure similar to cuttings or in vitro micropropagated shoots. The system allows multiple treatments with chemicals as well as the evaluation of different environmental conditions on a large number of explants. It is therefore suitable for high throughput chemical screenings and experiments that require numerous data points for statistical analysis. Using this assay, the adventitious root induction capacity of classical auxins was evaluated and a differential response to the different auxins could be demonstrated. NAA, IBA and IAA stimulated adventitious rooting on the stem segment, whereas 2,4-D and picloram did not. Light conditions profoundly influenced the root induction capacity of the auxins. Additionally to the environmental control of adventitious root formation, we also investigated the spatial and temporal aspects of stem-based adventitious root organogenesis. To determine the cells involved in de novo root initiation on the adult stems, we adopted scanning electron microscopy, which allows the visualization of the auxin responsive stem tissue. Using this technique, direct (without callus interface) and indirect (with intermediate callus phase) organogenesis was readily distinguished. The described micro-stem segment system is also suitable for other non-woody species and it is a valuable tool to perform fast evaluations of different treatments to study adventitious root induction.

Key words: Adventitious roots, *Arabidopsis thaliana*, Micropropagation, Auxin

Ive De Smet (ed.), *Plant Organogenesis: Methods and Protocols*, Methods in Molecular Biology, vol. 959, DOI 10.1007/978-1-62703-221-6_10, © Springer Science+Business Media New York 2013

1. Introduction

Adventitious roots (ARs) are defined as roots that develop on non-root tissue or non-root explants, such as leaves, hypocotyls, stems, and shoots (1, 2). The AR formation process is distinct from other organogenesis processes as it involves the de novo initiation of a meristem. In contrast to animals, plants develop, through the reiteration of post-embryonic organ initiation occuring in lateral or axillary meristematic zones (2, 3). True leaves and reproductive organs are not preformed during embryogenesis. They develop at later stages during adult plant growth. Therefore, a plant has to maintain a certain level of plasticity to be able to engage in the formation of new body shapes and organs (4). Through this plasticity, a plant has the capacity to tolerate changing environmental conditions via morphological adaptations and also the physiological age of a plant ensues morphological changes. The extent of this plasticity is demonstrated by the initiation of roots on stem cuttings. This process is rather rare in the absence of human intervention, but it is a commonly used industrial handling to reproduce elite cultivars. Although this technology is widely spread, there are still many plant species that do not evoke the same adaptive response and these species show recalcitrance to form roots on stem cuttings. So far, we have little knowledge about the mechanisms that underpin the capacity to form de novo organs. We know that a subpopulation of cells in the explants maintain the unique capacity to react to root-inducing signals including both endogenous factors such as phytohormones and exogenous environmental factors such as temperature and light (2). To study the processes involved in AR induction, we need a system that is amenable for chemical and genetic experimentation.

In contrast to the detailed studies of lateral root (root-borne roots) organogenesis (5, 6), the factors that regulate AR development are still poorly understood. In commercial plant propagation, poorly rooted cuttings drop out during the acclimatization phase, which is the last step in the production process and hence represents a substantial loss of investment (7). Some elite varieties of woody species, like apple, eucalypt and pine, are difficult-to-root and their interesting elite genotypes can their interesting elite genotypes can therefore not been established via clonal propagation. Because of commercial interest, several studies on apple (1, 8, 9), eucalypt (10–12) and loblolly pine (13–15) have been conducted. However, the number of molecular studies of the rooting process in these species is very limited (1, 9, 16–19). By analyzing the genes involved in adventitious rooting, we may get a better understanding of the causes for the rooting-recalcitrance.

Due to the availability of the genome sequence of apple and poplar (20, 21), we anticipate that in the future more research will develop in this direction.

In conjunction with AR research in crop plants, *Arabidopsis thaliana* is emerging as a model for AR formation. Hence, several protocols for AR induction in *Arabidopsis* have been developed. Falasca et al. (22) reported a technique of de novo AR formation from thin cell layers (TCLs). These cell layers consist of epidermal and cortical cells, lacking any cambial cells. Consequently, the AR formation process can be studied from those cells without interference of other tissues. Other systems implement de-rooted plants and detached leaves to evaluate adventitious root-inducing treatments (23). In another system, segmented hypocotyls containing a pericycle-like cell layer (the cells competent for lateral root formation in the primary roots) and vasculature tissue are used. Both tissues often produce callus and/or roots in the presence of auxin (24, 25). As a consequence of the presence of a pericycle-like layer in hypocotyls the degree of similarity with the primary root is high. This could hamper the extension of conclusions from this model to AR induction in adult stems (25). A similar objection can be made for hypocotyl-based induction of ARs on intact dark-grown seedlings (26–28). Moreover, this last system does not involve wounding, which is often an important trigger for AR induction. Recent studies in the etiolated seedling system have demonstrated that the auxin-controlled signalling pathway, which leads to adventitious rooting, implicates other auxin response factors (ARFs) than those involved in lateral root formation (26, 28).

In order to compare AR formation in *Arabidopsis* with AR formation in in vitro shoots or cuttings, we developed an adventitious rooting test-system based on stem segments. This system is more reminiscent to agricultural practices and at the same time it is amenable for molecular research. Because of the structural similarity between herbaceous (e.g., *Arabidopsis thaliana*) and woody species, we anticipate that some developmental aspects of the induction and formation of ARs may be common (29). The inflorescence stem of *Arabidopsis* is structurally very similar to a vegetative shoot of other annual dicotyledonous plants, with an average of 5–10 vascular bundles embedded in the cortical tissue (30). The vascular bundles are arranged in a circle and separate the cortex from the central pith. Centrally in a vascular bundle, a meristematic zone is maintained: the vascular cambium. Cell divisions of cambial cells give rise to xylem (inside) and phloem (outside). Xylem tissue conducts water and mineral nutrients upwards in the plants. The phloem flow on the other hand is downward and contains photoassimilates. The secondary lignified wall of the xylem vessels is also important for the support of the phototropic growing stems. The parenchyma in the central pith of the stem consists of bigger cells, which could be used as storage cells for photosynthesis products. In the cortical region, the parenchyma cells are smaller and contain chlorophyll to carry out photosynthesis. The outermost layer of a stem is the epidermis, a compact continuously arranged cell layer, which mainly functions to protect the inside tissues. At the base of the *Arabidopsis* inflorescent stem, secondary growth forms a vascular ring, which resembles the structure of the vasculature in woody species (31, 32). At the onset of secondary growth, the parenchyma cells adjacent to the cambium differentiate to form interfascicular cambium. The ring of interfascicular and regular cambium divides to generate secondary xylem inwardly and secondary phloem outwardly. As a result of continued cell division of the expanded cambium, the primary phloem is pushed outwards and collapses. The primary xylem remains in the center of the stem and shows extensive lignifications.

However, when considering regeneration capacity of species, *Arabidopsis* might not be the preferred option because it only shows limited regenerative capacity in tissue culture in comparison with for example tobacco. For instance, *Arabidopsis* inflorescence stem cuttings do not form adventitious roots spontaneously and in vitro stem segments require the presence of auxin in the medium to induce ARs (33). In laboratories, *Columbia* (*Col*) is the most frequently grown *Arabidopsis* ecotype. Despite the common use, this accession is rather recalcitrant in in vitro root and leaf explant regeneration experiments and also the division capacity of protoplasts is less in comparison with other accessions (34–36). In line with these findings, *Arabidopsis* accessions are transformed with variable success (37, 38). This natural variation in regeneration capacity can be exploited to find *Arabidopsis thaliana* genes involved in this trait and to uncover the underlying molecular processes and genetic elements involved in AR formation. This approach has been followed for instance to unravel the genetic control of other traits such as seed size and light sensitivity (39, 40). The natural variation in the physiological responses observed may also be underpinned by transcriptional variation as demonstrated for genes controlled by auxin (41). In light of these findings, it is clear that we need to determine whether the adventitious rooting trait is also variable between different *Arabidopsis thaliana* ecotypes. As demonstrated in a recent publication by da Rocha Correa et al. (23) the appropriate selection of both the ecotype and the explant is important when generating data to study the rooting response in *Arabidopsis thaliana*. Another major advantage of this species is the availability of mutants with altered root formation (27, 29, 42). Until we have more model species with an annotated genome available, *Arabidopsis thaliana* will remain the predominant model species for molecular adventitious root research.

One of the major factors known to play a role in adventitious root formation is the plant growth regulator auxin (1, 27). Auxin is generally described as the rooting hormone (7, 43), but also other phytohormones promote or accelerate rooting, mostly as a pre- or a post-auxin treatment (43). Auxin plays a central role in the determination of rooting capacity (12, 43). Besides auxin, also ethylene stimulates adventitious root formation (44, 45). There is a direct correlation between the reduced number of AR and the inhibition of ethylene biosynthesis or perception. Ethylene may also enhance the sensitivity to auxins, which results in a stimulation of AR initiation. Furthermore, it induces enzymes, which facilitate root emergence through stem tissues. Last but not least, ethylene would promote rooting by stimulating cytokinin catabolism ((44) and references therein). Cytokinin has been shown to act as a rooting inhibitor (2, 29). Therefore, downregulation of the endogenous cytokinin content via catabolism promotes rooting. A role for other phytohormones in AR induction and formation has been reported: gibberellins (46), abscisic acid (47–49), jasmonate (50), alkamides and nitric oxide (18, 51–53), polyamines (43, 54) and flavonoids (55).

Here, we focus on auxin as the most important growth regulator triggering AR initiation. A high concentration of auxin is required during the induction phase to allow the reprogramming of cells, whereas during the formative phase, the auxin concentration is less critical and may even need to be reduced to allow organogenesis (1, 12, 43). In *Arabidopsis thaliana* however, constitutive application of rather high doses of auxin (10 μM) does not prevent the emergence of roots. We determined the induction capacity of five commercially available auxins: indole-3-acetic acid (IAA), indole-3-butyric acid (IBA), 2,4-dichlorophenoxyacetic acid (2,4-D), α-naphthalene acetic acid (NAA) and picloram. We observed differential responses of the explants to the auxins applied. NAA, IBA, and IAA induced AR formation, whereas 2,4-D and picloram, applied in the same concentration, induced callus. NAA had the strongest root inducing activity. Structurally, the binding of NAA to the auxin receptor TIR1 is very similar to that of IAA, involving the same binding cavity and associated amino acids (56, 57). Nonetheless, NAA shows a lower binding affinity with TIR1 in comparison with the natural auxin IAA (58–61). The lower binding affinity does not correlate with the activity of this synthetic auxin compound. This suggests that the observed differences between the various auxins are most likely due to induction of a different signal transduction. The morphogenetic response to auxins therefore seems to depend on other factors such as transport, metabolization and conjugation, which determine the concentration of active auxin reaching the target cells. Moreover, the endogenous auxin synthesis and the physiological status or receptivity of the target cells will also play a role in the organogenic response. The physiological status of the tissue is dependent on environmental factors as humidity, nutrition, temperature and light, which are described to have a strong influence on AR formation (2, 10, 12, 26, 27, 42, 62, 63). In vitro light quality and temperature are important preconditioning factors for rooting (64). But also the imprinted status of the explants dependent on epigenetic changes induced by environmental stresses (via siRNAs (65, 66) and miRNAs (67)) and the ontogenetic and physiological age determine the organogenic capacity of the cells (15, 68–71).

During normal plant development, specific cells in the primary root pericycle are predetermined by auxin pulses in the root meristem to start dividing actively and form lateral root primordia. Externally applied auxin enhances this activation and results in the formation of more primordia (72–74). At the onset of adventitious root formation however, there are no such predestined cells present. Adventitious root formation in in vitro conditions thus requires cellular dedifferentiation in order to acquire the competence to proliferate and to regenerate organs (42). The developing root primordia can either emerge from undifferentiated callus or directly from reprogrammed cells. Hence, the cellular origin of ARs is less predictable than lateral roots (33). Moreover, the distinction between direct and indirect (intermediate callus phase) organogenesis is not always very clear because callus formation and organogenesis may occur simultaneously on the same explant. Detailed microscopic observation and histological sections are required to determine the origin and implanting of the newly formed ARs on the explants. Limited callus growth takes place during early phases of adventitious rooting. We have used scanning electron microscopy (SEM) to visualize the AR induction process in the auxin-responsive stems. The analysis of the stem tissues indicated that initially root organogenesis in *Arabidopsis thaliana* stem segments is direct. The AR emerged from cells that are located at the center of the stem structure. Histological sections pointed to cambial/phloem cells that start dividing upon auxin application. At later time points, callus is formed and this callus may produce secondary ARs via the indirect pathway. The SEM technique could indicate whether direct organogenesis took place or not, a parameter that is very important for future success of the explants during acclimatization and ex vitro growth.

2. Materials

Use purified, deionized water to prepare all solutions. Please check the waste disposal regulations when disposing waste solutions and pay attention to the appropriate safety guidelines when handling certain liquids.

2.1. Plant Growth

1. Seed sterilization:
   - Commercial bleach.
   - Hypochloric acid.

2. Kl germination medium: 0.5 MS (Murashige and Skoog, basal salt mixture), 1% D-sucrose, 100 mg/L myo-inositol, 0.5 g/L MES, and 8% agar (IDG, Plant tissue culture agar MC29).
3. In vivo growth: The plants are grown in individual containers of the arasystem-design (http://www.arasystem.com). This way, each tray contains 51 plantlets that grow individually and that are not intertangled while, at the same time, the space is optimally used and the control of pests and watering is still easy.

2.2. Explant Sterilization

1. 70% desinfectol solution: Desinfectol is a denatured ethanol with up to 5% ether. A 70% (v/v) solution of desinfectol in water is made to pretreat the explants during the sterilization.
2. HAZTABS solution (see Note 2): Haztabs are chlorine-releasing tablets that are used for disinfecting treatments. Per tablet there is 2.5 g available chlorine. For this sterilizing solution, one tablet is dissolved in one liter of water, giving rise to a solution containing 2,500 ppm chlorine.

2.3. Inductive Media

1. Auxin stock solutions: 1 mg/mL concentration
   Auxins are acidic compounds, and are therefore soluble in 1 N NaOH. When making the stock solutions 50 mg of the auxins is dissolved in 5 mL of an alkaline 1 N NaOH solution. Afterwards, water is added to a total volume of 50 mL (see Note 3).

Auxin	Volume from stock needed to have 10 μM concentration in 100 mL medium
IAA	175.18 μL
IBA	203.24 μL
2,4-D	221.04 μL (see Note 3)
NAA	186.21 μL
Picloram	241.46 μL

2. The medium that is used for the adventitious root inducing test is liquid medium: full-strength Murashige and Skoog (MS, basal salt mixture), 1% D-sucrose, 100 mg/L myo-inositol, and 0.5 g/L MES. The auxins are filter sterilized and added to the medium after the autoclaving step (30 min, 120°C). The volumes added are mentioned in the table above.
3. The medium is pipetted in a 24-well plate. Each well contains 1 mL of medium. It is advisable to prepare multiple replicates of the same medium and the basal medium to compare within one plate.

2.4. SEM Preparation Samples

1. 0.5 M Phosphate buffer, pH 7.2: The buffer is prepared using monopotassium phosphate and its conjugate base dipotassium phosphate. The volumes needed for the 0.5 M phosphate buffer solution are calculated using the Henderson–Hasselbalch equation: pH (7.2) = pKa(6.86) + log ([base]/[acid]), so the ratio [base]/[acid] = 2.1878. The molarity of the buffer is the sum of the molarities of the acid and base or 0.5 M = [acid] + [base]. Substitution in the ratio gives a solution [base] = 0.3432 moles/L and [acid] = 0.1568 moles/L. Therefore, 21.3373 g of KH_2PO_4 and 78.3285 g of $K_2HPO_4{\cdot}3H_2O$ are dissolved in somewhat less than 1 L of water. The pH is checked using a pH meter and adjusted when necessary using 1 M HCl or 1 M KOH. Once the desired pH is reached, water is added to exactly 1 L. This buffer was autoclaved and stored at 4°C.
2. The fixation solution: 2% glutaraldehyde is commonly applied in histological research. The stock solution of glutaraldehyde is a 50% v/v aqueous solution (Sigma, Grade 1). The 2% solution is prepared by mixing 4 mL of the 50% stock solution in 96 mL of phosphate buffer (see Notes 4, 5).
3. Dehydration steps in ethanol: Water from the samples is removed through the application of a series of ethanol dehydration steps with increasing percent volumes of ethanol (stock solution 100%). A 30%, 50%, 70% and 90% ethanol solution is prepared using pure water to dilute the ethanol stock solution.
4. Critical point drying device: CPD20 Balzers Union, CO_2 with critical point 31.1°C and 72.9 atm.

2.5. Scanning Electron Microscopy

Table top Hitachi TM 1000 and associated software.

3. Methods

3.1. Plants

1. Perform surface seed sterilization of *Arabidopsis thaliana* seeds using chlorine gas.
2. Place seeds in an open Eppendorf tube in a desiccator, which is placed inside a fume hood.
3. Put a beaker containing 100 mL of commercial bleach next to the seeds and add 3 mL of concentrated hypochloric acid (see Note 1).
4. Seal the tubes following incubation in chlorine gas for 4–6 h and transfer them to a laminar flow to be sown on K1-plates.
5. Spread the sterile seeds over this medium and seal the plates with Micropore surgical tape.

6. To have a synchronized germination of the seeds, they undergo a stratification treatment by storing the plates at 4°C for 4 days.
7. Afterwards, the plates are transferred into a growth room (25°C, 70 μmol/m²s, 16h light/8h dark).
8. Evaluate 6-day-old seedlings and transfer them from the plates into water-saturated soil in a tray.
9. Put the trays in a growth chamber sustaining vegetative growth of the plantlets (12h light/12h dark, 21°C).
10. Water the plants every 2 days and add 2 mL/L of Wuxal (liquid NPK fertilizer, Aglukon) to the water once a week.
11. Change the light regime to stimulate flowering (16-h light/8-h dark, 21°C) 4–5 weeks afterwards, at the onset of bolting.
12. Cut the main inflorescence stem (of ±20 cm) 1 week after flower induction and use this stem for further treatments.

3.2. Explant Excision and Sterilization (Fig. 1)

1. The excised inflorescence stem contains internodal segments and nodes carrying leaves, siliques, or flowers. First, cut out the internodes.
2. Maintain only the lower, in acropetal direction, 3–4 internodes for further treatments. Always exclude the internodes between siliques from the treatments, because they are too thin.
3. Further treatments of these internode segments require aseptic conditions. Therefore, the segments have to be sterilized. The sterilization is done with multiple segments in a 50 mL falcon tube. Add firstly 70% EtOH while turning the falcon over during 5 s.
4. Decant the EtOH solution and pour secondly 1% HAZTABS solution into the falcon (see Note 2). Add 1 drop of commercial detergent to improve the surface contact between explants and the sterilizing agent.

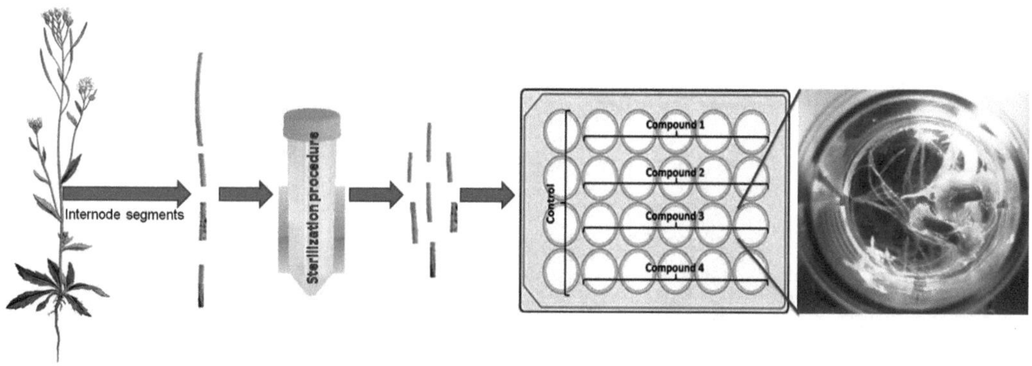

Fig. 1. Schematic overview of the manipulations of the stem segments prior to incubation in the adventitious root inducing medium (the compounds mentioned could be auxin or other treatments).

5. Mix the falcon gently during 5 min incubation.
6. Remove the HAZTAB solution and rinse the segments with sterile demineralized water.
7. Repeat this rinsing step 5 times until foam does not form any longer.
8. Do not allow the segments to dry and end the sterilization by adding water to the falcon tube.
9. The internode segments are now surface-sterilized. The applied sterilization procedure is rather harsh and a slight decolorization at the cut ends of the segments can be noticed.
10. Consequently, remove, at both cut ends, 2–4 mm and cut the remaining segment in fragments of approximately 0.5 cm.
11. Incubate these 0.5 cm segments subsequently in root inductive medium in 24-multiwell plates.

3.3 Incubation

1. We evaluated the adventitious root induction capacity of five commercially available auxins: IAA, IBA, 2,4-D, NAA, and picloram.
2. Auxins were added to the medium at a final concentration of 10 μM and pictures to evaluate the effect of the treatment were taken 14 days after the start of the incubation.

Figure 2 shows the effect of the different auxins. NAA, IBA, and IAA clearly stimulated root induction. Among the three auxins tested, NAA was the most effective auxin. It induced adventitious roots at both ends of the cutting, although predominantly more

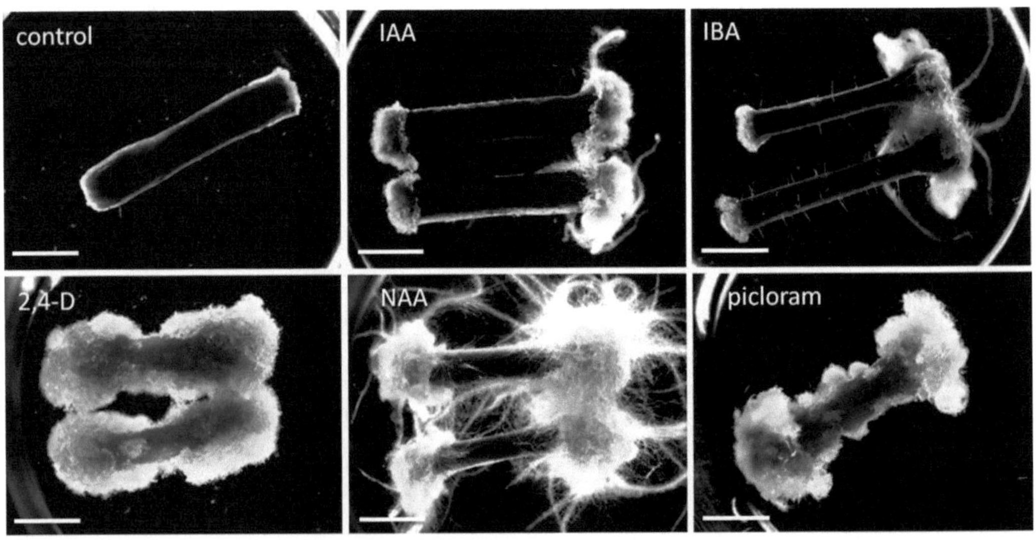

Fig. 2. Auxin response in *Arabidopsis thaliana* inflorescence stem segments. The phenotype shown is the result of 14-day incubation in liquid medium containing 10 μM of the respective auxins. The pictures are taken with a 16× magnification at a stereomicroscope (Leica MZ95, Leica DC300 software). The scale bar represents 0.1 cm for all pictures.

at the basal end. IAA and IBA also induced adventitious roots but in treatments with these auxins, the adventitious roots were uniquely formed at the basal end. 2,4-D and picloram, on the other hand, strongly induced callus when they were applied in the same concentration.

3.4 Critical Point Drying

1. Fix the stem samples at the end of incubation in a 2% glutaraldehyde solution. Therefore, the samples are immersed in a large volume of fixation solution (see Notes 4, 5).
2. To enhance the penetration of fixative, put the vessels containing fixative and samples in a vacuum desiccator for 1 h.
3. Afterwards, maintain the samples in the fixative for 24 h at 4°C.
4. Remove the fixation solution and rinse the samples three times with phosphate buffer.
5. Incubate the samples in increasing concentration series of EtOH. The incubation for each drying step takes 20 min. Treat the samples subsequently with a 30% EtOH, 50% EtOH, 75% EtOH and 95% EtOH solution (see Note 6).
6. Transfer the stem segments to a 100% EtOH solution and incubate during 1 h under vacuum. The aim of this step is to physically replace the water in the samples with the organic solvent ethanol.
7. Next, during critical point drying, ethanol is exchanged by fluid CO_2. CO_2 is fluid at a temperature of 31°C and at higher temperatures (and pressure) CO_2 undergoes a transition to the gas state. Gaseous CO_2 can be blown and the samples are then dry. This method is applied to dry tissues without deforming the cellular structure (or causing collapsing of the cells) and it is generally a part of the sample preparation process for SEM.
8. Bring the samples in the critical point drying apparatus. In this apparatus, the final substitution with liquid CO_2 and the drying will take place (see Note 7).
9. Release the pressure after the drying run and coat the dried tissue with metal (mostly gold) before transfer into the SEM. However, the metallization step is not necessary for the imaging described here.

3.5 Scanning Electron Microscope Analysis

As described in Subheading 3.3 the auxins predominantly induced adventitious roots at the basal side of the cuttings. Figure 3a, b depicts the basal end of a stem segment. The detail picture 3B shows the implanting of the root on the stem. This picture indicates a direct organogenesis of AR on the auxin-treated stem segments because there are no callus cells visible here. However, additional sections of the stems are necessary to exclude internal callus formation upon auxin treatment. It is generally assumed that

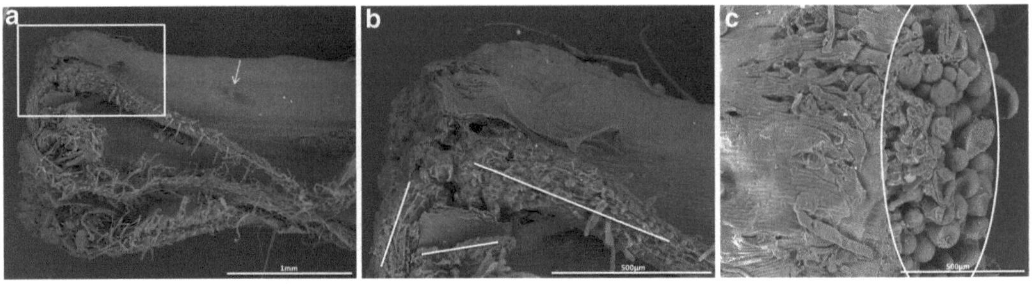

Fig. 3. (**a**) 10 μM NAA-treated stem segments, picture taken 14 days after starting incubation. The adventitious roots are predominantly formed at the basal end of a cutting. The *arrow* marks a trichome on the epidermal surface of the stem segment. The square area indicated in (**a**) is shown in more detail in (**b**). This demonstrates the direct implanting of the newly formed adventitious roots on the stem (the orientation of the roots is delineated). (**c**) Shows one of the ends of a stem segment, incubated for 14 days in 10 μM 2,4-D. At this end, only hypertropic callus cells are visible.

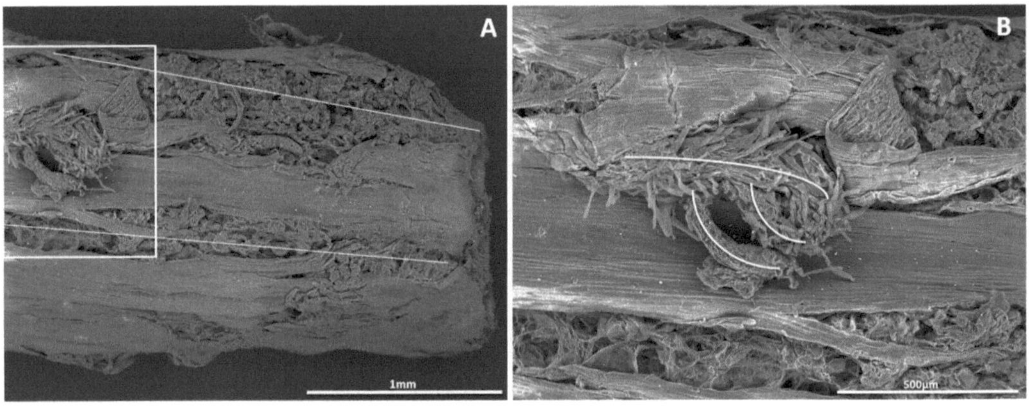

Fig. 4. (**a**) At the apical pointing end of the stem cutting, there are no adventitious roots induced. However, at this end, it is clear that the epidermis of the stem cracks almost at a regular spacing. In one of this open cracks adventitious roots can emerge (as indicated in **b**).

direct adventitious roots provide a better connectivity to the main stem. Therefore, direct roots would better ensure the survival of the cuttings after transfer into soil and during the acclimatization phase. During vegetative propagation of plants direct rooting is important for the survival of the plant and hence economically relevant. As a comparison, Fig. 3c shows the basal end of a 2,4-D-treated stem segment with hypertropic callus cells visible and here there is no adventitious root formation.

At the positions where ARs are formed, a crack is formed that splits the epidermal layer. As shown in Fig. 4a, these cracks are regularly spaced (line pattern) and coincide with the position of newly formed roots generating multiple rays of ARs. Figure 4b depicts adventitious roots being formed in one of these cracks. These roots are completely covered with root hairs.

4. Notes

1. Care needs to be taken upon addition of HCl, as the chlorine gas starts to form immediately. The desiccator should be closed quickly. The chlorine gas, formed when hypochloric acid is added to the commercial bleach solution, is irritating for the respiratory system. Only perform this procedure in a good-functioning fume hood in a well-ventilated room. It is also a strong oxidizer and reacts with flammable materials.
2. The HAZTABS solution (see Subheading 2.2, step 2) is light sensitive and should not be stored longer than 1 month. Protect the bottle from the light and avoid high temperatures. Use gloves when you handle this solution.
3. 2,4-D does not dissolve as easily as the other auxins. To make the stock solution it could be worth putting the vessel in a sonicator bath (or to heat the solution a little bit) after adding the alkaline solution but before adding water.
4. When using the glutaraldehyde fixation solution, attention must be paid to the quality of the glutaraldeyde solution (50% v/v water). The optimal solution is EM grade (Grade 1), because that solution mostly contains monomer and low polymers that are small enough to penetrate the tissue fairly quickly. The size of the mono- or polymers also determines the penetration depth of the solution and the reaction with the proteins in the samples.
5. Glutaraldehyde is toxic and hazardous in case of skin contact, eye contact, ingestion, or inhalation. Therefore, wear suitable protective gloves when using this solution.
6. The replacement of water in the samples with ethanol is an essential step preceding critical point drying. This replacement has to be done in order to carry out the critical drying at a convenient temperature and pressure. Moreover, CO_2 is not mixable with water and so the pretreatment with another fluid is necessary to ensure the later CO_2 interaction.
7. To ensure that there is no recondensation of the liquid CO_2, the temperature should be around 5°C above the critical point (31°C) during the drying phase.

Acknowledgments

This research was supported by the IWT (The Flemish institute for Innovation by Science and Technology). We are grateful to Marjolein Couvreur (Faculty of Sciences, Department Biology) for the critical point drying of the samples and to Dr. Daniel Van Damme for the introduction of the SEM.

References

1. De Klerk GJ, Van der Krieken W, De Jong JC (1999) Review—the formation of adventitious roots: new concepts, new possibilities. *In Vitro* Cell Dev Biol Plant 35:189–199
2. Geiss G, Gutierrez L, Bellini C (2009) Adventitious root formation: new insights and perspectives. In: Beeckman T (ed) Root development. Wiley–Blackwell, Hoboken, pp 127–156
3. Golz JF, Hudson A (2002) Signalling in plant lateral organ development. Plant Cell 14:S277–S288
4. Sultan SE (2000) Phenotypic plasticity for plant development, function and life history. Trends Plant Sci 5:537–542
5. De Smet I, Vanneste S, Inze D, Beeckman T (2006) Lateral root initiation or the birth of a new meristem. Plant Mol Biol 60:871–887
6. Peret B, De Rybel B, Casimiro I, Benkova E, Swarup R, Laplaze L, Beeckman T, Bennett MJ (2009) *Arabidopsis* lateral root development: an emerging story. Trends Plant Sci 14:399–408
7. Haissig BE, Davis TD, Riemenschneider DE (1992) Researching the controls of adventitious rooting. Physiol Plant 84:310–317
8. De Klerk GJ, Keppel M, Terbrugge J, Meekes H (1995) Timing of the phases in adventitious root formation in apple microcuttings. J Exp Bot 46:965–972
9. Smolka A, Welander M, Olsson P, Holefors A, Zhu LH (2009) Involvement of the *ARRO-1* gene in adventitious root formation in apple. Plant Sci 177:710–715
10. Fett-Neto AG, Fett JP, Goulart LWV, Pasquali G, Termignon RR, Ferreira AG (2001) Distinct effects of auxin and light on adventitious root development in *Eucalyptus saligna* and *Eucalyptus globulus*. Tree Physiol 21:457–464
11. da Rocha Correa LD, Fett-Neto AG (2004) Effects of temperature on adventitious root development in microcuttings of *Eucalyptus saligna* Smith and *Eucalyptus globulus* Labill. J Therm Biol 29:315–324
12. Fogaça CM, Fett-Neto AG (2005) Role of auxin and its modulators in the adventitious rooting of *Eucalyptus* species differing in recalcitrance. Plant Growth Regul 45:1–10
13. Greenwood MS, Weir RJ (1995) Genetic variation in rooting ability of Loblolly pine cuttings—effects of auxin and family on rooting by hypocotyl cuttings. Tree Physiol 15:41–45
14. Hamann A (1998) Adventitious root formation in cuttings of loblolly pine (*Pinus taeda* L.): developmental sequence and effects of maturation. Trees-Struct Funct 12:175–180
15. Rasmussen A, Hunt MA (2010) Ageing delays the cellular stages of adventitious root formation in pine. Aust For 73:41–46
16. Butler ED, Gallagher TF (1999) Isolation and characterization of a cDNA encoding a novel 2-oxoacid-dependent dioxygenase which is up-regulated during adventitious root formation in apple (*Malus domestica* "Jork 9") stem discs. J Exp Bot 50:551–552
17. Brinker M, van Zyl L, Liu WB, Craig D, Sederoff RR, Clapham DH, von Arnold S (2004) Microarray analyses of gene expression during adventitious root development in *Pinus contorta*. Plant Physiol 135:1526–1539
18. Pagnussat GC, Lanteri ML, Lombardo MC, Lamattina L (2004) Nitric oxide mediates the indole acetic acid induction activation of a mitogen-activated protein kinase cascade involved in adventitious root development. Plant Physiol 135:279–286
19. Ricci A, Rolli E, Dramis L, Diaz-Sala C (2008) N,N′-bis-(2,3-methylenedioxyphenyl)urea and N,N′-bis-(3,4-methylenedioxyphenyl)urea enhance adventitious rooting in *Pinus radiata* and affect expression of genes induced during adventitious rooting in the presence of exogenous auxin. Plant Sci 175:356–363
20. Velasco R, Zharkikh A, Affourtit J et al (2010) The genome of the domesticated apple (*Malus×domestica* Borkh). Nat Genet 42:833–839

21. Tuskan GA, DiFazio S, Jansson S et al (2006) The genome of black cottonwood, *Populus trichocarpa* (Torr. & Gray). Science 313: 1596–1604
22. Falasca G, Zaghi D, Possenti M, Altamura MM (2004) Adventitious root formation in *Arabidopsis thaliana* thin cell layers. Plant Cell Rep 23:17–25
23. da Rocha Correa L, Troleis J, Mastroberti AA, Mariath JEA, Fett-Netto AG (2012) Distinct modes of adventitious rooting in *Arabidopsis thaliana*. Plant Biol 14:100–109
24. Ozawa S, Yasutani I, Fukuda H, Komamine A, Sugiyama M (1998) Organogenic responses in tissue culture of *srd* mutants of *Arabidopsis thaliana*. Development 125:135–142
25. Falasca G, Altamura MM (2003) Histological analysis of adventitious rooting in *Arabidopsis thaliana* (L.) Heynh seedlings. Plant Biosyst 137:265–273
26. Sorin C, Bussell JD, Camus I, Ljung K, Kowalczyk M, Geiss G, McKhann H, Garcion C, Vaucheret H, Sandberg G, Bellini C (2005) Auxin and light control of adventitious rooting in *Arabidopsis* require ARGONAUTE1. Plant Cell 17:1343–1359
27. Sorin C, Negroni L, Balliau T, Corti H, Jacquemot MP, Davanture M, Sandberg G, Zivy M, Bellini C (2006) Proteomic analysis of different mutant genotypes of *Arabidopsis* led to the identification of 11 proteins correlating with adventitious root development. Plant Physiol 140:349–364
28. Gutierrez L, Bussell JD, Pacurar DI, Schwambach J, Pacurar M, Bellini C (2009) Phenotypic plasticity of adventitious rooting in *Arabidopsis* is controlled by complex regulation of auxin response factor transcripts and microRNA abundance. Plant Cell 21:3119–3132
29. King JJ, Stimart DP (1998) Genetic analysis of variation for auxin-induced adventitious root formation among eighteen ecotypes of *Arabidopsis thaliana* L-Heynh. J Hered 89:481–487
30. Altamura MM, Possenti M, Matteucci A, Baima S, Ruberti I, Morelli G (2001) Development of the vascular system in the inflorescence stem of *Arabidopsis*. New Phytol 151:381–389
31. Oh S, Park S, Han KH (2003) Transcriptional regulation of secondary growth in *Arabidopsis* thaliana. J Exp Bot 54:2709–2722
32. Ko JH, Han KH, Park S, Yang JM (2004) Plant body weight-induced secondary growth in *Arabidopsis* and its transcription phenotype revealed by whole-transcriptome profiling. Plant Physiol 135:1069–1083
33. Ludwig-Müller J, Vertocnik A, Town CD (2005) Analysis of indole-3-butyric acid-induced adventitious root formation on *Arabidopsis* stem segments. J Exp Bot 56: 2095–2105
34. Dovzhenko A, Dal Bosco C, Meurer J, Koop HU (2003) Efficient regeneration from cotyledon protoplasts in *Arabidopsis thaliana*. Protoplasma 222:107–111
35. Velazquez I, Valencia S, Lopez-Lera A, de la Pena A, Candela M (2004) Analysis of natural allelic variation in *in vitro* organogenesis of *Arabidopsis thaliana*. Euphytica 137:73–79
36. Passardi F, Dobias J, Valerio L, Guimil S, Penel C, Dunand C (2007) Morphological and physiological traits of three major *Arabidopsis thaliana* accessions. J Plant Physiol 164: 980–992
37. Chateau S, Sangwan RS, Sangwan-Norreel BS (2000) Competence of *Arabidopsis thaliana* genotypes and mutants for *Agrobacterium tumefaciens*-mediated gene transfer: role of phytohormones. J Exp Bot 51:1961–1968
38. Marion J, Bach L, Bellec Y, Meyer C, Gissot L, Faure JD (2008) Systematic analysis of protein subcellular localization and interaction using high-throughput transient transformation of *Arabidopsis* seedlings. Plant J 56:169–179
39. Alonso-Blanco C, Blankestijn-de Vries H, Hanhart CJ, Koornneef M (1999) Natural allelic variation at seed size loci in relation to other life history traits of *Arabidopsis thaliana*. Proc Natl Acad Sci USA 96: 4710–4717
40. Maloof JN, Borevitz JO, Dabi T, Lutes J, Nehring RB, Redfern JL, Trainer GT, Wilson JM, Asami T, Berry CC, Weigel D, Chory J (2001) Natural variation in light sensitivity of *Arabidopsis*. Nat Genet 29:441–446
41. Delker C, Poschl Y, Raschke A, Ullrich K, Ettingshausen S, Hauptmann V, Grosse I, Quint M (2010) Natural variation of transcriptional auxin response networks in *Arabidopsis thaliana*. Plant Cell 22:2184–2200
42. Konishi M, Sugiyama M (2003) Genetic analysis of adventitious root formation with a novel series of temperature-sensitive mutants of Arabidopsis *thaliana*. Development 130:5637–5647
43. Kevers C, Hausman JF, Faivre-Rampant O, Dommes J, Gaspar T (2009). What we have learned about the physiology of in vitro adventitious rooting of woody plants and how it relates to improvements in the practice. In: Niemi K and Scagel C (Eds) Adventitious root formation of forest trees and horticultural plants - from genes to applications. Research Signpost, Kerala, pp 209–225

44. Kuroha T, Satoh S (2007) Involvement of cytokinins in adventitious and lateral root formation. Plant Root 1:27–33
45. Negi S, Sukumar P, Liu X, Cohen JD, Muday GK (2010) Genetic dissection of the role of ethylene in regulating auxin-dependent lateral and adventitious root formation in tomato. Plant J 61:3–15
46. Gou JQ, Strauss SH, Tsai CJ, Fang K, Chen YR, Jiang XN, Busov VB (2010) Gibberellins regulate lateral root formation in *Populus* through interactions with auxin and other hormones. Plant Cell 22:623–639
47. Signora L, De Smet I, Foyer CH, Zhang HM (2001) ABA plays a central role in mediating them regulatory effects of nitrate on root branching in *Arabidopsis*. Plant J 28:655–662
48. De Smet I, Signora L, Beeckman T, Inze D, Foyer CH, Zhang HM (2003) An abscisic acid-sensitive checkpoint in lateral root development of *Arabidopsis*. Plant J 33:543–555
49. De Smet I, Zhang HM, Inze D, Beeckman T (2006) A novel role for abscisic acid emerges from underground. Trends Plant Sci 11:434–439
50. Fattorini L, Falasca G, Kevers C, Rocca LM, Zadra C, Altamura MM (2009) Adventitious rooting is enhanced by methyl jasmonate in tobacco thin cell layers. Planta 231:155–168
51. Pagnussat GC, Lanteri ML, Lamattina L (2003) Nitric oxide and cyclic GMP are messengers in the indole acetic acid-induced adventitious rooting process. Plant Physiol 132:1241–1248
52. Campos-Cuevas JC, Pelagio-Flores R, Raya-Gonzalez J, Mendez-Bravo A, Ortiz-Castro R, Lopez-Bucio J (2008) Tissue culture of *Arabidopsis thaliana* explants reveals a stimulatory effect of alkamides on adventitious root formation and nitric oxide accumulation. Plant Sci 174:165–173
53. Li SW, Xue LG, Xu SJ, Feng HY, An LZ (2009) Mediators, genes and signaling in adventitious rooting. Bot Rev 75:230–247
54. Nag S, Saha K, Choudhuri MA (2001) Role of auxin and polyamines in adventitious root formation in relation to changes in compounds involved in rooting. J Plant Growth Regul 20:182–194
55. Curir P, Vansumere CF, Termini A, Barthe P, Marchesini A, Dolci M (1990) Flavonoid accumulation is correlated with adventitious roots formation in *Eucalyptus* Gunni Hook micropropagated through axillary bud stimulation. Plant Physiol 92:1148–1153
56. Tan X, Calderon-Villalobos LIA, Sharon M, Zheng CX, Robinson CV, Estelle M, Zheng N (2007) Mechanism of auxin perception by the TIR1 ubiquitin ligase. Nature 446:640–645
57. Calderon-Villalobos LI, Tan X, Zheng N, Estelle M (2010) Auxin perception-structural insights. Cold Spring Harb Perspect Biol 2(7):a005546
58. Dharmasiri N, Dharmasiri S, Estelle M (2005) The F-box protein TIR1 is an auxin receptor. Nature 435:441–445
59. Kepinski S, Leyser O (2005) The *Arabidopsis* F-box protein TIR1 is an auxin receptor. Nature 435:446–451
60. Badescu GO, Napier RM (2006) Receptors for auxin: will it all end in TIRs? Trends Plant Sci 11:217–223
61. Spartz AK, Gray WM (2008) Plant hormone receptors: new perceptions. Genes Dev 22:2139–2148
62. Hansen J (1975) Light dependent promotion and inhibition of adventitious root formatioin by gibberellic acid. Planta 123: 203–205
63. Niemi K, Julkunen-Tiitto R, Tegelberg R, Haggman H (2005) Light sources with different spectra affect root and mycorrhiza formation in Scots pine *in vitro*. Tree Physiol 25:123–128
64. Kiss JZ, Miller KM, Ogden LA, Roth KK (2002) Phototropism and gravitropism in lateral roots of *Arabidopsis*. Plant Cell Physiol 43:35–43
65. Chen XM (2009) Small RNAs and their roles in plant development. In: Schekman R (Ed) Annual review of cell and developmental biology, vol 25. Annual Reviews, Palo Alto, pp 21–44
66. He GM, Elling AA, Deng XW (2011) The epigenome and plant development. In: Merchant SS, Briggs WR, Ort D (eds) Annual review of plant biology, vol 62. Annual Reviews, Palo Alto, pp 411–435
67. Rubio-Somoza I, Weigel D (2011) MicroRNA networks and developmental plasticity in plants. Trends Plant Sci 16:258–264
68. Diaz-Sala C, Garrido G, Sabater B (2002) Age-related loss of rooting capability in *Arabidopsis thaliana* and its reversal by peptides containing the Arg-Gly-Asp (RGD) motif. Physiol Plant 114:601–607
69. Diaz-Sala C, Hutchison KW, Goldfarb B, Greenwood MS (1996) Maturation-related loss in rooting competence by loblolly pine stem cuttings: the role of auxin transport,

metabolism and tissue sensitivity. Physiol Plant 97:481–490

70. Poethig RS (2003) Phase change and the regulation of developmental timing in plants. Science 301:334–336
71. Vidal N, Arellano G, San-Jose MC, Vieitez AM, Ballester A (2003) Developmental stages during the rooting of *in-vitro*-cultured *Quercus robur* shoots from material of juvenile and mature origin. Tree Physiol 23:1247–1254
72. Boerjan W, Cervera MT, Delarue M, Beeckman T, Dewitte W, Bellini C, Caboche M, Vanonckelen H, Vanmontagu M, Inze D (1995) Superroot, a recessive mutation in *Arabidopsis*, confers auxin overproduction. Plant Cell 7:1405–1419
73. Ruegger M, Dewey E, Gray WM, Hobbie L, Turner J, Estelle M (1998) The TIRl protein of *Arabidopsis* functions in auxin response and is related to human SKP2 and yeast Grr1p. Genes Dev 12:198–207
74. Casimiro I, Marchant A, Bhalerao RP, Beeckman T, Dhooge S, Swarup R, Graham N, Inze D, Sandberg G, Casero PJ, Bennett M (2001) Auxin transport promotes *Arabidopsis* lateral root initiation. Plant Cell 13:843–852

Chapter 11

High-Throughput, Noninvasive Imaging of Root Systems

Anjali S. Iyer-Pascuzzi, Paul R. Zurek, and Philip N. Benfey

Abstract

Root systems are complex structures key to plant health. The three-dimensional distribution of the root system, known as the root architecture, is important for optimal uptake of water and nutrients, as well as anchorage in the soil. Despite the importance of root systems, little is known about the genes that control root architecture, in part because of the difficulty of non-destructively viewing root systems. The Benfey lab has developed a gel-based imaging method to non-invasively examine root system architecture over time. Root systems of a variety of plant species can be quickly imaged daily or weekly. The platform is relatively inexpensive, high-throughput, and is ideally suited for researchers aiming to understand the genetic control of root architecture. Here we describe the application of this method to rice and maize root systems.

Key words: Root system architecture, Imaging, Rice, Maize, Gel

1. Introduction

Root systems are central to plant fitness. They are necessary for water and nutrient acquisition, and provide anchorage in the soil. The three-dimensional spatial distribution of the different types and ages of roots, known as the root system architecture (RSA), is a key factor in determining plant water and nutrient acquisition (1). Recent work has suggested that modification of RSA could contribute to increases in yield and/or stress tolerance (2, 3). Identifying the genes that control RSA and understanding how root systems develop over time are important areas of biological research. Although RSA within an inbred variety is generally very reproducible, RSA among different varieties within a species can be highly variable (4). Dicots such as Arabidopsis typically have a single taproot with lateral roots branching from this root, while

Ive De Smet (ed.), *Plant Organogenesis: Methods and Protocols*, Methods in Molecular Biology, vol. 959,
DOI 10.1007/978-1-62703-221-6_11, © Springer Science+Business Media New York 2013

monocots such as rice and corn have much more complex root systems consisting of multiple types of roots. Given the importance of RSA to plant fitness, surprisingly little is known about the genetic control of root architecture, particularly for crop plants such as rice and maize. This is in part due to the challenge of observing roots nondestructively. Traditional methods of examining root systems involve soil excavation or soil cores, but these destroy both the fine roots and the spatial distribution of the whole root system. Several new advances in the imaging field, such as X-ray micro CT (5), or combined PET-MRI (6), now provide researchers with multiple options for root imaging (7). Each of these has advantages and disadvantages. Which method to use ultimately depends on the researcher's question and goals.

The Benfey lab has developed a nondestructive platform to image roots in real time over several days or weeks. This platform involves growing roots in gellan gum, and is ideal for experiments in which high-throughput phenotyping is required, such as QTL mapping and gene isolation. In addition, the protocol described below is relatively inexpensive and the technology is accessible. We describe the imaging method we use for rice and maize, although this can be adapted to suit any plant species. This method can be very basic, with a simple digital camera, commercial turntable, and incandescent lights, or it can be scaled-up using sophisticated cameras, high precision turntables, and LED lighting for optimized images. The simpler setup will yield good images from which many accurate features can be extracted, especially if care is taken with the lighting, but they may not be of the quality necessary to extract all the features of the root system. We have tried to provide an outline of what is needed at both ends of the spectrum. What the researcher chooses to use will depend on experimental goals, time frame, and money.

2. Materials

2.1. Media Preparation

1. Maize-Hoagland's solution: $Ca(NO_3)_2{\cdot}H_2O$, NH_4NO_3, KNO_3, $MgSO_4{\cdot}7H_2O$, $MgSO_4{\cdot}7H_2O$, $ZnSO_4{\cdot}7H_2O$, $MnSO_4{\cdot}7H_2O$, $CuSO_4{\cdot}5H_2O$, H_3BO_3, $Na_2MoO_3{\cdot}H_2O$, KCl, $NH_4H_2PO_4$, FeDPTA (Sequestrene 330).
2. Rice-Yoshida's solution: NH_4NO_3, $NaH_2PO_4{\cdot}2H_2O$, K_2SO_4, $CaCl_2$, $MgSO_4{\cdot}7H_2O$, $MnCl_2{\cdot}4H_2O$, $(NH_4)_6MO_7O_{24}{\cdot}4H_2O$, H_3BO_3, $ZnSO_4{\cdot}7H_2O$, $CuSO_4{\cdot}5H_2O$, $FeCl_3{\cdot}6H_2O$, citric acid (monohydrate), MES.
3. Gelzan (Caisson laboratories, see Note 1).
4. Large ~28 L jars for maize work (see Note 2); 2 L ungraduated cylinders for rice (Fisher).

5. 1 ml glass pipettes.
6. Sylgard 184 Silicone Elastomer (Dow Corning).
7. Millipore tape.
8. Laser Level, square.
9. Large stir plates (for maize work).
10. Acetic acid.

2.2. Seed Preparation and Transplanting

1. Hydrogen Peroxide.
2. Sterile water.
3. Petri plates (rice).
4. Long tweezers.
5. Ethanol.

2.3. Imaging

1. Digital camera and Lens (see Note 3).
2. Turntable (such as from Ortery Technologies or, for better control, a high-end turntable from LinearX Systems Inc, see Note 4).
3. Light source (such as a light box from VWR, or, for better contrast, LED lights, see Note 3).
4. Water tank (see Fig. 2).
5. Tripod or table for stable imaging (see Figs. 1 and 2 and Note 5).
6. Imaging software (such as from Ortery Technologies or custom made, see Note 4).
7. Data storage.

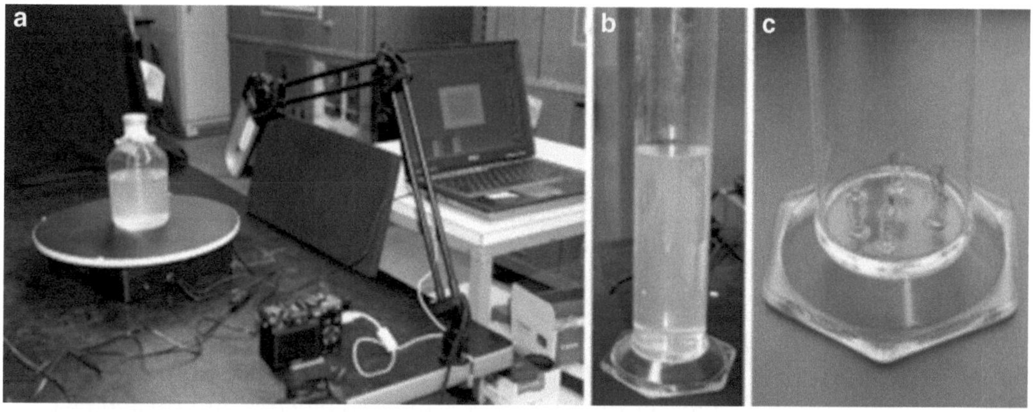

Fig. 1. Simple (1st generation) imaging setup. (**a**) Shows the bare minimum needed for an imaging setup including turntable, camera, light source, and a laptop. Additionally a (**b**) simple ungraduated cylinder is more than sufficient for any plant with a compact root system such as rice. (**c**) Shows pipette fragments embedded in the bottom of a cylinder using Sylgard 184 silicone to prevent the gel column from spinning.

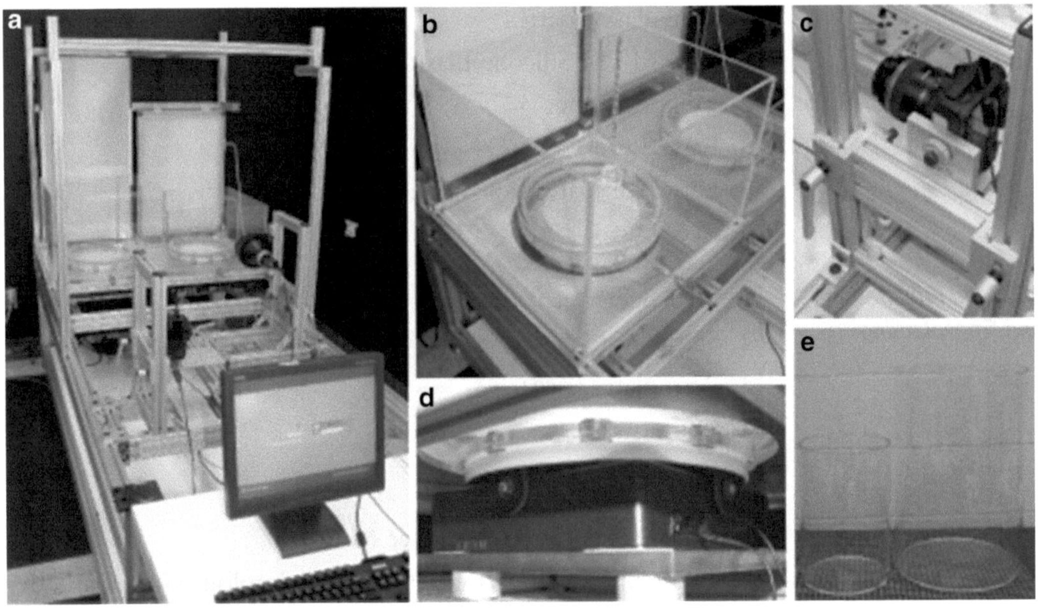

Fig. 2. Dedicated (4th generation) imaging table. (**a**) Shows the complete table, with two separate imaging stations, each of which is customized for a specific experiment. Each station is composed of a (**b**) backlight (fluorescence or LED) and an acrylic water tank, (**c**) a digital camera positioned in front of the tank on an easily adjustable camera tower, and a (**d**) turntable resting underneath the water tank. Additionally a computer (**a**) is used to control both the turntable and the camera. (**e**) Shows different custom-sized cylinders designed for plants of different root system sizes.

8. Optional (but highly encouraged): Dedicated room for imaging, next to growth space if possible.
9. Optional (encouraged for large projects): Barcoding/barcode scanner/barcode software.

2.4. Analysis Software

1. Adobe Photoshop/ImageJ/Fiji/GIMP for manual extraction of some features.
2. Gia-roots for automatic extraction (9).

2.5. Post-experiment Analysis

1. Digital camera.

3. Methods

3.1. Cylinder Preparation (Rice Only)

1. Since the gel strength we use is quite low, and we use cylinders, the gels will move in the cylinder when the turntable spins. To prevent this, we break 1 ml glass pipettes into small 0.5–1 cm. pieces and use Sylgard silicone elastomer to attach them to the bottom of the cylinder (Fig. 1c). Contact between the gel and the pipette pieces prevents movement of the gel within the

cylinder as the turntable spins. Pipette 25 ml of the silicone into the bottom of the 2 L cylinder and then place the small glass pieces directly on the bottom.

3.2. Media Preparation

1. Prior to making gels, autoclave cylinders with foil or mesh over the top to maintain sterility.
2. For rice, we use Yoshida's media. First make the macronutrient stock solutions: Make 5 macronutrient stock solutions in water (per liter):
   (a) 91.4 g NH_4NO_3
   (b) 40.3 g $NaH_2PO_4{\cdot}2H_2O$
   (c) 71.4 g K_2SO_4
   (d) 88.6 g $CaCl_2$
   (e) 324 g $MgSO_4{\cdot}7H_2O$.
3. Next, make the micronutrient stock solution.
   (a) Per liter, dissolve the following separately in 50 ml ddH_2O: 1.5 g $MnCl_2{\cdot}4H_2O$, 74 mg $(NH_4)_6MO_7O_{24}{\cdot}4H_2O$, 934 mg H_3BO_3, 35 mg $ZnSO_4{\cdot}7H_2O$, 31 mg $CuSO_4{\cdot}5H_2O$, 7.7 g $FeCl_3{\cdot}6H_2O$, 11.9 g citric acid (monohydrate).
   (b) Add 50 ml concentrated H_2SO_4 to 550 ml ddH_2O.
   (c) Combine with the above micronutrients. Bring final solution to 1 L with ddH_2O.
4. Use 1× Yoshida's for rice: Add 1.25 ml of each of the stock solutions per liter of distilled water. Add 0.55 g MES hydrate per liter. pH to 5.8 with NaOH.
5. To make rice gels: Add 1 L of 1× Yoshida's solution to a 2 L beaker with a stir bar. While stirring, add 2.5 g Gelzan (0.25% final concentration). Autoclave for 20 min. When cool enough to handle, pour 1 L into sterile, 2 L ungraduated cylinders and cover again with foil/mesh. Let cool in cylinders overnight. Save a small amount of leftover gel solution and pour into small sterile Petri dishes (60 × 15 mm) for pre-germination step below.
6. To make modified Hoagland's for maize: First, make stock solutions:
   (a) Solution A (for 1 L of solution): 298.25 g $Ca(NO_3)2{\cdot}4H_2O$, 20.57 g NH_4NO_3.
   (b) Solution B (for 1 L of solution): 152 g KNO_3, 126.7 g $MgSO_4{\cdot}7H_2O$, 0.286 g $ZnSO_4{\cdot}7H_2O$, 0.195 g $MnSO_4{\cdot}7H_2O$, 0.021 g $CuSO_4{\cdot}5H_2O$, 0.725 g H_3BO_3, $Na_2MoO_3{\cdot}2H_2O$, 3.154 g KCl, 59.030 $NH_4H_2PO_4$ (see Note 6).

(c) FeDPTA stock (for 1 L solution): 37.5 g FeDPTA (commercial name: Sequesterene 330, see Note 7).

(d) Autoclave or filter sterilize solution A and B; filter sterilize FeDPTA. see Note 7 regarding storage.

7. To make maize gels: Make sure that you have a stir plate large and strong enough to hold the large jar. Place a stir bar in each jar, and fill each with 14 L of ddH_2O water. Add 28 mL of each of Solution A and B to each jar. Add 2.8 mL of FeDTPA to each jar. While mixing, adjust the pH to 6.0 using 5 M KOH (should start at ~4.8). Use less concentrated KOH and HCl to fine-tune if needed (see Note 8). Add 21 g (0.15%) of Gelzan to each jar and let dissolve into the solution. Use wide heavy-duty aluminum foil to seal the top of the jars. Tape in at least several places, but leave some space for steam to escape. Autoclave on the liquid cycle for 90 min. Transfer the jar back to the stir plate and let mix for a few minutes. Set aside to cool and solidify (12 h), see Note 9.

3.3. Seed Preparation

1. Before rice or maize seeds can be planted they need to be (a) surface sterilized and (b) pre-germinated. For rice sterilization: Dehull seeds by manually removing the outer hull. Place in a 15 ml conical tube (labeled with the genotype) and sterilize with 10% H_2O_2 for 10 min, 70% ethanol for 1 min, and rinse 3× with sterile water. We usually sterilize three times as many seeds as we need. If you wish to weigh each seed and keep track of each, see Note 10. For rice, continue with step 3.

2. For maize sterilization: Place the kernels in a 50 mL conical tube. Add ~40 mL of 35% H_2O_2, and shake for 20 min. At this stage the seed may turn white. Decant the H_2O_2.Wash 5 times with sterile water. Place in sterile ddH_2O in a Petri dish O/N at 28°C or 30°C. Use enough water so that the seeds are in water, but not completely submerged. The following day, repeat the entire process, but only keep seeds in the H_2O_2 for 10 min.

3. Pre-germinate rice seeds in the dark on Yoshida's media poured into small (60×15 mm) petri dishes made for 2 days at 28°C.

4. Maize pre-germination (done after the 2nd sterilization step): Add ~3–4 mL sterile water to as many 60×15 mm petri dish plates as you have seeds. Place an individual seed in each plate. Because it is very hard to get to a point where all the seeds are sterile, this will prevent contamination from spreading, which happens if a single plate with many seeds is used. Seal the plates with Millipore tape and transfer to a dark, 28°C or 30°C incubator. The seeds should be ready for planting in about 2–3 days after beginning the protocol. Ideally they should be at a stage where the primary root has emerged and is ~2–3 cm in length, while the coleoptile is around 1 cm long, and there are no seminal roots.

3.4. Planting and Growth Conditions

1. Rice: When radicles have just emerged (at about 2 days), plant rice seeds approximately 1 cm deep into the gel in a sterile hood. Try to choose seeds that are similar in developmental stage (see Note 11). Cover the tops of cylinders with either aluminum foil or Millipore tape, or leave open (see Note 12). If desired, cover the roots of cylinders with aluminum foil or other material (see Note 13). We grow the rice in growth chambers set to 12-h day/night, 28°C day, and 25°C night. Our light intensity is 680–730 umol/m^2/s at the top of the cylinders.
2. Maize: Transfer a jar to a hood, and spray the outer glass surface with 70% EtOH. Gently remove the aluminum foil cap from the jar. Using the pipette aid and a 25 mL pipette, dig a small circular hole in the middle of the gel. It should be around 1 cm deep and 1 cm in diameter. Flame a pair of large tweezers and transfer the seedling to the jar such that the kernel is resting in the hole with the coleoptile facing up, and the primary root embedded into the side of the hole. Place the aluminum cap back on top of the jar. Seal the cap to the jar completely using Millipore tape. Transfer the jars to the growth chamber, see Note 14.
3. Maize growing conditions: 28°C day/25°C night, 800 μmol light, 65% humidity at day/80% humidity at night, 16-h/8h day/night cycle, see Note 15.

3.5. Imaging

1. Wipe down sides of jars and place cylinders in water tank (see Note 16). Be very careful that the jars are level (use a level). If using Ortery software and turntable, follow the directions in the software (custom turntables and software can also be used). Using this software, it takes about 5–6 min to produce a set of 40 images of a single cylinder. Factoring in additional time to move jars in and out of the tank, one person can image about 30 plants a day with a simple setup, and double that with a dedicated setup as show in Fig. 2. Plants can be imaged as many times during a day or week as desired (see Note 17).
2. Review images. Make sure that the lighting is providing good contrast. Test your images by thresholding (i.e., in Photoshop, ImageJ, GIMP). All, or nearly all, of the root system should remain visible once thresholded, with few to no extra pixels in areas of the image without a root. Upload to a server for storage (see Notes 18 and 19).

3.6. End of Experiment Plant Analysis (If Desired)

1. Remove plants from gel (simply pull out of gel—they come out easily) and place in tray containing clean water.
2. Take images of the entire plant using digital camera or scanner.

3. Cut roots from shoots and determine fresh weight of each.
4. Dry roots and shoots in an oven for several days; weigh each when dry (see Note 20).

3.7. Analysis

1. Images can be cropped and thresholded using Adobe Photoshop or the open-source equivalent, GIMP.
2. Parameters can be measured automatically with GIA-roots (traits as in Subheading 4, see Note 21, and reference (8) or RootReader (9)).
3. Roots can be reconstructed in 3D using several different off-the-shelf reconstruction software tools, although this may require additional steps not covered here. Additional software for 3D root reconstruction is found in ref. (9).
4. Quality control: Image a standard of known dimensions (i.e., a simple object like a ring or a tuning fork, or more complex such as a model root). Measure this object in the images and compare to the known dimensions to assess the accuracy of your system.

4. Notes

1. Different types of gelling agents can be used, but they differ in clarity, hardness, and nutrient concentration. We have had good success with Gelzan. The concentration used may differ for different species and is something that the researcher will need to test. Note that different agents can be used to change the gel conditions—adding salt, for example—but these can change the hardness and the clarity of the gel.
2. Jars the size needed for maize are custom made. We found a glass manufacturer and ordered the required size. Note that when they contain the hardened gel, these jars weigh approximately 40 lbs/18 kg.
3. Image quality is impacted by a number of factors, including lighting, camera and lens quality, and camera placement. We have found that a standard digital SLR camera with high resolution works well; however, for optimal images, a more specialized CCD camera (such as Allied Vision Tech Stingray F-046) will probably be desired. For lighting, good contrast is essential to identify fine roots in the images. Standard incandescent lightbulbs or a lightbox work well, but for even better contrast and uniformity LED lights may be warranted. What you choose depends on your goals, money, and how high- throughput you want the system to be.

4. The turntable from Ortery Technologies is useful, but not as accurate (degrees turned per step can be inconsistent) as high-end tables like those from LinearX Systems Inc. Also, the researcher cannot customize the software from Ortery, as there is no API provided.
5. We designed a custom table with tracks for moving the camera up, down, forward, and backward. However, our first imaging setup (see Fig. 1) was a simple table. Any flat-level surface will work well; you can customize the table to fit your needs.
6. For Hoagland's Solution B, we have found that it is best to start with almost all the water necessary. Also, when making Hoagland's, note that the yellow color disappears after exposure to light.
7. http://www.phytotechlab.com is a good source for FeDPTA; store prepared stock solution in the dark.
8. 14 L should require about 15 drops of 5 M KOH to get to pH 6.0. If it takes significantly more or less, something might be wrong. Do not mix A + B directly. It will result in a precipitate that will not dissolve even after water is added.
9. The media that you use, as well as the growing conditions, will need to be optimized for your particular species. For example, the media for maize is 0.5× Hoagland's but with only ¼ of the Fe because excess iron causes a yellow precipitate to form that interferes with the image processing.
10. If you wish to keep track of each seed, use this method of seed sterilization. The sterilization is described for corn, but you can use the same pillbox method with the rice sterilization method above for rice. Pill organizers (e.g., http://www.organize.com/wepiorl.html) with each of the organizers having 28 individual chambers. Drill a hole in the top cover over each chamber, and another two holes in the bottom of each chamber, as well as several larger holes in the side. Weigh each seed, and place it in its own chamber in the pill organizer. Keep track of which seed is where and its weight. Fill a plastic container with 35% H_2O_2, and place the pill organizer with the seed in the plastic container. If needed, use a heavy weight to keep the pill organizer from floating. Shake for 20 min. At this stage the seed might turn white. Decant the H_2O_2.Wash the seed 5 times with sterile water. (Leave the organizer in the plastic container. Fill the container with sterile water, shake, and decant.) Place each seed in an individual Petri plate in sterile ddH_2O and seal with Milipore tape. Make sure that the seed is not completely submerged (~3–4 mL of water for 60 mm plates). Place in a dark, 28°C or 30°C, incubator overnight. Keep track of the weight of each individual seed. The following day, repeat the process but only sterilize with H_2O_2 for 10 min. Pre-germinate as stated above for maize or rice.

11. The length of the root is important for transplantation success. For rice, make sure that the radicle has emerged, but only to about 1 cm or it will break when planting. For maize: The radicle should be about 2–3 cm in length, but do not plant the seed if the seminal roots are already out. Also make sure that when planting, the coleoptile is above the level of the gel. Dig a hole in the gel prior to planting (about 1.5–2 cm). For both rice and maize, the whole seed should be in the gel; otherwise some roots will not penetrate the gel and will grow along the top of the gel. For all species when planting, it is important to use seeds that are at the same stages of growth so that results are comparable.
12. For covering the top of the cylinders, we use aluminum foil (maize) or Millipore tape (rice). You can also leave the cylinders open if they are small and the experiment is short. If left open, contamination is a risk. Also, when plants are large and cylinders are left open, large volumes of the gel can be lost due to transpiration.
13. We have found that the roots of some rice varieties are sensitive to light. Roots can be covered during growth with aluminum foil (and uncovered only during imaging) if this is a problem.
14. Because the media used here is only 0.15%, transferring the jars directly to the growth chamber can result in the hole filling up with water from condensation and from the gel becoming semiliquid (syneresis). In a lot of the cases this prevents the seed from growing. The best way to avoid this is to either transfer the jars to the chamber during the night part of the day/night cycle or keep the cylinder in cooler conditions until the coleoptile has had a chance to grow ~1 cm above the level of the gel.
15. The amount of growth space required is an important consideration and depends on the size of jars used and experimental goals. We grow our plants in growth chambers, but a greenhouse could also be used.
16. Make sure that water in tank is DI water as this will reduce the number of bubbles. Any bubbles in your image will reduce your ability to phenotype well. Empty tank daily or every other day for cleanliness.
17. How long to do an experiment depends on both your research questions/goals and container size. For accurate root architecture assessment, it is extremely important that the roots do not hit the sides of the jars, as this alters root growth (and thus RSA). We image rice on days 12, 14, and 16, but could go longer or shorter. Maize is imaged on days 3, 6, 9, and 12.
18. Consider whether to take images as JPEGs or RAW files. Lossy compression used with most JPEG formats discards some data

and can introduce compression artifacts, but produces much smaller files, while lossless formats such as RAW will include all the data, but will also produce much larger files and will require longer processing times at the analysis stage.

19. Two major challenges with imaging are data storage and labeling. For large projects, barcoding cylinders increases efficiency. Uploading directly to a server and to an external hard drive for data storage instead of storing on the computer is necessary for large datasets.
20. To be sure that the roots are dry, weigh them first after a few days in the oven when they appear dry. Put them back in the oven and weigh again the next day. If the weight has not changed, they are dry.
21. To access software for automatic phenotyping used in ref. (4) (in beta-testing for use by multiple users as of this writing), contact gia-roots@biology.gatech.edu and see (8). Some traits can be measured manually using ImageJ/Fiji/Photoshop, especially if using a small number of images or traits.

References

1. Lynch J (1995) Root architecture and plant productivity. Plant Physiol 109:7–13
2. Steele KA, Virk DS, Kumar R, Prasad SC, Witcombe JR (2007) Field evaluation of upland rice lines selected for QTLs controlling root traits. Field Crops Res 101:180–186
3. Tuberosa R, Salvi S, Sanguineti MC et al (2002) Mapping QTLs regulating morpho-physiological traits and yield: case studies, shortcomings, and perspectives in drought-stressed maize. Ann Bot 89:941–963
4. Iyer-Pascuzzi AS, Symonova O, Mileyko Y et al (2010) Imaging and analysis platform for automatic phenotyping and trait ranking of plant root systems. Plant Physiol 152:1148–1157
5. Hargreaves CE, Gregory PJ, Bengough AG (2009) Measuring root traits in barley (Hordeum vulgare ssp vulgare and ssp spontaneum) seedlings using gel chambers, soil sacs, and X-ray microtomography. Plant Soil 316:285–297
6. Jahnke S, Menzel MI, van Dusschoten D et al (2009) Combined MRI-PET dissects dynamic changes in plant structures and functions. Plant J 59:634–644
7. de Dorlodot S, Forster B, Pages L et al (2007) Root system architecture: opportunities and constraints for genetic improvement of crops. Trends Plant Sci 12:474–481
8. Galkovskyi T, Mileyko Y, Bucksch A et al (2012) GiA Roots: software for the high throughput analysis of plant root system architecture. BMC Plant Bio 12:116–127
9. Clark R, MacCurdy R, Jung J et al (2011) Three-dimensional root phenotyping with a novel imaging and software platform. Plant Physiol 156:455–465

Chapter 12

Small-Molecule Screens to Study Lateral Root Development

Dominique Audenaert, Bert De Rybel, Long Nguyen, and Tom Beeckman

Abstract

Development of the root system is essential for proper plant growth and development. Extension of the root system is achieved by the continuous establishment of new meristems in existing parental root tissues, which leads to the development of lateral roots. This process of lateral root organogenesis consists of different developmental stages, which are all controlled by the plant hormone auxin. In this chapter, we describe a screening method in *Arabidopsis thaliana* to identify small synthetic molecules that interfere with the process of lateral root development during specific developmental stages.

Key words: Lateral root, Synthetic molecules, High-throughput screening

1. Introduction

Roots are of high importance for proper growth and development of plants. They not only provide water and nutrients to the plant but are also required for stable anchorage in the soil. In addition, roots are the first line of defense against biotic and abiotic stresses. A typical property associated with the root system is its extreme flexibility and plasticity. Because of the sessile nature of plants, root plasticity is essential to retrieve new sources of nutrients and water. This is achieved by the ability of roots to continuously develop new branches from the existing root tissue, which eventually leads to an extended network of root branches.

In dicotyledons, a single primary root develops from the radicle during the process of germination. During root growth and development, root branching occurs by the addition of lateral roots that originate from xylem-pole pericycle cells of the primary root. Lateral root development can be divided into distinct developmental stages (1). At the primary root tip, more specifically in

Ive De Smet (ed.), *Plant Organogenesis: Methods and Protocols*, Methods in Molecular Biology, vol. 959,
DOI 10.1007/978-1-62703-221-6_12, © Springer Science+Business Media New York 2013

the region of the basal meristem, pre-initiation events specify founder pericycle cells from which a lateral root will develop. Functional studies have shown that the control of lateral root founder cell identity is mediated by the GATA23 transcription factor (2). Higher up in the root, cell cycle is activated in the pericycle founder cells, and the first cell divisions occur that lead to the formation of a lateral root primordium (3). Subsequently, the activated pericycle cells proliferate through the overlying cell layers, i.e., endodermis, cortex, and epidermis, and eventually emerge from the primary root (4).

The plant hormone auxin is a main regulator of lateral root development during every developmental stage (1). As such, the application of auxin, auxin-analogues, and auxin antagonists makes it extremely difficult to pinpoint and study a particular lateral root developmental stage (5). The availability of small synthetic molecules that inhibit and/or activate a specific developmental step would aid tremendously in this respect. However, in order to obtain these specific molecules, several thousands of compounds need to be screened. Inevitably, this requires a high-throughput screening assay that highlights a lateral root developmental stage of interest. Here, we describe a miniaturized expression-based screening approach in transgenic *Arabidopsis thaliana* seedlings, based upon a lateral root-inducing system (6), that is compatible with high-throughput screening technology. Dependent on the marker gene, the method allows identifying small molecules that interfere with a specific process during lateral root development.

2. Materials

2.1. Plant Material

Transgenic *A. thaliana* seeds that are homozygous for a marker construct (see Notes 1 and 2).

2.2. Seed Sterilization

1. Sterile distilled water.
2. 70% (v/v) ethanol solution.
3. 5% (v/v) NaOCl/0.05% (v/v) Tween 20 solution.

2.3. Primary and Confirmation Screening

1. 96-Well filter plates with plastic lids.
2. Porous tape for air-permeable sealing.
3. Sterile distilled water.
4. Half-strength Murashige and Skoog (MS) growth medium: 0.215 g/L MS salts supplemented with 10 g/L sucrose, 0.1 g/L myo-inositol, and 0.5 g/L 2-(*N*-morpholino)ethanesulfonic acid (MES) monohydrate in distilled water. Adjust the

final pH to 5.7 by adding 1 M KOH. Autoclave the medium at 1 bar overpressure for 20 min. Growth medium can be stored at 4°C.

5. 50 mM (14.57 mg/ml) *N*-(1-naphthyl)phthalamic acid (NPA) in 100% dimethyl sulfoxide (DMSO) (see Note 3).
6. 50 mM (9.31 mg/ml) 1-naphthaleneacetic acid (NAA) in 100% DMSO (see Note 3).
7. Collection of small molecules, dissolved in 100% DMSO at a concentration of 5 mM, pre-plated in 96-well plates (Corning Life Sciences) and stored at −20°C (see Notes 4 and 5).
8. Continuous light growth chamber (110 μE/m^{2}s photosynthetically active radiation, supplied by cool-white fluorescent tungsten tubes, Osram) at 22°C.
9. Orbital shaker.
10. Vacuum manifold.
11. Vacuum/pressure pump.

2.4. Analysis

1. 90% (v/v) acetone solution.
2. 200 mM 5-bromo-4-chloro-3-indolyl β-D-glucopyranosid, sodium salt (X-Glc) in 100% dimethyl formamide (DMF) (see Note 3).
3. Triton X-100.
4. 500 mM ethylenediaminetetraacetic acid (EDTA) in distilled water. Adjust the final pH to 8 by adding 10 N NaOH. EDTA solution can be stored at room temperature.
5. Phosphate buffer: Mix 615 ml of a 500 mM Na_2HPO_4 solution and 385 ml of a 500 mM NaH_2PO_4 solution. Phosphate buffer can be stored at room temperature.
6. 100 mM potassium ferricyanide ($K_3Fe(CN)_6$) in distilled water. Ferricyanide solution can be stored at −20°C.
7. 100 mM potassium ferrocyanide ($K_4Fe(CN)_6$) in distilled water. Ferricyanide solution can be stored at −20°C.
8. Dissecting microscope.

3. Methods

3.1. Seed Sterilization

1. Add dry transgenic *A. thaliana* seeds in a sterile 2 ml microcentrifuge tube (see Note 6).
2. Add 70% ethanol for 2 min.
3. Remove the 70% ethanol solution and add 5% NaOCl/0.05% Tween 20 solution for 15 min.

4. Remove the 5% NaOCl/0.05% Tween 20 solution and wash five times with sterile distilled water. Leave water in the tubes after the last wash and put at 4°C for at least 2 days.

3.2. Preparation of Assay Plates

1. Prepare the growth medium by adding NPA to half-strength MS medium to a final concentration of 10 μM (see Note 7).
2. Add 150 μl of the growth medium to 96-well filter plates by using a multichannel pipette.
3. Add between three and ten sterilized seeds per well by using a 200 μl pipette with tips of which the top is cut to increase the opening size (see Note 8).
4. Seal the plates with plastic lids and air-permeable tape.
5. Put the plates in the continuous light growth chamber under continuous shaking (150 rpm).
6. Incubate for 5 days.

3.3. Distribution of Compounds

1. Remove the NPA-containing growth medium by vacuum filtration (see Note 9).
2. Add 147 μl of half-strength MS medium to the plates by using a multichannel pipette.
3. Add 1.5 μl of the compound screening collection (5 mM, 100% DMSO) to the plates (columns 2–11) by using a multichannel pipette (see Note 10).
4. Add 1.5 μl of a 100% DMSO solution to columns 1 and 12.
5. Seal the plates with plastic lids and air-permeable tape.
6. Put the plates in the continuous light growth chamber under continuous shaking (150 rpm).
7. Incubate for 2 h (see Note 11).
8. Prepare a 1 mM NAA solution (2% DMSO) by diluting a 50 mM NAA solution (100% DMSO) with sterile distilled water.
9. Prepare a 1 mM NPA solution (2% DMSO) by diluting a 50 mM NPA solution (100% DMSO) with sterile distilled water.
10. Add 1.5 μl of the 1 mM NAA solution (2% DMSO) to columns 1–11, and add 1.5 μl of the 1 mM NPA solution (2% DMSO) to column 12 (see Notes 12 and 13).
11. Seal the plates with plastic lids and air-permeable tape.
12. Put the plates in the continuous light growth chamber under continuous shaking (150 rpm).
13. Incubate for 24 h (see Note 14).

3.4. Analysis

1. Remove the growth medium by vacuum filtration.
2. Add 150 μl of a 90% acetone solution to the plates by using a multichannel pipette.
3. Incubate for 30 min at 4°C (see Note 15).
4. Remove the 90% acetone solution by vacuum filtration.
5. Wash three times with phosphate buffer.
6. Prepare the GUS staining solution by adding 1.5 ml of a 200 mM X-Glc solution in 100% DMF, 1.5 ml Triton X-100, 0.6 ml of a 500 mM EDTA solution, 1.5 ml of a 100 mM potassium ferricyanide solution, and 1.5 ml of a 100 mM potassium ferrocyanide to a final volume of 300 ml phosphate buffer.
7. Remove the phosphate buffer by vacuum filtration.
8. Add 150 μl of the GUS staining solution to the plates by using a multichannel pipette (see Note 16).
9. Incubate for several hours at 37°C (see Note 17).
10. Remove the GUS staining solution by vacuum filtration.
11. Wash three times with phosphate buffer (see Note 18).
12. Analyze the effect of the compounds on the expression pattern of the marker gene by using a dissecting microscope (see Note 19).
13. Identify wells with compounds that interfere with the expression pattern of the marker gene under study (see Note 20).

4. Notes

1. Typically, transgenic *A. thaliana* lines contain a construct consisting of the promoter of a marker gene (gene-of-interest) fused to the reporter gene β-glucuronidase (GUS) to visualize expression of the marker gene.
2. Dependent on the lateral root developmental stage of interest, different marker genes can be considered: for example *GATA23* for lateral root founder cell specification in the pre-initiation stage (2), *CYCB1;1* for the lateral root initiation stage (6), *IAA3* for lateral root protrusion through the endodermis (7), and *LAX3* for lateral root protrusion through the cortex (7).
3. DMSO- and DMF-dissolved chemical reagents should be freshly prepared from dry powder.
4. DMSO is a hygroscopic solvent. The absorption of water leads to compound degradation, reduced solubility, and changes in compound concentration and as such affects the overall quality

of the chemical library. This can be circumvented by storing the DMSO-dissolved compound collections in low-humidity environments such as under nitrogen in enclosed containers. In addition, because collections are stored in frozen condition (−20°C), solutions need to be thawed to access them and refrozen for further storage. Freeze/thaw cycles should be kept to a minimum because they have dramatic effects on compound integrity because of precipitation and/or sample degradation (8).

5. Several companies supply pre-plated diverse sets of DMSO-dissolved small molecules that can be used for screening purposes. These companies include ChemBridge Corporation, Enamine, Life Chemicals, Maybridge, Asinex, etc.
6. Screening of 20 96-well plates requires about 500 μl of dry seeds.
7. The addition of NPA inhibits auxin transport in the developing primary root during the germination process. Consequently, all seedlings have a primary root without any additional lateral roots. This allows synchronization of the lateral root development process over all seedlings, which results in a more robust screening assay.
8. The number of seeds per well does not need to be exactly the same in each well because the analysis is qualitative (assessment of changes in expression pattern) rather than quantitative.
9. After each vacuum filtration step, tap the bottom of the plates with paper tissue to remove the remaining drops.
10. Screening collections are mostly delivered in 96-well plates with columns 1 and 12 left empty, which allows adding controls in each assay plate.
11. A pre-incubation step will allow the synthetic molecules to penetrate the root cells previous to the activation of lateral root development by NAA treatment.
12. This approach aims at the identification of small molecules that inhibit lateral root development at the developmental stage reported by the marker gene. Removing the step of NAA addition to the compound-treated wells will lead to the identification of small molecules that activate lateral root development.
13. Column 1 will serve as a positive control (enhanced lateral root development), and column 12 will serve as a negative control (absence of lateral root development).
14. Because pre-initiation events occur as early as 2 h after NAA treatment and cell cycle activation (lateral root initiation) occurs at 6 h after NAA treatment, shorter incubation times can be considered to identify compounds that interfere with these early lateral root developmental stages.

15. Longer incubation times should be avoided as this will affect the filters at the bottom of the plate, which could eventually result in leakage of the plates in subsequent steps.
16. To analyze 20 plates, at least 300 ml of GUS staining solution is required.
17. The expression pattern of both positive and negative controls should be checked regularly during the incubation. When the difference between positive and negative controls is easily scoreable, GUS staining can be stopped. Total incubation time is dependent on the marker gene and the *A. thaliana* GUS line. Typically, incubation times range between 2 and 24 h.
18. At this stage, plates can be stored at 4°C for several weeks prior to analysis of the expression pattern.
19. On average, scoring of an entire 96-well plate takes about 15 min.
20. Compounds that are identified as hits should be retested by using the same protocol to confirm their effect.

References

1. Péret B, De Rybel B, Casimiro I, Benková E, Swarup R, Laplaze L, Beeckman T, Bennett MJ (2009) Arabidopsis lateral root development: an emerging story. Trends Plant Sci 14:399–408
2. De Rybel B, Vassileva V, Parizot B, Demeulenaere M, Grunewald W, Audenaert D, Van Campenhout J, Overvoorde P, Jansen L, Vanneste S, Möller B, Wilson M, Holman T, Van Isterdael G, Brunoud G, Vuylsteke M, Vernoux T, De Veylder L, Inzé D, Weijers D, Bennett MJ, Beeckman T (2010) A novel aux/IAA28 signaling cascade activates GATA23-dependent specification of lateral root founder cell identity. Curr Biol 20:1697–1706
3. Vanneste S, De Rybel B, Beemster GT, Ljung K, De Smet I, Van Isterdael G, Naudts M, Iida R, Gruissem W, Tasaka M, Inzé D, Fukaki H, Beeckman T (2005) Cell cycle progression in the pericycle is not sufficient for SOLITARY ROOT/IAA14-mediated lateral root initiation in *Arabidopsis thaliana*. Plant Cell 17:3035–3050
4. Malamy JE, Benfey PN (1997) Organization and cell differentiation in lateral roots of *Arabidopsis thaliana*. Development 124:33–44
5. De Rybel B, Audenaert D, Beeckman T, Kepinski S (2010) The past, present, and future of chemical biology in auxin research. ACS Chem Biol 4:987–998
6. Himanen K, Vuylsteke M, Vanneste S, Vercruysse S, Boucheron E, Alard P, Chriqui D, Van Montagu M, Inzé D, Beeckman T (2004) Transcript profiling of early lateral root initiation. Proc Natl Acad Sci USA 101:5146–5151
7. Swarup K, Benková E, Swarup R, Casimiro I, Péret B, Yang Y, Parry G, Nielsen E, De Smet I, Vanneste S, Levesque MP, Carrier D, James N, Calvo V, Ljung K, Kramer E, Roberts R, Graham N, Marillonnet S, Patel K, Jones JD, Taylor CG, Schachtman DP, May S, Sandberg G, Benfey P, Friml J, Kerr I, Beeckman T, Laplaze L, Bennett MJ (2008) The auxin influx carrier LAX3 promotes lateral root emergence. Nat Cell Biol 10:946–954
8. Kozikowski BA, Burt TM, Tirey DA, Williams LE, Kuzmak BR, Stanton DT, Morand KL, Nelson SL (2003) The effect of freeze/thaw cycles on the stability of compounds in DMSO. J Biomol Screen 8:210–215

Chapter 13

Cell Lineage Analyses in Living Tissues

John Runions and Smita Kurup

Abstract

Developmental biologists require methods for marking cell lineages as they arise in living tissues. Traditionally, lineages have been traced in fixed tissues but these observations are difficult to verify. We present a method by which a progenitor cell and all of its lineage become marked by a nuclear-localised fluorescent protein. This allows rapid estimation of the effects of genetic or physical manipulation of developing tissues. Heat shock is used to activate *YFP* expression in single progenitor cells which is heritable by all daughter cells in subsequent rounds of mitosis. Heat shock can be applied to specimens generally using an incubator to generate random lineage patterns or more specifically to single cells or small regions using laser activation of the lineage marking system.

Key words: Lineage, GFP, Clonal, Chimera, Heat-shock promoter, Transposon

1. Introduction

Cell lineage refers to the relationships between cells that results because of cell division during development. Cell lineages in multicellular organisms can be invariant, in which case the specification of cell fate is correlated to the cell division pattern. However, in many organisms, we can observe indeterminate divisions, in which the fate of a cell is determined by a cell's interaction with its environment. In order to understand the developmental processes, it is important to both have genealogical information and an understanding of the influence of neighboring cells.

To trace cell lineage, various methods have been employed over the years in both plants and animals. The earliest observations were based on microscopical studies. In plants these anatomical studies gave the first understanding of lineages based on patterns of primary cell wall deposition after cell division, but were not entirely

Ive De Smet (ed.), *Plant Organogenesis: Methods and Protocols*, Methods in Molecular Biology, vol. 959,
DOI 10.1007/978-1-62703-221-6_13, © Springer Science+Business Media New York 2013

satisfying. Clonal analysis and genetic mosaics furthered the study of genealogy. These chimeras contain cells that are differentiated due to polyploidy or albinism. They allow the generation of fate maps which have proved useful in determining the patterns of development (1–4). X-ray-induced sectoring (5–7) and microinjection (8) are other tools that have been employed in Arabidopsis and maize to study cell fate in the apical meristem. In recent years, several other techniques have been utilized. These include use of retroviral vectors (9), quail-chick chimeras (10, 11), and vital dye staining (reviewed in ref. 12). The use of DNA recombination events (13) to activate the expression of a marker gene increased the ease of studying lineages. The Cre–loxP recombination system was used in mice and Arabidopsis (14–17), and FLP recombinase was used in tobacco, *Arabidopsis* (18, 19), and *Drosophila* (20–22).

The use of inducible promoters further enhanced the use of recombinant systems. These allow expression of the transgene to be induced at the desired time point in development. The most popular inducible promoter has proved to be the heat-shock promoter from *Drosophila* in animals and from soybean in plants (18, 23, 24).

Use of heat-shock promoters has some advantages. The promoter can be induced at a particular time in development and duration of the treatment can be altered to vary the expression of the transgene. However, there may be problems associated with heat shocking, e.g., lethality. The use of lasers to induce the heat shock has been reported in *C. elegans* and *Drosophila* (23–25). These studies were done primarily to create mosaics. Reporter genes used have been largely *β-gal* (animals) or *uidA* (plants), both of which are detected using destructive histochemical assays. There have been reports of limited use of GFP as a reporter gene (20, 21).

The maize Ac/Ds transposition system has been introduced into Arabidopsis (26, 27) and used successfully to create mosaics. By placing the Ac transposase under the control of a heat-shock promoter from soybean, the activity of the transposons could be induced at high temperatures (27).

The last several years have also seen the development of image registration algorithms that allow visualization of cell division in real time (28, 29).

1.1. Visualization of Lineages by Fluorescent Protein Expression in Living Plant Cells

We have generated a novel system to mark clonal sectors in living tissues such that all daughter cells of an induced progenitor are fluorescently marked (Fig. 1). We demonstrated the first application of using the heat-inducible Ac/Ds transposon-based recombinant system linked to a live cell marker, YFP, to follow lineages in live cells in Arabidopsis (30). In addition, expression of the promoter can also be induced by using a laser to target specific cell types of interest.

In this system, heat shock at 42°C results in expression of an Ac transposase gene which, in turn, causes excision of a Ds1 element that these plants carry between a 35S promoter and the

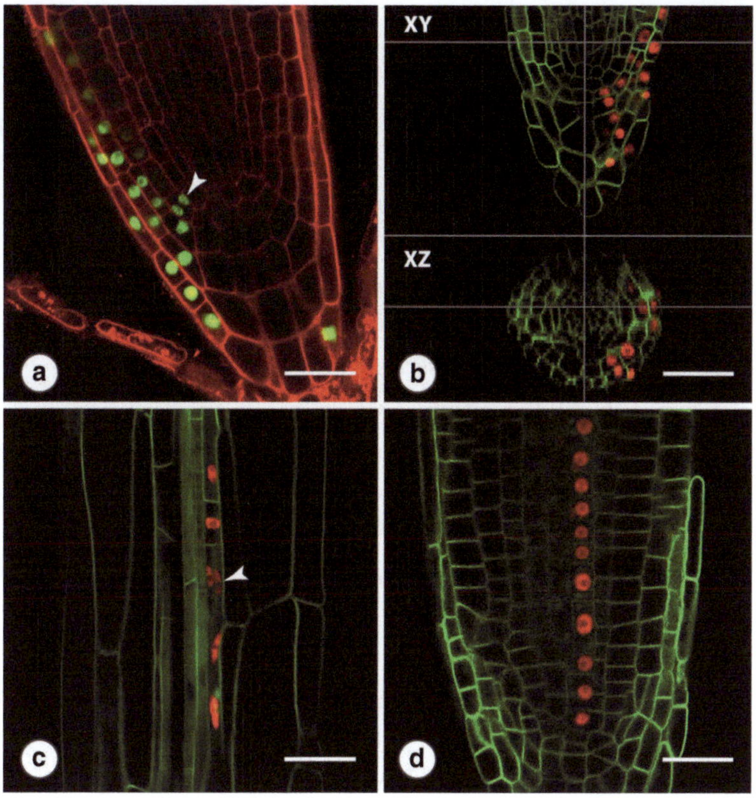

Fig. 1. Fluorescently marked lineages in living roots that result from heat-shock-activated expression of nuclear fluorescence in a single progenitor cell followed by repeated rounds of mitosis in daughter cells. (**a**) Main root tip. An epidermis/lateral root cap lineage resulting from activation of the single progenitor initial cell for these tissues at the location of the *arrowhead*. A single activated columella cell is visible in the *lower right part* of the image—it has not divided post activation. H2B-YFP marks nuclei in this lineage (*green*). Cells are outlined by propidium iodide staining (*red*). Bar = 20 μm. (**b–d**) Lateral roots. H2B-YFP marks nuclei (*red*) and plasma membranes are marked with LTI6b-GFP (*green*). (**b**) An epidermis/lateral root cap lineage as in (a) is shown in the *XY* view. The *XZ* view is useful because it demonstrates the way that this lineage spreads in 3D. Activation of a single initial cell results in marking of the epidermis and lateral root cap around ¼ of the circumference of the developing root. Bar = 50 μm. (**c**) The pericycle of the main root is marked because of activation in one of the pericycle initial cells. Early divisions leading to formation of a lateral root are anticlinal to the axis of main root growth and the first periclinal division that will lead to outward proliferation of the lateral root is marked by an *arrowhead*. Bar = 100 μm. (**d**) A uniseriate lineage of endodermal cells is marked. This lineage will have resulted from activation of a differentiated endodermal cell in the cell division zone. Had the cortex/endodermis initial cell been activated, both tissues would be marked. Bar = 20 μm.

H2B-YFP nuclear-marking fluorescent protein fusion (see Fig. 1 in ref. (30)). This, however, is a relatively rare event after a brief heat shock and *YFP* expression is therefore induced in only a very few cells in a developing meristem. This is desirable as the goal is to allow development to continue for a period of time, e.g., 12–48 h, after heat shock in order to generate a lineage pattern. These patterns result because expression of the 35S::H2B-YFP construct that results after Ds1 excision is stably inherited by all daughter cells that result from division of the originally marked cell.

2. Materials

1. Arabidopsis lines carrying the Ds1 construct (Ds-H2B-YFP).
2. Arabidopsis lines carrying the hs-Ac construct.
3. Arabidopsis lines carrying the EGFP:LTI6b membrane marker.
4. Propidium iodide (5 μg/ml solution in water made fresh); keep away from light.
5. Microscope slides.
6. #0 Coverslips.
7. 0.8% Low-melting-point (LMP) agarose (keep warm at 35°C).
8. Nunc coverglass-bottomed 2-cell chambers.
9. Sodium hypochlorite 10% solution with 0.05% Triton X-100.
10. Distilled water.
11. Microfuge tubes to sterilize seed.
12. ½-Strength medium: ½-Strength MS basal salts + Gamborg's B5 vitamins, 0.5 g/l MES, 0.7% agar, pH 5.6.
13. Gas-permeable tape to seal Petri dishes and chambers.
14. Pipettes and pipette tips for sterilizing and plating out seed.
15. Forceps.
16. Water bath to keep agarose at 35°C.
17. Confocal laser scanning microscope.
18. Epifluorescence stereomicroscope with YFP/GFP and red filters.
19. Growth room to allow plant growth at 18–22°C with appropriate light.
20. Incubator set at 42°C.

3. Methods

3.1. Generation of Plant Material

1. Arabidopsis lines carrying the Dsl construct (Ds-H2B-YFP) were crossed to Arabidopsis lines carrying the hs-Ac construct to generate hs-Ac/Ds-H2B-YFP plants. Seeds from these plants were used for further experiments.
2. If possible, introduce the EGFP:LTI6b marker in these lines by floral dip or crossing.

3. Arabidopsis lines carrying the hs-Ac construct hs-Ac/ Ds-H2B-YFP seedlings were surface sterilized by immersion in the sodium hypochlorite solution for 10 min with agitation followed by three washes with sterile distilled water.
4. Sterile seeds were plated on ½-strength media plates (as described in Subheading 2, item 12), sealed with gas-permeable tape and placed in growth room under continuous light at 18–22°C.

3.2. Heat Shock Conditions

1. Plates with seedlings at the appropriate age were heat shocked at 42°C in an incubator for 30 min. It is useful to optimize the time of incubation for the tissue of choice (see Note 1).
2. After heat shock, seedlings were transferred back to the growth room for 1 to several days. It is useful to screen seedlings for expression of YFP in cell types of interest at intervals post incubation to optimize detection (see Note 2).

3.3. Analysis of Lineage Patterns

1. Seedlings should be screened initially for YFP fluorescence using a stereomicroscope equipped for epifluorescence (see Note 3).
2. Further imaging of detected lineages is best done using a confocal laser microscope.
3. If the EGFP:LTI6b membrane marker is not present, cells can be outlined by staining seedlings with propidium iodide (5 μg/ml) for 1.5 min, rinsing and mounting in water (see Note 4).
4. To prevent seedling movement during high-magnification confocal microscopy, seedlings can be mounted between microscope slides and #0 coverslips in warm (35°C) LMP agarose (see Note 5).
5. Excitation of YFP in nuclei and EGFP in plasma membranes is via an argon ion laser.
   - Use the laser set to 50% power output. EGFP is excited using the 488 nm line at 15% transmission.
   - EGFP fluorescence is detected using an emission band of 505–530 nm.
   - YFP is excited using the 514 nm line at 50% transmission.
   - YFP fluorescence is detected using an emission band of 527–553 nm.
   - Propidium iodide is excited with the 488 nm argon ion laser line.
   - Propidium iodide fluorescence is detected between 606 and 635 nm.
6. Lineage patterns have been observed in all tissue types (see Note 6).

3.4. Laser Induction of Ds1 Transposition

1. Grow seedlings in coverglass-bottomed 2-cell chambers (e.g., Nunc chambers from Fisher Scientific) containing media as described above (see Notes 7 and 8).
2. At 3–5 days post germination, roots will have grown through the agarose to the bottom of the chamber and then continued to grow along the coverglass (see Note 9).
3. Observation and heat shocking are done with a confocal microscope system equipped with an inverted microscope (see Note 10).
4. Prior to heat shocking, tissues of interest, e.g., main root tips, should be imaged as a control using EGFP and YFP settings similar to those described above.
5. Heat shocking specific cells within the root tip is accomplished by repeated scanning of the 488 nm argon ion laser line within a region-of-interest drawn over the cell region (see Note 11).
6. A 63× 1.4 NA oil-immersion objective lens gives the best results when the laser is set to maximum output power with 100% transmission.
7. Region-of-interest scans can be preformed either continuously or at 5-s intervals using the time-series function of the confocal software for various amounts of time from 1 to 15 min (see Note 12).
8. The most efficient protocol was found to be scanning at 2–5-s intervals for 2–4 min. This needs to be optimized for the system being used, laser output, and tissue of interest.
9. Immediately post heat shock, laser power should be reduced to the levels described above and EGFP and YFP fluorescence imaged as a time zero control.
10. Chambers containing heat-shocked seedlings can then be placed back into the growth room.
11. Root tips should be imaged periodically during the next 72 h and lineage pattern development recorded (see Note 13).

4. Notes

1. A heat-shock promoter from barley used to generate transgenic wheat was tested for optimization of heat shock conditions using a range of temperatures (31). Optimal condition for heat shock treatment determined by the authors is 38–40°C for 1–2 h.
2. Estimate the efficiency of heat-shock activation by looking at the number of individual lineages that result, for example, in a root tip. If the line is too sensitive or seedlings were exposed to heat

for too long, many lineage patterns may converge making the job of describing individual lineages difficult. It takes a good deal of practice to distinguish individual complex lineages from those that are the result of two or more simple patterns converging. As an example, the lateral root cap/epidermis lineage in the main root tip is complex looking and at first glance appears to be the result of more than a single transposition event.

3. If a fluorescence stereoscope is not available, screen using a compound microscope with the lowest magnification objective available.
4. Propidium iodide outlines living cells because it is excluded from entering the symplast by an intact plasma membrane. Nuclei of dead cells will stain strongly. Use caution when working with propidium iodide; always wear gloves.
5. LMP agarose remains liquid at lower temperature than does normal agarose and so is suitable for mounting seedlings without exposing them to excess heating. To prepare LMP agarose for this purpose, boil it but be sure to cool it to 35°C or lower before putting living seedlings into it. Keep the flask containing agarose warm by placing it into a larger water dish on a warming plate. When seedlings are put into the liquid agarose on a microscope slide, apply the coverslip and press down on it for a few seconds as the agarose solidifies.
6. We have observed lineage patterns in all root tissues, hypocotyl epidermis, and leaf epidermis and guard cells. In principle, the technique should mark lineages in all tissues but we have not investigated laser activation of lineages in tissues other than roots.
7. Coverglass-bottomed dishes are necessary for this task because high-resolution microscope objectives will not focus through a thicker bottom. Be sure you are not using microscope slide-bottomed dishes.
8. Place chambers inside larger Petri dishes to prevent desiccation of seedlings grown in continuous light at 18–22°C in a growth room.
9. Several seeds should be started in each well of the chambered slide to guarantee that you will find enough root tips in the proper orientation for imaging. Often, root tips grow slightly angled downwards and this makes it difficult to get nice images of the median root section.
10. An inverted microscope is required because, in the case of laser induction, we need to keep the seedlings *in situ* during heat shock and for several days subsequently. Seedlings that are removed from media and mounted on microscope slides will not thrive during this type of observation. In any case, the seedlings need to remain fixed in the location they were in

during heat shocking so that the activated area can be observed later.

11. To avoid heating which might activate the heat-shock promoter and to avoid photobleaching, it is desirable to use fast scanning for the initial setup of the region-of-interest, followed by a slow scan to obtain an image with high resolution.
12. We have observed that continuous scanning heats the tissue too much and results in quantitative excision of Ds1 elements within all cells of the root tip. This means that individual lineages are not visible because all cells become fluorescent.
13. It takes a good deal of practice to recognize the area that was heat shocked if, for example, you are observing root tips which grow quickly. It helps to observe particles and other irregularities within the medium surrounding the tissue at time zero and to use these as reference points.

References

1. Doring HP, Lin J, Uhrig H et al (1999) Clonal analysis of the development of the barley (*Hordeum vulgare* L.) leaf using periclinal chlorophyll chimeras. Planta 207:335–342
2. Marcotrigiano M, Bernatzky R (1995) Arrangement of cell layers in the shoot apical meristems of periclinal chimeras influences cell fate. Plant J 7:193–202
3. Pyke K, Zubko MK, Day A (2000) Marking cell layers with spectinomycin provides a new tool for monitoring cell fate during leaf development. J Exp Bot 51:1713–1720
4. Szymkowiak EJ, Sussex IM (1996) What chimeras can tell us about plant development. Annu Rev Plant Physiol Plant Mol Biol 47:351–376
5. Furner IJ, Pumfrey JE (1992) Cell fate in the shoot apical meristem of *Arabidopsis thaliana*. Development 115:755–764
6. Furner IJ, Pumfrey JE (1993) Cell fate in the inflorescence meristem and floral buttress of *Arabidopsis thaliana*. Plant J 4:917–931
7. Irish VF, Sussex IM (1992) A fate map of the *Arabidopsis* embryonic shoot apical meristem. Development 115:745–753
8. Lusardi MC, Neuhaus-Url G, Potrykus I et al (1994) An approach towards genetically engineered cell fate mapping in maize using the *Lc* gene as a visible marker: transactivation capacity of *Lc* vectors in differentiated maize cells and microinjection of *Lc* vectors into somatic embryos and shoot apical meristems. Plant J 5:571–582
9. Cepko CL, Ryder E, Austin C et al (1998) Lineage analysis using retroviral vectors. Methods 14:393–406
10. Le Douarin NA (1973) Biological cell labeling technique and its use in experimental embryology. Dev Biol 30:217–222
11. Itasaki N, Bel-Vialar S, Krumlauf R (1999) 'Shocking' developments in chick embryology: electroporation and *in ovo* gene expression. Nat Cell Biol 1:E203–E207
12. Stern CD, Fraser SE (2001) Tracing the lineage of tracing cell lineages. Nat Cell Biol 3:E216–E218
13. Bonnerot C, Nicolas JF (1993) Clonal analysis in the intact mouse embryo by intragenic homologous recombination. C R Acad Sci III 316:1207–1217
14. Kilby NJ, Snaith MR, Murray JA (1993) Site-specific recombinases: tools for genome engineering. Trends Genet 9:413–421
15. Yamauchi Y, Abe K, Mantani A et al (1999) A novel transgenic technique that allows specific marking of the neural crest cell lineage in mice. Dev Biol 212:191–203
16. Sieburth LE, Drews GN, Meyerowitz EM (1998) Non-autonomy of AGAMOUS function in flower development: use of a Cre/loxP method for mosaic analysis in *Arabidopsis*. Development 125:4303–4312
17. Dor Y, Brown J, Martinez OI et al (2004) Adult pancreatic beta cells are formed by self-duplication rather than stem-cell differentiation. Nature 429:41–46

18. Kilby NJ, Davies GJ, Snaith MR et al (1995) FLP recombinase in transgenic plants: constitutive activity in stably transformed tobacco and generation of marked cell clones in *Arabidopsis*. Plant J 8:637–652
19. Kilby NJ, Fyvie MJ, Sessions RA et al (2000) Controlled induction of GUS marked clonal sectors in *Arabidopsis*. J Exp Bot 51:853–863
20. Weigmann K, Cohen SM (1999) Lineage-tracing cells born in different domains along the PD axis of the developing *Drosophila* leg. Development 126:3823–3830
21. Besse F, Pret AM (2003) Apoptosis-mediated cell death within the ovarian polar cell lineage of *Drosophila melanogaster*. Development 130:1017–1027
22. Gorfinkiel N, Sanchez L, Guerrero I (2003) Development of the Drosophila genital disc requires interactions between its segmental primordia. Development 130:295–305
23. Stringham EG, Candido EP (1993) Targeted single-cell induction of gene products in *Caenorhabditis elegans*: a new tool for developmental studies. J Exp Zool 266:227–233
24. Halfon MS, Kose H, Chiba A et al (1997) Targeted gene expression without a tissue-specific promoter: creating mosaic embryos using laser-induced single-cell heat shock. Proc Natl Acad Sci USA 94:6255–6260
25. Harris J, Honigberg L, Robinson N et al (1996) Neuronal cell migration in *C. elegans*: regulation of Hox gene expression and cell position. Development 122:3117–3131
26. Dean C, Sjodin C, Page T et al (1992) Behaviour of the maize transposable element Ac in *Arabidopsis thaliana*. Plant J 2:69–81
27. Balcells L, Sundberg E, Coupland G (1994) A heat-shock promoter fusion to the *Ac* transposase gene drives inducible transposition of a *Ds* element during *Arabidopsis* embryo development. Plant J 5:755–764
28. Reddy GV, Heisler MG, Ehrhardt DW et al (2004) Real-time lineage analysis reveals oriented cell divisions associated with morphogenesis at the shoot apex of *Arabidopsis thaliana*. Development 131:4225–4237
29. Liu M, Yadav RM, Roy-Chowdhury A et al (2010) Automated tracking of stem cell lineages of Arabidopsis shoot apex using local graph matching. Plant J 62:135–147
30. Kurup S, Runions J, Kohler U et al (2005) Marking cell lineages in living tissues. Plant J 42:444–453
31. Freeman J, Sparks CA, West J et al (2011) Temporal and spatial control of transgene expression using a heat-inducible promoter in transgenic wheat. Plant Biotech J 1–9. doi:10.1111/j.1467-7652.2011.00588.x

Chapter 14

Protein Immunolocalization in Maize Tissues

Cristian Forestan, Nicola Carraro, and Serena Varotto

Abstract

The analysis of gene expression at transcript and at protein level is of outstanding importance in plant developmental biology. Proteins can be localized with subcellular resolution by immunolocalization using specific antibodies or generating reporter lines carrying the specific protein fused to a fluorescent protein. Immunolocalization is particularly suitable to confirm the expression pattern of transgenic reporter lines. It also represents a valid alternative, especially for plants such as maize, for which transformation is time consuming and still often unsuccessful, by-passing also some side-effects of fusion proteins. Indeed, the availability of specific antibodies for immunolocalizations and observations of maize tissues under a confocal microscope is a powerful tool for studying protein targeting to different cellular compartments.

In the following chapter we describe the complete procedure for the localization of proteins in different maize tissues both at cellular and sub-cellular level, using polyclonal or monoclonal antibodies. Tissues can be embedded in different substrates, such as paraplast, PEG400 and agarose, depending on the tissue and the desired use: epifluorescence or confocal microscope observations. An additional protocol for the analysis of the nuclear distribution of modified histones in squashed maize root apexes is also presented.

Key words: Immunolocalization, Antibody, Maize, Embedding, Fluorescence and confocal microscopy

1. Introduction

A crucial step in the determination of gene function is the analysis of its expression domains at both the transcript and protein level. More specifically, the localization pattern of a gene product at the tissue, cellular, or subcellular level can be indicative of its function. In many cases, the transcript and the protein coded by a single gene have been shown to co-localize in plant tissues (the so-called cell-autonomous function). However proteins

Ive De Smet (ed.), *Plant Organogenesis: Methods and Protocols*, Methods in Molecular Biology, vol. 959,
DOI 10.1007/978-1-62703-221-6_14, © Springer Science+Business Media New York 2013

encoded by some plant MADS transcription factors can also be detected in wider cellular domains than their transcripts (cell non-autonomous function), as they move between cells through plasmodesmata (1, 2). Within this framework, the localization of proteins at cellular and subcellular level is an important aspect of elucidating their function (3). Furthermore, a single protein can have different accumulation or targeting sites in the cell and consequently it can take part in different biochemical pathways or exert other different functions. In plants, two techniques can be used for protein localization studies at the cellular and subcellular levels: fluorescent proteins (FPs) can be fused in frame with specific gene coding sequences and promoter regions to be used as passenger proteins to assess their in vivo targeting (4). In spite of the availability of fluorescent protein-tagged lines (http://maize.jcvi.org) this approach is not widely diffused in maize developmental biology studies mainly due to the difficult and long process needed to achieve stable transformation. This also requires that primary transformants (T0 generation) undergo several backcrosses to a highly homozygous inbred line (generally B73).

On the other hand immunolocalization assays require the availability of specific antibodies. Maize is considered to be an ancient tetraploid species, which has reverted to diploid behavior. Thus its genome contains multiple copies of the same genes, characterized by sub-functionalization as well as redundancy. The consequences of gene redundancy and high sequence similarity make it difficult to produce antibodies capable of binding to single specific proteins. Nevertheless protein localization through immunolocalization has proven to be a valid tool in assessing protein accumulation sites and in defining targeting of maize-redundant gene products to the plasma membrane or to different cellular compartments (5–7).

For protein immunolocalization maize tissue can be embedded in different substrates, such as paraplast, PEG 400, and agarose. While paraplast is commonly used for observations under a light or fluorescence microscope, depending on the choice of the secondary antibody, PEG 400-embedded tissues are more suitable for observations of plant cells with a confocal microscope, allowing a better resolution at cellular and subcellular level. Agarose embedding can be used for roots or other tissues that require higher hydration and prevents collapsing of the cell walls, still allowing subcellular resolution imaging. Based on the three alternative embedding media, different protocols for immunolocalization are presented. An additional protocol can be used to analyze nuclear histone modification distribution in non-embedded, squashed, root apexes (8).

2. Materials

All solutions are prepared using milliQ water or bidest sterile water. If not otherwise stated, solution and buffers can be stored at room temperature.

2.1. Tissue Fixation

1. 1× phosphate-buffered saline (PBS) buffer: 137 mM NaCl, 2.7 mM KCl, 10 mM Na_2HPO_4, 2 mM KH_2PO_4, pH 7.4.
2. Paraformaldehyde.
3. Formalin–acetic-alcohol (FAA) fixative: 50% absolute ethanol; 5% acetic acid; 4% formaldehyde 40%.

2.2. Tissue Embedding

2.2.1. Embedding in Paraplast

1. 1× PBS.
2. 100% Ethanol.
3. Xylene.
4. Paraplast plus embedding media.

2.2.2. Embedding in PEG 400

1. 1× PBS.
2. Ethanol.
3. PEG 400 distearate.
4. 1-Hexadecanol.

2.2.3. Embedding in Agarose

1. 1× PBS.
2. Electrophoresis grade agarose.

2.3. Root Apex Squashes

1. Pectinase.
2. Cellulase Onozuka R-10.
3. Pectolyase Y-23.

2.4. Section Pretreatment, Antibodies, Labeling, and Mounting

1. Xylene.
2. 100% Ethanol.
3. 1× PBS.
4. Bovine serum albumin (BSA) 100 mg/ml in distilled water (aliquots can be stored at −20°C).
5. 4-Nitro blue tetrazolium chloride (NBT): 75 mg/ml dissolved in 70% dimethylformamide.
6. 5-Bromo-4-chloro-3-indolyl-phosphate, 4-toluidine salt (BCIP): 50 mg/ml dissolved in dimethylformamide.
7. Buffer 1: 100 mM Tris-HCl pH 7.5, 150 mM NaCl.
8. Buffer 2: 100 mM Tris-HCl pH 7.5, 150 mM NaCl, 3% BSA, 0.3% Triton X.

9. Buffer 3: 100 mM Tris-HCl pH 9.5, 100 mM NaCl, 50 mM $MgCl_2$.
10. Buffer 4: 100 mM Tris-HCl pH 9.5, 100 mM NaCl, 50 mM $MgCl_2$, plus 2 μl/ml NBT and 1.5 μl/ml BCIP.
11. Primary antibody.
12. Secondary antibody against the host of the primary antibody conjugated to alkaline phosphatase or to Alexa Fluor® Dyes.
13. Vectashield® mounting medium containing anti-fade reagent, with or without DAPI to stain nuclei.
14. Distyrene-Plasticizer-Xylene (DPX) mounting medium.

2.5. Equipment

1. Fume hood.
2. Oven: temperature range from 37 to 58°C.
3. Slide warmers.
4. Microtome.
5. Vibratome.
6. Metal slide rack to fit in glass boxes.
7. Vertical glass slide boxes.
8. Humid chambers for slide incubation with the antibody.
9. Autoclave sterilized stir bars, spatulas, and glass cylinders.
10. Plastic and glass Pasteur pipettes.
11. Polysine® slides (Menzel-Gläser) and coverslips.
12. Nail polish.
13. Petri dishes or molds.

3. Methods

3.1. Tissue Preparation

Before starting the immunolocalization of proteins it is a good rule to verify the expression and localization of the mRNAs of interest in the different organs and tissues of the plant by RT-PCR and in situ hybridization, respectively. These experiments can give preliminary information on which plant organs, tissues, and developmental stages are ideal for fixation and embedding. A gene transcript, even if expressed at low levels in a few cell layers, can give a good result in in situ hybridization and, of course, in immunolocalization experiments. Knowing the mRNA localization patterns can also be useful to validate the immunolocalization results (see Note 1).

3.2. Fixation

1. Maize tissues are fixed with 4% paraformaldehyde or FAA in 1× PBS solution at 4°C.
2. Fixatives are prepared fresh a few hours before fixation and kept on ice.
3. Paraformaldehyde is dissolved heating the PBS solution to a maximum temperature of 70°C or increasing the pH of the solution to 9–10 by adding NaOH, under the fume hood.
4. The fixative is chilled on ice, filtered through 3MM filter paper, and the volume and pH (final pH = 6.7–7) are adjusted before use by adding H_2SO_4.
5. Collect desired plant materials and quickly submerge the samples into the fixative on ice, preferably in glass vials. Sample size should be about 4 mm × 4 mm to allow proper infiltration of fixative and following solutions.
6. Caryopses must be trimmed at the top or on both sides, depending on the cellular domain of interest, in order to optimize fixative infiltration. Larger maize inflorescences can be cut in half following their longitudinal or transversal axis.
7. Fix the samples in excess solution overnight at 4°C. Fixative must be replaced after a few hours when fixing maize caryopses containing starch. During fixation the sample must be completely submerged. If necessary, place open vials under vacuum to help sample submersion.
8. After fixation discard the sample floating on the solution.

3.3. Embedding

Machines for automatic paraplast or wax embedding can be properly set to speed up the procedures of infiltration. The embedding steps up to the 70% ethanol solution wash are performed in the cold room (see Note 2).

3.3.1. Embedding in Paraplast

1. For embedding in paraplast plus, a graded series of ethanol solutions (from 10 to 90% in water) in 1× PBS buffer are prepared before use and kept on ice. Perform the following dehydration steps keeping samples on ice:
   (a) Two washes of 15 min each in 1× PBS solution.
   (b) Two washes of 10 min 10% ethanol.
   (c) Two washes of 10 min 20% ethanol.
   (d) Two washes of 10 min 30% ethanol.
   (e) Two washes of 10 min 40% ethanol.
   (f) Two washes of 10 min 50% ethanol.
   (g) Two washes of 10 min 70% ethanol.
   (h) Two washes of 15 min 80% ethanol.
   (i) Two washes of 15 min 90% ethanol.
   (j) Two washes of 30 min 100% ethanol.

2. Plant material is transferred into glass tubes for the following washes performed at room temperature:
   (a) 60 min in ethanol:xylene (3:1).
   (b) 60 min in ethanol:xylene (1:1).
   (c) 60 min in ethanol:xylene (1:3).
   (d) 60 min in xylene 100%.
3. There are two ways to proceed with paraplast embedding:
   (a) Xylene (100%) is used to prepare saturated solutions of paraplast with different melting temperature: 37, 42, and 50°C. The solutions are kept at RT and used during the embedding steps described below.
   (b) During embedding, paraplast chips are added directly to the sample submerged in xylene to reach solution saturation at 37, 42, and 50°C. This second procedure is considered the standard.
4. Add ¼ volumes of paraplast chips to the sample in 100% xylene and incubate overnight at room temperature in a fume hood.
5. The next day, place the tubes at 37°C and when the paraplast chips are completely melted add an additional ¼ volume of chips.
6. After 8 h, place the tubes at 42°C for 2 h, then discard ¼ of the volume of the solution, add ¼ volumes of paraplast chips to saturate the solution, and leave the tube overnight at 42°C.
7. During the night put some paraplast chips in molds or Petri dishes to melt.
8. The next day raise the temperature to 48°C, remove ¼ volume of solution, and add new chips.
9. After about 4 h move the tubes to 52°C, remove ¼ of the solution, and add an equal volume of chips.
10. After about 4 h raise the temperature to 58°C and when the chips are completely melted, remove the solution from the tubes and place the plant samples in molds or Petri dishes (kept warm) containing freshly melted paraplast.
11. During the night one or more beakers of paraplast chips should be placed at 58°C to melt and be ready for the subsequent changes.
12. Change the melted paraplast three times a day for 2 days, always leaving the molds open in the oven at 58°C.
13. After the last paraplast change, arrange the tissue sample to desired orientation for sectioning and let the molds reach room temperature.
14. Embedded samples can be safely stored in a sealed container at RT for a long period.

3.3.2. Embedding in Wax (PEG:Hexadecanol, 9:1)

For embedding in wax, PEG 400 distearate is melted at 60°C in a water bath and 1-hexadecanol is added: 1 part hexadecanol+9 parts PEG 400 (w/w). The solution is stirred for 4 h, poured into plastic Petri dishes or into Falcon tubes, and cooled to room temperature. Petri dishes or tubes containing the wax are stored at RT in the dark and can be conserved for several months.

1. Sample dehydration is performed on ice with gentle shaking; ethanol solutions are freshly prepared in 1× PBS buffer and kept on ice:
   (a) 30 min in 30% ethanol.
   (b) 30 min in 50% ethanol.
   (c) 30 min in 70% ethanol.
   (d) 30 min in 90% ethanol.
   (e) 30 min in 100% ethanol at room temperature.
2. Replace the 100% ethanol with 4 ml 100% ethanol and transfer the samples to 42°C for 5 min.
3. Slowly add 4 ml of liquid wax to the samples and leave them at 42°C overnight. The wax-infiltrated samples will be leaning against the bottom of the vial the next morning, while the ethanol phase will lie on top.
4. Pour off the upper ethanol phase and replace wax three times at 2-h intervals. Carefully avoid the samples coming out of the wax; always leave a sufficient amount to cover the tissues.
5. After the final substitution, there should be only pure wax.
6. Properly orient the samples in molds or Petri dishes and let the wax solidify at RT.
7. Store embeddings at 4°C.

3.3.3. Embedding in Agarose (4–8% Agarose in Distilled Water)

1. For agarose embedding, prepare a 4–8% electrophoresis grade agarose solution in distilled water.
2. Quickly pour warm agarose into 15 ml falcon tubes or desired molds and insert the sample (rinsed three times in 1× PBS after fixation) in the desired orientation as soon as the solution becomes viscous and starts to solidify.
3. Let the blocks solidify at room temperature and place them at 4°C for 30–45 min (4). Samples should be stored in sealed humid boxes at 4°C to prevent dehydration of the agarose block and subsequent shrinkage (see Note 3).

3.4. Sectioning

3.4.1. Paraplast-Embedded Samples

1. Prepare the paraplast-embedded samples for sectioning by trimming the blocks into a trapezoid shape and leaving about 2–3 mm of wax around the tissues.

2. It is helpful to keep the samples at 4°C for half an hour before placing the blocks on the microtome. Cooling the cutting block in the freezer for 20–30 min also helps, especially for PEG-embedded samples.
3. Section the region of interest such that the longer of the two parallel faces is at the bottom.
4. Sections of 8 μm are collected in long ribbons when sectioning paraplast-embedded tissues.
5. After sectioning, pre-warm the slides on a heating plate at 42°C. Use Polysine® slides.
6. Several drops of sterile water are applied on the slides that can be marked with a pencil to prevent the marks from disappearing during later incubations.
7. Long ribbons of sections are gently laid to float on the water and allowed to flatten out.
8. Quickly remove the excess water with a sterile Pasteur pipette.
9. Let the slides dry at 40°C overnight on the slide warmer.
10. Store the slides in sealed plastic boxes at room temperature.

3.4.2. Wax-Embedded Samples

1. When sectioning wax-embedded tissues, keep the slides at room temperature (about 22°C).
2. Apply a few drops of sterile water on Polysine® slides.
3. Wax sections of 10–12 μm do not form long ribbons during sectioning. Collect the single sections with a paint brush, place them on the water over the slides, and let the slides dry at room temperature overnight.
4. Store the slides in sealed boxes at room temperature.

3.4.3. Agarose-Embedded Samples

1. When sectioning agarose-embedded tissues, mount the blocks in the desired orientation on a vibratome.
2. Cut 80–100 μm sections. The sections are single and do not form ribbons.
3. Collect sections from the water-filled lower compartment of the vibratome and place into 96-well Elisa plates or Eppendorf tubes filled with PBS.
4. All washes are subsequently performed changing the solutions with a plastic Pasteur pipette.

3.5. Immuno-localization

In immunolocalization experiments the primary antibody binds to the target protein in the plant cells. Subsequently, a secondary antibody that recognizes an epitope carried by the primary antibody is applied. The secondary antibody can be conjugated either to alkaline phosphatases or to a fluorophore molecule. When the secondary antibody is conjugated to alkaline phosphatases, two substrates (NBT and BCIP) are added to the detecting solution in order to produce a purple insoluble precipitate. The purple precipitate can be visualized by light microscopy. The fluorophore can be directly visualized using both fluorescence and confocal microscopes carrying a UV lamp and the appropriate laser lines and filters (see Note 4).

Before the incubations of the tissue sections with the primary and secondary antibodies, slides are pretreated to remove the paraplast/wax from the tissues and maximize the specificity of the reactions. This way the signal of the secondary antibody is enhanced and the formation of excessive background signal is reduced. Agarose-embedded sections do not need treatment to eliminate the embedding matrix as this will detach spontaneously after sectioning (one or two washes in 1× PBS can help to physically detach the matrix, which can be removed with forceps).

1. First, tissue sections are de-waxed and rehydrated through an ethanol series rinse. De-waxing and dehydration are performed at RT in vertical glass boxes with 50 ml of solution: the sections are always completely submerged.
2. Normally all solutions are prepared before starting.
3. The solutions are subsequently poured into the vertical boxes, and the slides are submerged and after incubation briefly dried on a paper towel and moved from one solution to the next fresh one.
4. Tissue sections are usually inspected after rehydration to be sure that the paraplast/wax is completely removed and that the tissues have a well-preserved cell structure.
5. At this time a selection of the best slides should be performed in order to choose the most informative ones with well-preserved tissue sections.
6. Subsequently, the sections are incubated in solutions containing BSA to neutralize possible nonspecific antibody reactions (blocking). Normally blocking is performed with 1–3% BSA in PBS for 45 min at RT and in humid boxes. For thicker sections, 0.2–0.3% (v/v) Tween20 or TritonX can be added to the blocking solution.
7. The incubations with the primary and secondary antibodies are carried out in a humid chamber.
8. The tissue sections are covered with 200–400 μl of 1× PBS containing the primary antibody and left in the humid chamber overnight at 4°C. The incubation time should be determined empirically for each antibody. In our case, positive results were obtained also incubating the sections with the primary antibody for 2–4 h at 37°C.

9. After washing away the primary antibody with appropriate solutions (see below) in vertical glass boxes, the incubation with the secondary antibody is carried out in the humid chamber for 1–3 h at RT. The secondary antibody dilution and incubation time can vary and need to be adjusted for each primary and secondary antibody combination.
10. In order to verify the specificity of primary and secondary antibody reactions, negative or positive controls should be used in the procedure. The best negative control for the primary reaction should imply the use of a mutant that lacks the protein of interest or alternatively a tissue that does not express the target gene product, although immunolocalization might not work as well in different tissues. Negative controls that include no primary antibody (only the secondary antibody is used) should be performed to verify that the secondary antibody does not stain membranes and cell walls aspecifically, thus generating a false positive signal. Positive controls can be performed using an antibody which binds to a broadly expressed protein such as actin or tubulin, to verify that the immunolocalization assay works correctly. The immunolocalization experiments should always be prepared after in situ hybridization experiments or quantitative real-time PCR (qRT-PCR), which allow to determine the transcript expression pattern in the different tissues, so that the expression data at transcript and protein level can be compared. When available, pre-immune serum can be used on some sections instead of primary antibody to verify the presence of an aspecific signal.

3.5.1. Immunolocalizations on Paraplast-Embedded Tissues

All steps are performed at RT in the fume hood.

1. Select the slides and use a pencil to label each slide with the name of the antibodies to be applied. For each solution use a clean vertical glass box.
2. De-wax the slides as follows:
   (a) Two washes for 10 min in 100% xylene.
   (b) Two washes for 10 min in 100% ethanol.
   (c) 5 min in 70% ethanol.
   (d) 5 min in 50% ethanol.
   (e) 5 min in 30% ethanol.
   (f) 5 min in PBS.
3. After de-waxing the slides are equilibrated in Buffer 1, pre-treated with Buffer 2 (a blocking buffer containing 1% BSA), in vertical glass boxes and then incubated with the primary antibody diluted in Buffer 2 in a humid chamber, as follows:
   (a) Three washes for 5 min in Buffer 1.

(b) 60 min in Buffer 2.

4. Briefly dry the slides keeping them in a vertical position on a paper towel.
5. Prepare the humid chamber and apply 200 μl of Buffer 2 containing the primary antibody to the slides. Antibody dilution has to be empirically determined but in our experience 1:200–1:500 v/v is a good starting point.
6. After incubation wash the slides in vertical glass boxes three times for 5 min in Buffer 1.
7. Again, briefly dry the slides keeping them in a vertical position on a paper towel.
8. Prepare the humid chamber and apply 200 μl of Buffer 2 containing the secondary antibody conjugated to alkaline phosphatase to the slides.
9. After incubation wash the slides in Buffer 1 in vertical glass boxes and submerge them in Buffer 3 as follows:
   (a) Three washes for 5 min in Buffer 1.
   (b) 10 min in Buffer 3.
10. Transfer the slides in Buffer 4 and let the colorimetric reaction go on for at least 30 min.
11. To avoid excessive color formation, monitor the staining on slides used as negative control (no primary antibody added).
12. Block the reaction in water for 5 min.
13. Air-dry slides in a fume hood.
14. Add DPX mounting medium and cover with coverslip.

3.5.2. Immunolocalizations on Wax-Embedded Tissues

All de-waxing steps are performed at RT in vertical glass boxes.

1. Select the slides and use a pencil to label each slide with the name of the antibodies to be applied.
   (a) Three washes for 20 min in 97% ethanol in 1× PBS.
   (b) 20 min 90% ethanol in 1× PBS.
   (c) 20 min 50% ethanol in 1× PBS.
   (d) Two washes for 5 min in 1× PBS.
2. After de-waxing, set up the humid chamber and prepare the antibody solution adding 10 μl of BSA from the stock solution (100 mg/ml) to 1 ml of 1× PBS.
3. Then add the necessary amount of antibody (typical dilutions range from 1:200 to 1:500, v/v).
4. It is generally recommended to perform an incubation step with a blocking buffer before the incubation with the primary antibody. The necessity of this step has to be determined

empirically for each antibody. Blocking can be performed for 1 h at room temperature with 1% BSA in 1× PBS.

5. Briefly dry the slides keeping them in a vertical position on a paper towel and apply 200–400 μl of primary antibody solution to the slides in the humid chamber. The incubation in the primary antibody solution is performed overnight at 4°C or 4–5 h at RT: in our hands, these conditions produced the same results.
6. After the incubation with the primary antibody, wash the slides twice for 5 min in 1× PBS.
7. Prepare the solution with the secondary antibody as follows: Add 10 μl of BSA from the stock solution (100 mg/ml) and the necessary amount of secondary antibody to 1 ml of 1× PBS.
8. Apply to the slides 200–400 μl of secondary antibody solution in the humid chamber and incubate for 1-3 hours at RT. In our hands, the Alexa Fluor® 568 and 488 (2 mg/ml)-conjugated secondary antibodies diluted 1:400 gave the best results. The choice of the secondary antibody mainly depends on the available filters or lasers in the fluorescence or confocal microscope.
9. Rinse the slides twice for 5 min with 1x PBS.
10. Briefly dry the slides keeping them in a vertical position on a paper towel.
11. Apply to the section two drops of Vectashield® anti-fade mounting medium containing DAPI if staining of the nucleus is desired.
12. Finally, cover the slides with coverslips, seal them with nail polish, and store at 4°C.

3.5.3. Immunolocalizations on Agarose-Embedded Tissues

1. The agarose is typically removed from the sections as soon as they are cut by means of mechanical vibrations. They can be collected from the lower compartment using fine point forceps, a brush, or a plastic Pasteur pipette and dropped into wells filled with 1× PBS.
2. All washes and reactions are performed in 96-well Elisa plates, Eppendorf tubes, or depression or well slides.
3. Blocking should be performed with BSA 4–6%, 0.2% (v/v) Tween 20 in 1× PBS for 1 h at RT (alternatively use 100 mM glycine + 0.2% (v/v) Tween 20 for 30 min and then the BSA + Tween 20 solution).
4. Rinse twice in 1× PBS changing the solution in the wells, add the primary antibody to the desired dilution, and leave overnight at 4°C slowly shaking. Alternatively the reaction can be performed for 2–3 h at 37°C.

5. Rinse twice with 1× PBS, add the secondary antibody (dilution 1:400 for Alexa Fluor® 568, 2 mg/ml), and incubate for 1 h at RT.
6. Rinse twice with 1× PBS.
7. Carefully transfer the sections to Polysine slides or leave in depression or well slides.
8. If using Polysine slides add glycerol solution (5% glycerol in water) or Vectashield® medium, seal slide covers with nail polish, let them dry, and observe with the fluorescent or confocal microscope.
9. Slides can be stored in the dark at 4°C for several months. Alternatively samples can be conserved for a short time in the wells filled with 1× PBS buffer.

3.6. Immunolocalizations on Squashed Root Apexes

Root apex squashes have long been used for cytological studies on chromosomes, often accompanied by colchicines treatments, but also for subcellular compartment observation and immunolocalization assays. This technique was successfully employed to visualize the nuclear distribution of histone modifications using specific antibodies against single epigenetic histone marks.

1. Germinate maize seeds and grow them vertically in wet, rolled paper towels for 2–3 days at 25°C.
2. Root apexes (about 2 mm long) are cut out and fixed in 4% paraformaldehyde in 1× PBS for 1 h on ice.
3. Wash the apexes three times for 15 min each in 1× PBS to remove paraformaldehyde residues.
4. Incubate them for 30 min at RT in a mixture of cell wall digesting enzymes (2.5%, w/v, pectinase; 2.5%, w/v, Cellulase Onozuka R-10; and 2.5%, w/v, Pectolyase Y-23) in 1× PBS to permeabilize the cell walls.
5. Wash again in 1× PBS for 15 min and using a forceps or a plastic Pasteur pipette transfer the apexes onto Polysine slides.
6. Cover with coverslips and gently squash the apexes applying moderate thumb pressure to flatten the cells.
7. Immediately freeze the slides in liquid nitrogen and remove the coverslip by flicking it.
8. Root cells are now ready for the antibody incubation with the primary antibody.
9. Nuclei are usually blocked in 8% BSA–0.1% Triton X-100 in 1× PBS, washed with 1× PBS and incubated with the appropriate primary antibody in a humid chamber overnight at 4°C.
10. After incubation with the primary antibody, wash slides in 1× PBS and incubate with a secondary antibody: goat-anti-rabbit-FITC (diluted 1:50; Sigma-Aldrich) or goat-anti-rabbit-rhodamine (1:100; Upstate-Millipore).

11. Counterstain nuclei with DAPI contained in the Vectashield® mounting medium.
12. Finally, cover the slides with a coverslip, seal them with nail polish, and store at 4°C.

3.7. Slide Observation

Slides should be observed immediately after mounting, in order to preserve the best signal (see Note 5). Slides properly stored can retain the signal for weeks but a partial loss of signal might occur, thus compromising the most informative slides. Avoiding repeated and extended exposure of the tissues to the confocal microscope lasers also prevents bleaching of the fluorophore and signal loss. The confocal microscopy settings (pinhole, gain, objective, scale) should be recorded in order to be able to repeat the observations and compare across successive experiments.

4. Notes

1. In the immunolocalization experiments performed on paraffin-embedded tissues the tissues can be used to obtain sections both for in situ hybridization and immunolocalization (as long as they have been kept free of RNAses). This is very important when the transcript and the protein are localized for the first time in a tissue and this first experiment can give important information about the mRNA and protein spatial and temporal expression, before setting up an immunolocalization with wax-embedded tissues. The immunolocalization procedure with paraplast-embedded tissues is suitable for sections that will be observed with the light or the fluorescence microscope. These sections are not suitable for observation under the confocal microscope since they present an excess of aspecific binding signal due to the nature of the paraplast matrix.
2. During dehydration of the samples for paraplast embedding the procedure can be safely stopped after the 70% ethanol washes and the samples can be stored for a long time at –20°C in ethanol solution. When resuming the dehydration series, put the samples on ice for 15–30 min, wash once again in 70% ethanol, and then continue with the protocol.
3. Sections obtained using the vibratome on agarose-embedded organs should be at least 80 μm thick and typically, one in every two sections is lost in the process. This type of sectioning is thus suitable only for thicker organs, not for fine structures such as the shoot apical meristem (SAM). Agarose embedding is particularly useful for root tissues in which the cell walls are prone to collapsing if embedded in harder and dryer matrices such as paraplast and wax (Fig. 1).

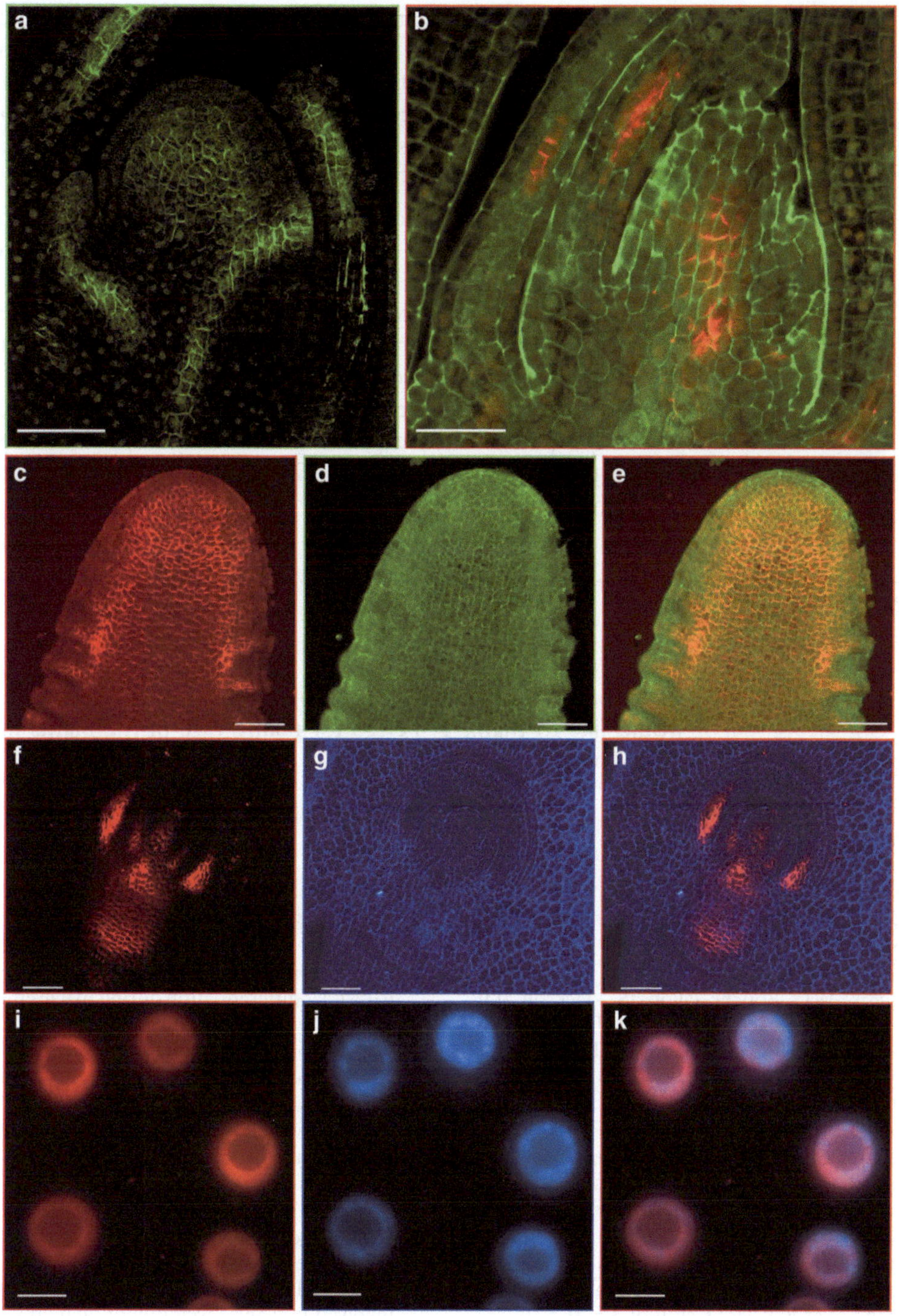

Fig. 1. Immunolocalization in maize tissues and nuclei. An anti-AtPIN1 monoclonal antibody labels plasma-membrane polarly localized proteins in maize shoot apical meristem (**a** and **b**), female inflorescences (**c** and **e**), and embryos (**f** and **h**). Panel **b** represents a confocal image in which green signal corresponds to the Alexa488 (Invitrogen) secondary antibody. Fluorescence signal is coded in *red* in images **b**, **c**, **e**, **f**, and **h** (Alexa 568 secondary antibody—Invitrogen), while *green* and *blue* represent tissue autofluorescence collected with GFP and DAPI filters, respectively. Panels **b**–**h** show epifluorescence images acquired with a Leica DC300F camera. Immunostaining with H3K4me2 (histone 3, dimethylated in lysine 4) antibody shows the distribution of this histone modification in maize nuclei (**i**). Nuclei DAPI staining is reported in panel **j** while a merged image is shown in **k**. Scale bars = 100 μm in **a**–**h**, 10 μm in **i**–**k**.

4. Maize tissues are characterized by strong autofluorescence in the yellow-green part of the light spectrum when excited with UV light: this characteristic must be taken into consideration when choosing the secondary antibody. It is suggested to use different fluorophores and choose the one giving the best signal/background results.
5. The results of immunolocalization experiments heavily depend on the quality of the tissue sections at the end of the procedure. Especially wide sections (i.e., maize inflorescences and maize kernels) can break and fall off the slides. When the tissues are damaged during the procedure, they can fold on the slides and become a sort of physical trap for the solutions and antibodies. In this case it is hard to correctly interpret the localization pattern of the proteins. It is very important to start with an excess of slides and discard the damaged sections at each step of the procedure.

Acknowledgments

This work was funded by PRIN Programmes of MIUR. The author would like to thank T. Pengo for taking care of the plants in the greenhouse.

References

1. Kim JY, Yuan Z, Jackson D (2003) Developmental regulation and significance of KNOX protein trafficking in arabidopsis. Development 130:4351–4362
2. Lucas WJ, Bouche-Pillon S, Jackson DP, Nguyen L, Baker L, Ding B, Hake S (1995) Selective trafficking of KNOTTED1 homeodomain protein and its mRNA through plasmodesmata. Science 270:1980–1983
3. Millar AH, Carrie C, Pogson B, Whelan J (2009) Exploring the function-location nexus: using multiple lines of evidence in defining the subcellular location of plant proteins. Plant Cell 21:1625–1631
4. Brandizzi F, Fricker M, Hawes C (2002) A greener world: the revolution in plant bioimaging. Nat Rev Mol Cell Biol 3:520–530
5. Varotto S, Locatelli S, Canova S, Pipal A, Motto M, Rossi V (2003) Expression profile and cellular localization of maize Rpd3-type histone deacetylases during plant development. Plant Physiol 133:606–617
6. Forestan C, Meda S, Varotto S (2010) ZmPIN1-mediated auxin transport is related to cellular differentiation during maize embryogenesis and endosperm development. Plant Physiol 152:1373–1390
7. Carraro N, Forestan C, Canova S, Traas J, Varotto S (2006) ZmPIN1a and ZmPIN1b encode two novel putative candidates for polar auxin transport and plant architecture determination of maize. Plant Physiol 142: 254–264
8. Rossi V, Locatelli S, Varotto S, Donn G, Pirona R, Henderson DA, Hartings H, Motto M (2007) Maize histone deacetylase hda101 is involved in plant development, gene transcription, and sequence-specific modulation of histone modification of genes and repeats. Plant Cell 19:1145–1162

Chapter 15

Auxin Immunolocalization in Plant Tissues

Cristian Forestan and Serena Varotto

Abstract

The major naturally occurring auxin, indol-3 acetic acid (IAA), coordinates many growth and differentiation processes by modulating gene expression during plant development. The sites of IAA biosynthesis and its polar transport (PAT) routes determine auxin accumulation and distribution during growth. From many studies on the model plant *Arabidopsis thaliana* over the last years, it has become evident that the expression and sub-cellular localization of multiple transport proteins are required to initiate and maintain directional auxin flows within plant organs and tissues, creating the auxin concentration gradients that regulate plant development. For this reason, the understanding of auxin dependent pattern formation also relies on the possibility to directly visualize auxin concentration and distribution in the tissues. The production and isolation of antibodies highly specific for IAA provide the means to detect and localize free IAA in different plant species during tissues differentiation and organ development. The immunolocalization protocol presented here uses a monoclonal anti-IAA specific antibody that can be used to visualize auxin accumulation in different organs and tissues during plant development. We successfully used this protocol to determine IAA maxima during kernel and inflorescence development in maize, highlighting also alterations in auxin accumulation patterns in a mutant with reduced auxin accumulation capacity and in plants treated with an inhibitor of auxin transport.

Key words: Indole-3-acetic acid (IAA), Auxin accumulation sites, Anti-IAA antibodies, Tissue sections

1. Introduction

Auxin and particularly its asymmetric distribution in plant cells and tissues regulate basic growth processes, such as cell division and elongation, and control a plethora of plant developmental aspects from embryogenesis to flowering. The current knowledge on auxin metabolism, transport, and signaling is mainly based on the research performed on the model species *Arabidopsis thaliana* in the last 20 years. In this model, many genes required for auxin transport

Ive De Smet (ed.), *Plant Organogenesis: Methods and Protocols*, Methods in Molecular Biology, vol. 959,
DOI 10.1007/978-1-62703-221-6_15, © Springer Science+Business Media New York 2013

and signaling have been identified using molecular genetics approaches and mutant characterization (1, 2). However, in spite of the great progress made in understanding how auxin is transported from cell to cell and its signal is perceived in the plant tissues, little is still known on the molecular mechanisms and the sites of auxin biosynthesis in plants (3).

In the last decade the need to transfer from model to crop species gave a new impulse to auxin research in monocots and maize in particular. So far the molecular mechanisms controlling auxin biosynthesis, transport, and signal transduction appear to be conserved between monocots and dicots, although differences and peculiarities, especially for what it concerns gene copy number and subfunctionalization in different tissues, have been demonstrated in monocots (4).

Along with the identification and characterization of auxin efflux carriers belonging to the *PIN-FORMED* (*PIN*) gene family of maize (5, 6) we became interested in understanding auxin-pattern formation in this species. There are essentially two ways for localizing auxin in plant tissues: the visualization of auxin-responsive promoter elements, the most used of which is the DR5 reporter (7), and the use of anti-indole-3 acetic acid (IAA is the major naturally occurring auxin)-specific antibodies. Auxin-responsive promoters and particularly DR5 have demonstrated to be good tools to visualize auxin signaling activities but they cannot localize both extra- and intracellular auxin (8). On the other hand, auxin antibodies might lack monospecificity and their use needs proper negative and positive control adoption during immunolocalization experiments.

Here we present a protocol for immunolocalization of IAA in different maize tissues with good specificity. To check the specificity of the anti-IAA antibody in maize kernels, we used tissue sections both from plants treated with NPA, an auxin transport inhibitor, and *defective endosperm-B18* (*de*-B18*) mutant defective in auxin accumulation (5, 9).

2. Materials

All solutions are prepared using milliQ water or bidest sterile water. Solutions and glassware are usually sterilized in autoclave for 20 min at 120°C. If not otherwise stated, solution and buffers can be stored at room temperature.

2.1. Tissue Fixation and Embedding

1. 1× phosphate-buffered saline (PBS) buffer: 137 mM NaCl, 2.7 mM KCl, 10 mM Na_2HPO_4, 2 mM KH_2PO_4, pH 7.4.
2. *N*-Ethyl-*N'*-(3-dimethylaminopropyl) carbodiimide hydrochloride (EDAC)—E1769, Sigma.
3. Paraformaldehyde.

2.2. Tissue Embedding

1. 1× PBS.
2. 100% Ethanol.
3. Xylene.
4. Paraplast plus embedding media.

2.3. Section Pretreatment, Antibody Labeling and Mounting

1. Xylene.
2. 100% Ethanol.
3. 1× PBS.
4. Bovine serum albumin (BSA).
5. 4-Nitro blue tetrazolium chloride (NBT): 75 mg/ml dissolved in 70% dimethylformamide.
6. 5-Bromo-4-chloro-3-indolyl-phosphate, 4-toluidine salt (BCIP): 50 mg/ml dissolved in dimethylformamide.
7. Fast Red TR/Naphtol AS—MX.
8. Blocking solution: 0.1% Tween 20, 1.5% glycine, 5% BSA, in 1× PBS, pH 7.
9. Washing solution 1: 0.88% NaCl, 0.1% Tween 20, 0.8% BSA, in 1× PBS.
10. Washing solution 2: 2.9% NaCl, 0.1% Tween 20, 0.1% BSA, in 1× PBS.
11. Washing solution 3: 0.8% BSA in 1× PBS.
12. Buffer 4: 100 mM Tris-HCl pH 9.5, 100 mM NaCl, 50 mM $MgCl_2$.
13. Buffer 5: 100 mM Tris-HCl pH 9.5, 100 mM NaCl, 50 mM $MgCl_2$, plus 2 μl/ml NBT and 1.5 μl/ml BCIP.
14. Primary antibody: IAA phytodetek monoclonal antibody—PDM09346-0096 (AGDIA—Biofords).
15. Secondary antibody, anti-mouse alkaline phosphatase-conjugated, IgG-AP.
16. Distyrene-Plasticizer-Xylene (DPX) mounting medium.
17. CC/Mount, mounting medium.

2.4. Equipment

1. Fume hood.
2. Oven: Temperature range from 37 to 58°C.
3. Slide warmers.
4. Microtome.
5. Metal slide rack to fit in glass boxes.
6. Vertical glass slide boxes.
7. Humid chambers for slide incubation with the antibody.

8. Stir bars, spatulas, and glass cylinders sterilized in autoclave.
9. Plastic and glass Pasteur pipettes.
10. Polysine® slides and coverslips.

3. Methods

3.1. Tissue Preparation

For immunolocalization of auxin, we omitted the pre-fixation step in EDAC in at least ten samples for each tissue analyzed. This not prefixed material allowed us to test the effect of pre-fixation in each IAA immunolocalization. As control for maize kernels we used the de*-B18 mutant (9), which was shown to contain very low levels of free and bound IAA, particularly in the endosperm. We also set up NPA treatments on maize female inflorescences, to check IAA localization in tissue in which polar auxin transport was altered (see Note 3).

3.2. Fixation

Maize tissues are prefixed in 3% EDAC in 1× PBS and post-fixed in 4% paraformaldehyde in 1× PBS. Fixatives are prepared fresh a few hours before fixation and kept on ice.

1. A stock of sterile solutions of 10× PBS are stored at room temperature and mixed to prepare a fresh appropriate volume of buffer 1× final concentration at pH 7.4.
2. Paraformaldehyde is dissolved in PBS 1× at about 70°C, under the fume hood.
3. The fixative is chilled on ice, filtered through 3MM filter paper, and the volume and pH are if necessary adjusted with H_2SO_4 before use.
4. Collect desired plant materials and quickly submerge the samples in EDAC solution (3% in PBS 1×) on ice for 1 h, preferably in glass vials.
5. After pre-fixation, fix the prefixed samples and fresh samples (no EDAC pre-fixation) in excess 4% paraformaldehyde solution overnight at 4°C. During fixation the sample must be completely submerged and only the samples lying at the bottom of the vials must be used for embedding.
6. Sample size should be about 3 mm × 3 mm to allow proper infiltration of pre-fixative and fixative. Caryopses are usually trimmed on both sides to optimize infiltration of fixative solutions into the kernel. It is necessary to empirically determine the fixation protocol for each maize tissue and for different growing conditions. A gentle vacuum incubation can help the fixative to reach the deeper tissues, but tissue over-fixation must be avoided: cells will look shrunken after embedding

and/or dewaxing and morphology will not be preserved. Over-fixed tissues cannot be used for immunolocalization (see Note 1).

3.3. Embedding

Machines for automatic paraplast embedding can be used for infiltration.

1. For embedding in paraplast plus a graded series of ethanol solutions (from 10 to 70%) in phosphate buffer pH 7.2 are freshly prepared before use.
2. Xylene (100%) is used to prepare saturated solutions of paraplast with different melting temperature: 37, 42, and 50°C. Paraplast is melted just before use in an oven at 58°C, avoiding higher temperature heating.
3. The embedding steps till 100% ethanol solution are done at 4°C and ethanol graded series solutions are prepared in PBS 1×. Perform the following dehydration steps:
   (a) Two washes of 15 min each in 1× PBS buffer pH 7.4.
   (b) Two washes of 10 min 10% ethanol.
   (c) Two washes of 10 min 20% ethanol.
   (d) Two washes of 10 min 30% ethanol.
   (e) Two washes of 10 min 40% ethanol.
   (f) Two washes of 10 min 50% ethanol.
   (g) Two washes of 10 min 70% ethanol.
   (h) Two washes of 15 min 80% ethanol.
   (i) Two washes of 15 min 90% ethanol.
   (j) Two washes of 30 min 100% ethanol.
4. Samples are transferred into glass tubes for the following washes performed at room temperature:
   (a) 60 min in ethanol:xylene (3:1).
   (b) 60 min in ethanol:xylene (1:1).
   (c) 60 min in ethanol:xylene (1:3).
   (d) 60 min in xylene 100%.
5. Add ¼ volume of paraplast chips to the tubes and incubate overnight.
6. Place the tubes at 37°C and when the paraplast chips are completely melted again add ¼ volume of chips.
7. After 4 h, place the tubes at 42°C for 2 h.
8. Discard ¼ volume of solution and add ¼ volume of paraplast chips to saturate the solution and leave the tube overnight at 42°C.
9. During the night melt the necessary amount of chips at 58°C.

10. The next day raise the temperature to 48°C, remove ¼ volume of solution, and add new chips.
11. After about 4 h move the tubes to 52°C, remove ¼ of the solution, and add an equal volume of chips.
12. After about 4 h raise the temperature to 58°C and when the chips are completely melted, remove the solution from the tubes and pour the plant materials into molds containing fresh melted paraplast.
13. Change the wax three times a day for 2 days, always leaving the molds open in the oven at 58°C. After the last wax change, arrange the tissue sample to obtain desired orientation for sectioning and let the molds reach room temperature. Embedded samples can be stored in a sealed container at RT for a long period.

3.4. Sectioning

1. Prepare the samples for sectioning by trimming the blocks into a trapezoid shape and leaving about 2–3 mm of wax around the tissues. It is helpful to keep the samples at 4°C for 30 min before placing the blocks on the microtome.
2. Section the region of interest such that the longer of the two parallel faces is at the bottom. Sections of 8–10 μm are collected in ribbons.
3. Prewarm the slides on the slide warmer at 42°C. Slides used are Polysine® (Menzel-Gläser).
4. Apply several drops of sterile water on the slides that can be marked with pencil, to avoid dissolving of marks during later incubations.
5. Long ribbons of sections are gently floated in the water and let flatten out.
6. Quickly remove the excess water with a sterile Pasteur pipette.
7. Let the slides dry overnight at 37°C on the slide warmer.
8. Store the slides in sealed plastic boxes at room temperature.

3.5. Immunolocalization

In IAA immunolocalization experiments, the primary antibody recognizes the free IAA that accumulates in plant cells. Subsequently, a secondary antibody conjugated to alkaline phosphatase is applied that recognizes and binds the primary antibody. Two different substrates NBT and BCIP or Fast red can be added to the detecting solution to produce either a purple or red precipitate by alkaline phosphatase. The precipitate can be visualized under the light microscope. Before the incubations of the tissue sections with the antibodies, paraplast is removed from the tissues and sections are treated to facilitate the antibody reactions and avoid excessive background.

1. Tissue sections are de-waxed and rehydrated through an ethanol series at RT in vertical glass boxes with 50 ml of fresh prepared solution: make sure that all the sections are completely submerged during the treatments.
2. After each incubation step, slides are briefly dried in paper towels and moved from one solution to the next fresh one.
3. After rehydration, check the slides to ascertain that paraplast has been completely removed from the tissues and that they have a well-preserved cell morphology.
4. Set up the humid chamber for incubation with the primary and secondary antibodies.
5. Select the best slides and incubate them firstly with a blocking solution and then with the anti-IAA antibody, in a solution containing BSA to neutralize possible nonspecific antibody reactions.
6. In the humid chamber, the tissue sections are covered with 400 μl of solution containing the primary antibody and left for 4–5 h at room temperature.
7. After incubation with the anti-IAA antibody, wash the primary antibody off the sections with washing solutions (see below) in vertical glass boxes.
8. The incubation with the secondary antibody conjugated to alkaline phosphatase is done in the humid chamber for 1–3 h at RT. The secondary antibody dilution and incubation time need to be adjusted for each of the maize tissues.
9. The specificity of primary and secondary antibodies is verified using either negative or positive controls in the procedure. Sections that are not prefixed with EDAC, treated with the primary or secondary antibodies only, are always included in the experiments to check the background staining (see Note 2).

3.5.1. Immuno-Procedure

All steps are performed at RT.

1. Select the slides and using a pencil label each with the antibodies to be applied. For each solution use a clean vertical glass box.
2. Dewax the slides as follows:
   (a) Two washes for 15 min in 100% xylene.
   (b) Two washes for 20 min in 100% ethanol.
   (c) 20 min in 90% ethanol in 1× PBS.
   (d) 20 min in 50% ethanol in 1× PBS.
   (e) Two washes for 5 min in 1× PBS.
3. Slides can be stored longer in 1× PBS if needed.

4. After dewaxing, check the slides at microscope and discard the ones in which sections are damaged.
5. In the meantime set the humid chamber and prepare the anti-IAA antibody solution.
6. Before the antibody treatment in the humid chamber the sections are covered with 400 μl/slides of blocking solution for 1 h and then washed as follows:
   (a) 5 min in Washing solution 1.
   (b) 5 min in 1× PBS 0.8% BSA (Washing solution 3).
7. After washing, sections are treated with a mouse anti-IAA monoclonal antibody AGDIA 0.1 mg/ml in 1× PBS 0.8% BSA for 5 h in the humid chamber at room temperature.
8. Wash the primary antibody off the slides with the following washes:
   (a) Two washes for 10 min in Washing solution 2.
   (b) 10 min in Washing solution 1.
   (c) Overnight in 1× PBS 0.8% BSA (Washing solution 3) at 4°C.
9. The next morning, let the slides warm for about half an hour at room temperature.
10. Apply 200 μl of secondary antibody solution to the slides and incubate for 3 h in the humid chamber at room temperature. We use a Donkey anti-Mouse alkaline phosphatase conjugated secondary antibody (IgG-AP-SC2097 Santa Cruz).
11. After the incubation with the secondary antibody, the slides are washed as follows:
    (a) Two washes for 15 min in Washing solution 1.
    (b) 15 min in water.
    (c) 5 min in Buffer 4.
12. Depending on the detection solution, slides are incubated in Buffer 5 containing NBT and BCIP or in a fresh solution containing Fast red TR/Naphtol AS-MX, prepared following the manufacturer's instructions.
13. The incubation time with NBT/BCIP lasts 30 min maximum, but the coloring reaction can be stopped earlier and must be determined empirically for each tissue. Incubation with Fast red lasts 15 min maximum, but can be stopped earlier as well, depending on the tissues. To determine the right time conditions for secondary antibody detection it is important to compare the staining between positive and negative control sections (see Note 3). This comparison monitors the staining due to unspecific binding of antibody, which depends mainly on the tissues and secondary antibody along with the procedure used.

14. For both detection methods, the slides are incubated in water for 5 min to block the reaction and then air-dried in a fume hood (see Note 4).

3.6. Slide Observation

Before observations, slides are mounted and sealed.

1. After detection with NBT/BCIP slides are mounted applying two drops of DPX mounting medium and gently placing the coverslip to avoid air bubble formation.
2. When Fast red is used for detection two drops of CC/Mount and coverslips are applied on the sections, again avoiding the formation of air bubbles.
3. Let the slide dry before observation under the light microscope. Properly stored slides can retain the signal for weeks.

4. Notes

1. A correct fixation of the tissues is the first crucial step in IAA immunolocalization. The EDAC pre-fixation is essential when using AGDIA anti-IAA-antibody. EDAC cross-links the carboxyl group of IAA to structural proteins in the plant tissues creating the epitope recognized by the AGDIA anti-IAA monoclonal antibody (10). Sections that are not prefixed with EDAC are good negative controls to monitor staining background and must be produced for each tissue of interest. We observed no color comparable with the prefixed maize section in our negative controls confirming both that pre-fixation with EDAC is essential for free IAA detection using this antibody and that it does not recognize other epitopes in our maize tissue sections (5). However, other authors reported that this cross-linking agent used to immobilize the auxin has drastic effects on F-actin in cells of maize root apices and might thus alter the localization of other antigens (8).
2. Confirmation of IAA localization data derived from immunolocalizations can be obtained using different and complementary strategies. Obviously, anti-IAA immunolocalization must be repeated several times in independent experiments to obtain a sufficient number of tissue sections and observations to elaborate an auxin accumulation pattern for each specific tissue. Furthermore, the use of either auxin transport inhibitors for treating the wild-type plants or mutants having altered auxin accumulation capacity can provide suitable tools to assess the results of anti-IAA immunolocalizations in wild-type tissue sections. Positive controls can be obtained by adding exogenous IAA to the tissues of growing plant in the presence of IAA. In

our lab, to confirm auxin localization data in maize wild-type kernels, IAA accumulation patterns were determined in kernel sections of plants treated with *N*-1-naphthylphthalamic acid (NPA), at two different concentrations 80 and 120 μM, via daily watering for 2 weeks starting from the tasseling stage (for the precise protocols of NPA treatments see (5, 11)). Anti-IAA immunolocalizations in NPA-treated plants showed that inhibition of auxin polar transport during kernel development determines a disruption of IAA accumulation pattern and particularly affects the formation of correct auxin gradients in maize embryo and endosperm. In NPA-treated plants, we observed that the defects in auxin accumulation concern those tissues where auxins are accumulated via polar transport, indicating that tissue sections obtained from NPA-treated plants are really good positive and negative controls for anti-IAA immunolocalizations.

The availability of a characterized maize mutant de*-B18 (9) recently mapped on maize chromosome 10 (12), whose kernels revealed levels of free and bound IAA at least 15 times lower than in the wild-type endosperm, allowed us to check the effect of auxin content reduction on IAA distribution by immunolocalization on kernel sections. Also in this case, immunolocalizations with the anti-IAA antibody showed alteration in auxin accumulation in the mutant compared to the wild type in specific cellular domains of the endosperm, indicating a correlation between auxin content and accumulation site and corroborating the efficiency of anti-IAA antibody in visualizing auxin accumulation sites.

3. The availability of a DR5 auxin-responsive promoter reporter line allows the auxin accumulation patterns of the line to be compared with the results obtained with anti-IAA immunolocalization experiments. In maize we observed very similar auxin accumulation patterns in male and female inflorescences comparing the DR5 reporter line from Jackson's lab (13) with the immunolocalization on inflorescence sections done in our lab (authors' unpublished results). Nevertheless, the DR5 promoter, being activated by brassinolides also in the absence of auxin level changes, is not completely specific for IAA (14). Similarly, other auxin-responsive promoters such as BA3 and GH3 seem to respond to signals different from auxin. All these IAA-responsive promoters, reflecting the activation of different signaling cascades, show different, often contrasting, expression patterns (8, 15). Taken together these observations suggest that a comparison of the results obtained from auxin reporter lines and the immunolocalizations with the anti-IAA-specific antibody is the best strategy to define auxin accumulation patterns in different maize tissues.

4. As in all experiments of immunolocalization, at the end of the procedure the quality of the sections is fundamental for a correct interpretation of the IAA localization pattern.

   Always start with an excess of slides and discard those bearing damaged sections, which can interfere with the interpretation of the results.

References

1. Zazimalova E, Murphy AS, Yang H, Hoyerova K, Hosek P (2010) Auxin transporters—why so many? Cold Spring Harb Perspect Biol 2:a001552
2. Vieten A, Sauer M, Brewer PB, Friml J (2007) Molecular and cellular aspects of auxin-transport-mediated development. Trends Plant Sci 12:160–168
3. Zhao Y (2010) Auxin biosynthesis and its role in plant development. Annu Rev Plant Biol 61:49–64
4. McSteen P (2010) Auxin and monocot development. Cold Spring Harb Perspect Biol 2:a001479
5. Forestan C, Meda S, Varotto S (2010) ZmPIN1-mediated auxin transport is related to cellular differentiation during maize embryogenesis and endosperm development. Plant Physiol 152:1373–1390
6. Carraro N, Forestan C, Canova S, Traas J, Varotto S (2006) ZmPIN1a and ZmPIN1b encode two novel putative candidates for polar auxin transport and plant architecture determination of maize. Plant Physiol 142: 254–264
7. Ulmasov T, Murfett J, Hagen G, Guilfoyle TJ (1997) Aux/IAA proteins repress expression of reporter genes containing natural and highly active synthetic auxin response elements. Plant Cell 9:1963–1971
8. Schlicht M, Strnad M, Scanlon MJ, Mancuso S, Hochholdinger F, Palme K, Volkmann D, Menzel D, Baluska F (2006) Auxin immunolocalization implicates vesicular neurotransmitter-like mode of polar auxin transport in root apices. Plant Sign Behav 1:122–133
9. Torti G, Manzocchi L, Salamini F (1986) Free and bound indole-acetic acid is low in the endosperm of the maize mutant *defective endosperm-B18*. Theor Appl Genet 72:602–605
10. Avsian-Kretchmer O, Cheng JC, Chen L, Moctezuma E, Sung ZR (2002) Indole acetic acid distribution coincides with vascular differentiation pattern during arabidopsis leaf ontogeny. Plant Physiol 130:199–209
11. Wu X, McSteen P (2007) The role of auxin transport during inflorescence development in maize (zea mays, poaceae). Am J Bot 94:1745–1755
12. Pasini L, Stile M, Puja E, Valsecchi R, Francia P, Carletti G, Salamini F, Marocco A (2008) The integration of mutant loci affecting maize endosperm development in a dense genetic map using an AFLP-based procedure. Mol Breed 22:527–541
13. Gallavotti A, Yang Y, Schmidt RJ, Jackson D (2008) The relationship between auxin transport and maize branching. Plant Physiol 147:1913–1923
14. Nakamura A, Higuchi K, Goda H, Fujiwara MT, Sawa S, Koshiba T, Shimada Y, Yoshida S (2003) Brassinolide induces IAA5, IAA19, and DR5, a synthetic auxin response element in arabidopsis, implying a cross talk point of brassinosteroid and auxin signaling. Plant Physiol 133:1843–1853
15. Jenik PD, Barton MK (2005) Surge and destroy: the role of auxin in plant embryogenesis. Development 132:3577–3585

Chapter 16

Gene Expression Analysis of Shoot Apical Meristem Cell Types

Ram Kishor Yadav, Stephen Snipes, Thomas Girke, and G. Venugopala Reddy

Abstract

Shoot apical meristems (SAMs) of higher plants harbor a set of stem-cells and provide cells for the development of all the above-ground biomass of plants. Most of the important pattern formation events such as maintenance of stem-cell identity, specification and differentiation of leaf/flower primordia, and the temporal control of the transition from vegetative to reproductive program are determined in SAMs. Genetic analysis has revealed molecular and hormonal pathways involved in stem-cell maintenance, organ differentiation and flowering time. However, limited information is available as to how different pathways interact with each other to function as a network in specifying different cell types and their function. Deciphering gene networks that underlie cell fate transitions requires new approaches aimed at assaying genome scale expression patterns of genes at a single cell type resolution. Here we provide details of experimental methods involved in protoplasting of SAM cells, generating cell type specific gene expression profiles and analysis platforms for identifying and inferring gene networks.

Key words: Stem cells, FACS, Protoplasts, Fluorescent reporter

1. Introduction

The process of pattern formation involves generation of specific cell types in a correct spatiotemporal sequence (1). It is a dynamic process which requires conversion of transient temporal signals into stable genetic circuits that involve feed-forward and feedback control of gene activation (1, 2). Therefore, deciphering gene networks that underlie cell fate transitions requires new approaches aimed at assaying genome-scale expression patterns of genes at a spatial resolution of single cell types (3–5). Shoot apical meristems (SAMs) of higher plants harbor a set of stem cells and provide cells for the development of all

Ive De Smet (ed.), *Plant Organogenesis: Methods and Protocols*, Methods in Molecular Biology, vol. 959,
DOI 10.1007/978-1-62703-221-6_16, © Springer Science+Business Media New York 2013

the above-ground biomass of plants (6). Traditionally, SAMs of higher plants have been divided into distinct domains of cells mostly based on their location within the SAMs and morphological criteria (6, 7). The SAM of *Arabidopsis* is a multilayered structure consisting of several hundred cells, organized into three clonally distinct layers of cells (6–8). The cells in the outermost L1 layer and the subepidermal L2 layer divide in anticlinal orientation (perpendicular to the SAM surface), while the underlying corpus forms a multilayered structure where cells divide in random orientations. Within this framework, the SAM is divided into cytological and functional zones (9). The central zone (CZ) is at the tip and harbors a set of stem-cells. Progeny of stem cells enter into differentiation pathways when they enter the surrounding regions—the flanking peripheral zone (PZ) where leaf and flower primordia are formed. The CZ also provides cells to the rib-meristem (RM) located beneath the CZ where cells of the stem form. The SAMs, thus, represent a dynamic continuum of different cell types that are organized both in radial dimension and also in clonally distinct cell layers. The challenge is to understand how the interconnected network of cells interprets complex spatiotemporal signals to regulate gene expression patterns to bring about cell fate transitions. This requires assaying gene expression profiles at a higher spatiotemporal resolution. Several methods have been described for isolating specific cell types for the purpose of gene expression profiling. Here we describe a method based on Fluorescence-Activated Cell Sorting (FACS)-based isolation of protoplasts from specific cell types of SAMs (3) which is adopted from studies on *Arabidopsis* root system (4). We describe methods for generating cell type-specific gene expression profiles and analysis platforms for identifying and inferring gene networks.

2. Materials

2.1. Equipment

New Brunswick Excella E24 Incubator Shaker.

Beckman Coulter Allegra X-15R Temperature Controlled Centrifuge.

Eppendorf Centrifuge 5415 D Tabletop Microcentrifuge.

Beckton Dickinson FACS-Aria.

Agilent 2100 Bioanalyzer.

Affymetrix Genechip.

Affymetrix Hybridization Oven 640.

Affymetrix Fluidics Station 450.

Affymetrix Scanner 3000, 7G.

Nanodrop ND-1000 Spectrophotometer.

Corning Inc. Petri Plate.

2.2. Solutions, Chemicals, and Buffers

Solution A (for 25 ml): KCl 1 M (250 μl), $MgCl_2$ 1 M (50 μl), $CaCl_2$ 1 M (50 μl), BSA (0.025 g), MES (0.0097 g), Mannitol (2.75 g), add H_2O up to 25 ml.

Solution B (for 20 ml): Solution A (20 ml), cellulase (Yakult, Japan, Cat #203039) (300 mg), pectolyase (Yakult, Japan, Cat #202047) (200 mg), hemicellulase (Sigma, USA, Cat #H2125) (200 mg).

RNA Isolation Kit.

RNeasy Plant Mini Kit (#74904 QIAGEN).

2.3. Plant Material

50–60 mg of *ap1-1;cal1-1* SAM tissue (about 200 shoots) expressing a fluorescent marker.

3. Methods

3.1. Protoplasting and Cell Sorting

This method has been developed based on two earlier studies (4, 10).

1. Grow *ap1-1;cal1-1* plants carrying a fluorescent cell type marker for up to 4 weeks.
2. Prepare 25 ml of solution A (pH 5.5) with 1 M Tris.
3. To prepare solution B, weigh the cell wall digesting enzymes, cellulose (300 mg), pectolyse (200 mg), and hemicellulose (200 mg), and dissolve them in 20 ml of freshly prepared buffer A. Mix vigorously by pipette until the solution becomes transparent.
4. In the meantime adjust the shaker temperature to 22 °C.
5. Harvest 50–60 mg of ap1;cal1 shoot tissue as quickly as possible.
6. Transfer the SAM tissue to a 70 μm nylon cell strainer (BD Falcon) and place the strainer in a Petri dish 35 mm × 10 mm (BD Falcon).
7. Add approximately 6–7 ml of solution B and place the Petri dish on a shaker set at 120 rpm at 22 °C. At every 10-min interval, gently rinse the SAM surface with a jet of Solution B using a pipette.
8. Set the centrifuge at 4 °C.
9. If protoplasting is efficient, the protoplasting solution should turn turbid by 45 min of treatment.
10. Pipetting and mixing of solution may result in loss of Solution B. Add Solution B at regular intervals.
11. Upon 1 h and 15 min of treatment, transfer the contents of Petri dish to a 15 ml Falcon tube and centrifuge at 500 × *g* at 4 °C for 10 min. Remove the supernatant without disturbing the pellet.
12. Dissolve the cell pellet in 500–600 μl of solution A by pipetting gently. Once the cell pellet is dissolved completely, place the tube on ice until FACS sorting.

3.2. Isolation of Total RNA from Sorted Cells

1. Collect fluorescently labeled protoplasts from FACS flow in 500 μl RLT buffer (QIAGEN RNeasy kit) and adjust the volume of cell suspension up to (3× of original) 3 ml by adding RLT buffer.
2. Add 30 μl of β-mercaptoethanol in it and one could either store it at −80 °C or proceed without freezing for RNA isolation.
3. Add 0.5 volume of chilled ethanol and mix gently.
4. Apply 700 μl of solution to RNeasy column and centrifuge at 10600 × *g* for 15 s. Add the solution to same sample column and repeat the same step again and again until you finish the remaining amount of precipitated solution.
5. Add 350 μl of buffer RW1 to spin column and centrifuge at 10600 × *g* for 15 s.
6. To perform on column DNA digestion, add 80 μl of DNA digesting solution to sample column at RT for 15 min (10 μl DNAase plus 70 μl RDD buffer).
7. Stop DNA digestion by adding 350 μl buffer RW1 and centrifuge at 10600 × *g* for 15 s.
8. Apply 500 μl RPE buffer provided with RNeasy kit after adding ethanol in it.
9. Centrifuge at 10600 × *g* for 2 min.
10. Air-dry the sample column for 2 min.
11. Elute RNA from column by applying 50 μl RNase-free water, centrifuge at 17900 × *g* for 1 min, and repeat the same step once more.
12. Precipitate RNA with 0.5 volume of 7.5 M ammonium acetate and add 2.5 volume ethanol along with 0.5 μl Glycoblue (15 mg/ml) overnight at −20 °C.
13. Centrifuge at 17900 × *g* for 30 min, remove supernatant, and wash RNA pellet with cold 70% ethanol.
14. Air-dry the pellet for ~5 min and resuspend RNA pellet in ~4–6 μl RNase-free water.
15. RNA integrity was determined using Agilent 6000 RNA Nano kit on Agilent 2100 Bioanalyzer according to the manufacturer's protocol.
16. The total RNA yield from 100,000 cells usually ranges between 200 and 300 ng and is insufficient for microarray experiments. To hybridize, it requires two-step amplification.

3.3. Two-Step RNA Amplification, Labeling and Target Probe Hybridization

Perform according to the manufacturer's protocol (Affymetrix).

Table 1
Overview of analysis steps

Preprocessing and quality control
affyPLM[1,2], arrayQualityMetrics[1]
Normalization and background correction
MAS5[1,2], RMA[1,2], GCRMA[1,2], VSN[1]
Identification of differentially expressed genes (DEG)
SAM[1,3], ANOVA[1,2,3], RankProd, LIMMA[1]
Clustering of gene expression profiles
Hierarchical Clustering[1,2,3], k-means[1,2,3], fuzzy k-means[1]
Functional categorization of gene expression sets
Gene Ontologies[1,2,3], KEGG[1,2,3]
Identification of regulatory motifs
Motif finding[1], motif discovery[1]

Examples of software tools providing the corresponding utilities are referenced with numbers: R/Bioconductor[1], BRB-ArrayTools[2], and MeV[3]

4. Data Analysis

4.1. Overview

This section gives a brief overview of the basic concepts used to analyze Affymetrix expression GeneChips® hybridized with cell type samples isolated from SAMs. Among the many software environments available for microarray analysis, we have chosen here to highlight three relatively popular representatives that are freely available in the public domain (Table 1). For more advanced users, the command-line-driven statistical environment R combined with its many Bioconductor packages is by far the most powerful solution (11, 12). BRB-ArrayTools (13) and MultiExperiment Viewer (MeV; 14) are software packages with graphical user interfaces that may be easier to learn, especially for occasional nonexpert users.

4.2. Data Preprocessing and Quality Control

- Organize the raw GeneChip® files (CEL files) in one directory and record all experimental variables in a tabular experiment definition file, containing CEL file names, replicate groups, cell types, and other variables.
- Import CEL files into chosen software environment (e.g., one of the tools mentioned in Table 1).

- Run array quality control steps provided by software. A simple approach for detecting outlier arrays can be the calculation of an all-against-all sample correlation matrix. The latter can be visualized with one of the common clustering methods, such as dendrograms generated by hierarchical clustering or scatter plots from multidimensional scaling (see cluster section for details). In these representations, samples from the same sources (e.g., tissue or cell type) should always cluster closer together than samples from distinct sources. Outlier CEL files may need to be removed from the downstream analysis steps.

4.3. Normalization and Background Correction

- Gene expression microarrays provide relative measurements that are only meaningful if their raw data have been properly normalized to account for technical variations among samples, such as variations occurring during sample preparation steps. Most software tools provide implementations of the normalization algorithms commonly used for GeneChip® data, including MAS5 (15), RMA (16), and GCRMA (17). For the quantile-based normalization methods, including RMA and GCRMA, it is important to normalize all CEL files (to be used for the downstream steps) in one batch, while the scaling-based MAS5 algorithm allows normalizing each file independently from the others. In addition, it is recommended to perform all downstream sample comparisons on data sets that were processed with the same normalization method. *Note*: Both RMA and GCRMA return $\log_2$ values, whereas MAS5 data are not on a log scale.
- If the above normalization methods are used with their default parameters then background correction is included in the normalization step.
- The MAS5 algorithm also provides present calls providing information as to whether a gene is expressed strongly enough to be detectable by the technology. The present calls are often used to estimate which and how many genes are reliably expressed in different cell types (18, 19).
- After the normalization, one can calculate mean expression values for the replicated samples. The averaged values are often useful for downstream gene clustering or various filtering steps.

4.4. Identification of Differentially Expressed Genes

- Differentially expressed gene (DEG) analysis allows determining genes with significant expression differences among cell types. To provide statistical support in form of confidence levels (*p*-values), it is absolutely crucial to have for all cell types at least two, preferentially three or more, biological replicates. Without replicates, this step will be restricted to a simple fold change analysis that lacks any information about the inherent variance of the biological system and to estimate false discovery rates (FDR).

- Common statistical methods used for identifying DEGs include two-sample *t*-tests, ANOVA, RankProd (20), SAM (21), and LIMMA. The *p*-values generated by these methods are usually adjusted to account for the multiple testing situations of microarrays measuring thousands of gene changes simultaneously. For instance, the Benjamini and Hochberg method is often used to adjust the *p*-values and to obtain FDRs (22).
- After the statistical DEG analysis, one can filter the data sets by adjusted *p*-values, fold changes, and present call information. For example, to identify all transcripts enriched in one cell type, one can filter for genes that are ≥2-fold higher expressed in one cell type than all the others while being supported by an adjusted *p*-value of ≤0.01 and ≥1 present calls (obtained in MAS5 step).
- To analyze the composition of genes in DEG sets or clusters, Venn diagram analyses can be often helpful. Support for this type of set comparisons is provided by some of the software tools listed in Table 1. There are also several online systems available, including Venny (http://bioinfogp.cnb.csic.es/tools/venny) and Venn Diagram Generator (http://www.pangloss.com/seidel/Protocols).

4.5. Clustering of Gene Expression Profiles

- Clustering is frequently used in the microarray field to partition expression profiles of genes or samples into similarity groups based on a chosen distance measure (23). Because genes are usually abundant on microarrays, but the number of measurements is often limited (here cell samples), it is important to emphasize that clustering of samples is almost always possible, whereas clustering of genes can only be performed if a reasonable number of sample measurements is available. The latter is required for the calculation of robust similarity measures of expression profiles among genes.
- To perform clustering, it is often useful to organize the normalized expression values averaged among the replicates in a table where the rows represent the genes and the columns the samples. Because clustering can be computationally very time consuming on large data sets, it is usually recommended to use only on a subset of the genes represented on a microarray. For instance, one can apply it just to the set of DEGs identified above.
- Subsequently, a distance matrix is computed in a separate step or it is generated as integral part of the clustering software. *Note*: Not all clustering methods require a distance matrix as input, but the ones introduced below use distance matrices in at least one part of their algorithm. Among the many distance measures available, the most commonly used ones in the microarray field are the Euclidean distance or distance-transformed correlation coefficients. Pearson's correlation coefficient

is probably the most popular method, but it is very sensitive to outliers (here extreme expression values). This limitation can be overcome by choosing Spearman's correlation coefficient instead, where a rank transformation step minimizes the influence of outliers. To convert correlation values (similarity metric) into a distance measure, they can be subtracted from 1. If one wants to treat negative correlations the same way as positive ones, one can use their absolute values.

- Among the vast number of algorithms available for clustering microarray data, we introduce here only two relatively popular methods: hierarchical clustering (24, 25) and fuzzy k-means clustering (26). The former was chosen as an example of strict clustering where each item can only be a member of one cluster, whereas the latter allows multiple cluster memberships. To perform hierarchical clustering, one can provide a distance matrix as input. Alternatively, the averaged expression values can be used, if the software takes care of the computation of the distance matrix internally. Most hierarchical clustering algorithms join the most to least similar items (here genes or samples) in an agglomerative manner using as cluster joining rule single, average, complete, or other methods. Average or complete linkages are often good choices here. The output of hierarchical clustering is a similarity tree (dendrogram) that can be used to sort the rows of the underlying expression table according to the gene order defined by the dendrogram. For visualization purposes, one can present the dendrogram next to a heat map where the expression values are represented by a color scheme. When clustering the samples as well, one can present in the same heat map the similarities among the cell types by sorting the columns according to their sample dendrogram.
- Fuzzy k-means clustering is another popular algorithm used in this area. This method performs a specific version of k-means clustering, called partitioning around medoids. It groups items (genes) into a predefined number of k clusters while allowing multiple memberships of items in clusters ("fuzzyness"). The method requires two user-definable parameters: the number of cluster k to consider during the clustering as well as a membership exponent controlling the fuzzyness of the result. Selecting these parameters in an objective manner can be difficult. However, permutation tests can be helpful to optimize them where one can monitor the enrichment of GO terms (see below) and then choose the settings that resulted in the highest enrichment.

4.6. Functional Categorization of Gene Expression Sets

- To functionally annotate gene sets obtained from DEG or cluster analyses, it can be very helpful to determine what functional annotation categories are enriched in these sets. Different functional annotation systems are available for this purpose, including Gene Ontologies (GO), KEGG, AraCyc, and Pfam domains. GO terms are most often used, because their annotations cover 95% of the genome, whereas pathway annotation systems provide much lower coverage (27, 28). As statistical enrichment tests, the hypergeometric distribution, Fisher's exact test, or Gene Set Enrichment Analyses (GSEA) are often used (29, 30). In principle, these methods determine whether any of the annotation categories are enriched in a sample gene set relative to random sampled gene sets of the same size.

- The software packages listed in Table 16.1 provide utilities for functionally annotating gene sets and statistical enrichment testing. In addition, there are many Web services available for these tasks (e.g., DAVID: http://david.abcc.ncifcrf.gov/).

4.7. Identification of Regulatory Motifs

- Both motif finding and de novo motif discovery methods can be used to identify regulatory elements (sequence motifs) enriched in the promoter regions of DEG or cluster sets. Often the sequence regions 500–2,000 bp upstream of each gene model are considered in these tests.
- For motif finding, one can use the *cis*-regulatory elements available in various databases (e.g., Agris: 31, PLACE: 32) and then map them to the upstream sequences of each gene set. Subsequently, one can test whether a motif is enriched in the promoter sequences relative to randomly selected gene sets. The hypergeometric distribution can be used to compute *p*-values for estimating the significance of this enrichment and the Bonferroni method to adjust the *p*-values for multiple testing (33).
- For de novo motif discovery, one can use one of the many motif discovery tools, such as MEME (http://meme.nbcr.net/meme4_6_1/intro.html), AlignAce (http://atlas.med.harvard.edu/), and Weeder (http://159.149.109.9/modtools/). A comprehensive comparison of these and many other tools is available in (34). Once novel motif candidates are identified, their enrichment in the promoter regions of interest needs to be tested with the same statistical methods as the known motifs.

Acknowledgments

Our genomics work is being carried out at the core facility of center for plant cell biology (CEPCEB) and institute of integrative genome biology (IIGB), University of California, Riverside. This work is funded by National Science Foundation grant (IOS-0820842) to GVR and TG.

References

1. Stathopoulos A, Levine M (2005) Genomic regulatory networks and animal development. Dev Cell 4:449–462
2. Barton MK (2010) Twenty years on: the inner workings of the shoot apical meristem, developmental dynamo. Dev Biol 341:95–113
3. Yadav RK, Girke T, Pasala S, Xie M, Reddy GV (2009) Gene expression map of the Arabidopsis shoot apical meristem stem cell niche. Proc Natl Acad Sci USA 106:4941–4946
4. Birnbaum K, Shasha DE, Wang JY, Jung JW, Lambert GM, Galbraith DW, Benfey PN (2003) A gene expression map of the Arabidopsis root. Science 302:1956–1960
5. Brady SM, Orlando DA, Lee JY, Wang JY, Koch J, Dinneny JR, Mace D, Ohler U, Benfey PN (2007) A high-resolution root spatiotemporal map reveals dominant expression patterns. Science 318:801–806
6. Steeves TA, Sussex IM (1989) Patterns in plant development. Cambridge University Press, New York, NY (shoot apical meristem mutants of Arabidopsis thaliana)
7. Meyerowitz EM (1997) Genetic control review of cell division patterns in developing plants. Cell 88:299–308
8. Reddy GV (2008) Live-imaging stem-cell homeostasis in the Arabidopsis shoot apex. Curr Opin Plant Biol 11:88–93
9. Baurle I, Laux T (2003) Apical meristems: the plant's fountain of youth. Bioessays 25: 961–970
10. Guangyu C, Conner AJ, Christey MC, Fautrier AG, Field RJ (1997) Protoplast isolation from shoots of Asparagus cultures. Int J Plant Sci 158:537–542
11. R Development Core Team (2010) R: A language and environment for statistical computing. URL: http://www.R-project.org. Vienna, Austria, ISBN: 3-900051-07-0
12. Gentleman R, Carey V, Dudoit S, Irizarry R, Huber W (2005) Bioinformatics and computational biology solutions using R and bioconductor. Springer, New York, NY
13. Simon R, Lam A, Li MC, Ngan M, Menenzes S, Zhao Y (2007) Analysis of gene expression data using BRB-ArrayTools. Cancer Inform 3:11–17
14. Saeed AI, Bhagabati NK, Braisted JC, Liang W, Sharov V, Howe EA et al (2006) TM4 microarray software suite. Methods Enzymol 411:134–193
15. Gautier L, Cope L, Bolstad BM, Irizarry RA (2004) Affy—analysis of Affymetrix GeneChip data at the probe level. Bioinformatics 20:307–315
16. Irizarry RA, Hobbs B, Collin F, Beazer-Barclay YD, Antonellis KJ, Scherf U, Speed TP (2003) Exploration, normalization, and summaries of high density oligonucleotide array probe level data. Biostatistics 4:249–264
17. Wu Z, Irizarry RA, Gentleman R, Martinez-Murillo F, Spencer F (2004) A model-based background adjustment for oligonucleotide expression arrays. J Am Stat Assoc 99:909–917
18. Liu WM, Mei R, Di X, Ryder TB, Hubbell E, Dee S, Webster TA, Harrington CA, Ho MH, Baid J et al (2002) Analysis of high density expression microarrays with signed-rank call algorithms. Bioinformatics 18:1593–1599
19. McClintick JN, Edenberg HJ (2006) Effects of filtering by present call on analysis of microarray experiments. BMC Bioinform 7:49
20. Hong F, Breitling R, McEntee CW, Wittner BS, Nemhauser JL, Chory J (2006) RankProd: a bioconductor package for detecting differentially expressed genes in meta-analysis. Bioinformatics 22:2825–2827
21. Tusher VG, Tibshirani R, Chu G (2001) Significance analysis of microarrays applied to the ionizing radiation response. Proc Natl Acad Sci USA 98:5116–5121
22. Benjamini Y, Hochberg Y (1995) Controlling the false discovery rate—a practical and powerful approach to multiple testing. J Roy Stat Soc Ser B Methodological 57:289–300
23. de Hoon MJ, Imoto S, Nolan J, Miyano S (2004) Open source clustering software. Bioinformatics 20:1453–1454
24. Murtagh F (1985) Multidimensional clustering algorithms. COMPSTAT Lectures 4. Physica-Verlag, Wuerzburg
25. Eisen MB, Spellman PT, Brown PO, Botstein D (1998) Cluster analysis and display of genome-wide expression patterns. Proc Natl Acad Sci USA 95:14863–14868
26. Krishnapuram R, Joshi A, Nasraoui O, Yi L (2001) Low-complexity fuzzy relational clustering algorithms for Web mining. IEEE Trans Fuzzy Syst 9:595–607
27. Mueller LA, Zhang P, Rhee SY (2003) AraCyc: a biochemical pathway database for Arabidopsis. Plant Physiol 132:453–460
28. Kanehisa M, Goto S, Hattori M, Aoki-Kinoshita KF, Itoh M, Kawashima S, Katayama T, Araki M, Hirakawa M (2006) From genomics to chemical genomics: new developments in KEGG. Nucleic Acids Res 34:D354–D357

29. Falcon S, Gentleman R (2007) Using GOstats to test gene lists for GO term association. Bioinformatics 23:257–258
30. Subramanian A, Tamayo P, Mootha VK, Mukherjee S, Ebert BL, Gillette MA, Paulovich A, Pomeroy SL, Golub TR, Lander ES, Mesirov JP (2005) Gene set enrichment analysis: a knowledge-based approach for interpreting genome-wide expression profiles. Proc Natl Acad Sci USA 102:15545–15550
31. Yilmaz A, Mejia-Guerra MK, Kurz K, Liang X, Welch L, Grotewold E (2011) AGRIS: Arabidopsis Gene Regulatory Information Server, an update. Nucleic Acids Res 39: D1118–D1122
32. Higo K, Ugawa Y, Iwamoto M, Korenaga T (1999) Plant cis-acting regulatory DNA elements (PLACE) database. Nucleic Acids Res 27:297–300
33. Sinha S, Ling X, Whitfield CW, Zhai C, Robinson GE (2006) Genome scan for cis-regulatory DNA motifs associated with social behavior in honey bees. Proc Natl Acad Sci USA 103:16352–16357
34. Tompa M, Li N, Bailey TL, Church GM, De Moor B, Eskin E, Favorov AV, Frith MC, Fu Y, Kent WJ, Makeev VJ, Mironov AA, Noble WS, Pavesi G, Pesole GR, Simonis N, Sinha S, Thijs G, van Helden J, Vandenbogaert M, Weng Z, Workman C, Ye C, Zhu Z (2005) Assessing computational tools for the discovery of transcription factor binding sites. Nat Biotechnol 23:137–144

Chapter 17

Kinematic Analysis of Cell Division in Leaves of Mono- and Dicotyledonous Species: A Basis for Understanding Growth and Developing Refined Molecular Sampling Strategies

Hilde Nelissen, Bart Rymen, Frederik Coppens, Stijn Dhondt, Fabio Fiorani, and Gerrit T.S. Beemster

Abstract

The cellular level processes cell division and cell expansion form a crucial level linking regulatory processes at the molecular level to whole plant growth rates and organ size and shape. With the rapid progress in molecular profiling, quantification of cellular activities becomes increasingly important to determine sampling strategies that are most informative to understand the molecular basis for organ and plant level phenotypes. Inversely, to understand phenotypes caused by genetic or environmental perturbations it is crucial to know how the cell division and expansion parameters are affected spatially and temporally.

Kinematic analyses provide a powerful and rigorous mathematical framework to quantify cell division and cell expansion rates. In dicotyledonous leaves, these processes are primarily changing over time, resulting in division, expansion, and mature phases of development. Monocotyledonous leaves have a persistent spatial gradient, with an intercalary meristem where division takes place, an expansion zone, and a mature part of the leaf. Here we describe in detail how to perform kinematic analyses in leaves of the model species *Arabidopsis thaliana* and in the leaves of the monocotyledonous crop species *Zea mays*. These methods can readily be used and adapted to suit other species using relatively standard equipment present in most laboratories. Importantly, the obtained results can be used to design sampling techniques for proliferating, expanding and mature cells.

Key words: Cell division, Cell expansion, Kinematics, Image analysis, Epidermal cells, Arabidopsis, Leaf growth, Maize

Ive De Smet (ed.), *Plant Organogenesis: Methods and Protocols*, Methods in Molecular Biology, vol. 959,
DOI 10.1007/978-1-62703-221-6_17, © Springer Science+Business Media New York 2013

1. Introduction

In plant sciences, growth analysis is widely used to describe genotypes and to characterize responses to environmental conditions. Manual measurements of growth at the whole plant or organ level often rely on relatively simple methodologies: they involve measuring size or weight at multiple times and calculation of the rate of increase of these parameters. In many cases however the purpose of experiments is to gain insight into the underlying mechanisms that regulate plant/organ growth. To this end two major categories of strategies have been used extensively, each focusing on different aspects of growth regulation. Firstly, "classical" growth analyses have been widely used to quantify the structural basis of observed growth phenotypes. Classic whole plant growth analysis essentially characterizes the exponential rate of growth on a mass basis, RGR (relative growth rate), based on allocation of photosynthesis-derived carbon into new photosynthetic area and into other parts of the plant (1). Cellular growth analysis considers the contributions of cell division and cell expansion into growth differences at the whole organ level. The kinematic analyses described in this chapter are an advanced version of cellular growth analysis. In recent years we witnessed considerable advances in molecular profiling techniques such as mRNA profiling by microarrays and RNA-sequencing, and metabolic and proteome profiling by mass spectroscopy analyses. Increasingly, these molecular profiling techniques are used to analyze and understand the molecular basis of phenotypes caused by genetic differences or experimental treatments. Although these approaches are extremely powerful, they restrict themselves to molecular level processes, which, indirectly through effects on cell division and expansion, result in the observed whole organ level differences.

We argue that to maximally understand growth regulatory mechanisms, growth and molecular analysis need to be combined. Particularly, kinematic analysis can instruct when and where differences at the cellular level are observed. Sampling of specific developmental stages and or zones for molecular analyses is likely to be significantly more powerful for unraveling the changes that underlie the growth phenomena than whole shoot or whole plant approaches.

To allow a broader use of kinematic approaches for quantification of how division and expansion contribute to differences in rates of growth at the whole organ level, we describe these methods here in detail for two different types of leaves (mono- and dicotyledonous). Kinematic methods have been pioneered halfway last century (2, 3) and a rigorous mathematical framework was developed based on laws of fluid dynamics a few decades later (4). Since then there has been a gradual increase in the experimental use of these methods, largely supported by the increasing power of PCs and the availability of powerful, easy to use, public domain image-analysis software like ImageJ (http://rsbweb.nih.gov/ij/). Here we describe the methods for two specific experimental systems: the first leaf pair of the dicotyledonous species *Arabidopsis thaliana* and the leaf of the monocotyledonous species *Zea mays*. We focus in detail on the practical implementation of these methods. Recent reviews are recommended for further reading about the conceptual basis and the derivation of the formulae used (4, 5). Together, these two experimental systems provide examples of the possibilities that this approach currently offers for representatives of monocotyledonous and dicotyledonous species, as the protocols can easily be adapted to suit most other species.

2. Materials

2.1. Analysis of Arabidopsis Leaves

1. 100 ml medium 0.5× MS salts, 10 g/l sucrose, 0.5 g/l MES, and 0.8 g/l plant tissue culture agar in nano-pure water. pH is adjusted to 5.8 before adding agar. The medium is autoclaved at 1 bar over-pressure for 20 min.
2. Petri dishes and porous tape to seal.
3. Bleach, 70% ethanol, and nano-pure water.
4. Lactic acid.
5. Hoyer medium: 80 g chloral hydrate in 30 ml water (6).
6. Mounting material: Object slides, cover slips.
7. A binocular microscope equipped with a camera and a microscope with 20× and 40× Plan DIC lenses equipped with a drawing tube.
8. Flatbed scanner.
9. Computer running image analysis software (e.g., ImageJ; Public domain image analysis software, freely available from http://rsbweb.nih.gov/ij/), a spreadsheet (e.g., MS Excel or OpenOffice; freely available from http://www.openoffice.org/).

2.2. Maize Leaf

1. Ruler.
2. Harvesting material: Razor blades, scalpel, forceps, and vials to store material.
3. Lactic acid, absolute ethanol, 3/1 (v/v) absolute ethanol:acetic acid.

4. Mounting material: Object slides and cover slips.
5. 4′,6-Diamidino-2-phenyindole (DAPI) solution 1 mg/ml, will be 1/1000 diluted in buffer for sample incubation.
6. Buffer solution: 50 mM NaCl, 5 mM EDTA, and 10 mM Tris–HCl.
7. Microscopy for cell length measurements: Microscope with a 20× PLAN DIC lens, ideally supplemented with a similar 10× and 40× lens. For meristem length measurements: an epifluorescence microscope with an excitation filter at wavelength 350 nm and emission filter at wavelength 420 nm, equipped with 20× and 40× lenses. Digital camera connected to a personal computer equipped with software which allows to measure cells on the "live" image of the microscope (Image J is compatible with many microscopes).
8. Computer running image analysis software (e.g., ImageJ or microscopy measuring software. We currently don't use ImageJ anymore but the built-in microscope software; Public domain image analysis software, freely available from http://rsbweb.nih.gov/ij/), a spreadsheet (e.g., MS Excel or OpenOffice; freely available from http://www.openoffice.org/), and R (Public domain statistical software, freely available from http://www.r-project.org/).

3. Methods

3.1. Analysis of Arabidopsis Leaves

After initiation at the shoot apex, leaves of dicotyledonous plants go through subsequent stages of cell division and expansion before they reach maturity. Although the transitions between these stages occur in a tip to base gradient (7) the approach outlined here calculates average rates of division and expansion across the entire leaf. The analysis is based on two sets of primary data: leaf blade area and cell area. For the first leaf pair of *Arabidopsis thaliana* ecotype Columbia-0, we measure these variables from 2 days after germination (DAG) until 22 DAG. This period spans from leaf emergence until maturity under our environmental conditions (growth chamber: 21°C, fluorescent (Cool White) light 80 μE/m^2/s, 16/8 h D/N, shelves cooled at 19°C to prevent condensation against the lids of the Petri dishes).

3.1.1. Plant Material

1. Seeds are sterilized for 15 min in 3.5% bleach solution in 1.5 ml Eppendorf tubes and washed three times with sterile water.
2. After the last wash the water is left in the tubes.
3. Seeds are sown using a 20 μl pipette with tips of which the top is cut 2 mm with a razor blade to increase the opening. About

a dozen seeds are aspirated with enough water to dispense them. The seeds are deposited individually on the agar. In this step it is important to gently touch the medium with the tip in order to break the surface of the agar layer, allowing the root to penetrate the medium.

4. The plates are left open in the flow for a few minutes to let excess water evaporate.
5. Then the Petri dishes (see Note 1) are closed using porous tape that allows gas exchange.

3.1.2. Sample Preparation and Mounting

1. For the earliest stages of development, whole plants are harvested. Once the first true leaves have formed a petiole (ca 11 days after plating, which is approximately 8 days after germination) they are dissected.
2. After harvesting, the leaves are placed overnight in 70% ethanol to remove chlorophyll.
3. Then the material is transferred to lactic acid for clearing and further storage. After one night, the cleared material can be mounted on a microscopic slide using lactic acid as mounting medium. When the clearing is not satisfactory and the leaves still look opaque, there is an alternative protocol (see Note 2).
4. For the cell analysis, the abaxial epidermis is used, therefore it is important to place the leaf on the microscopic slide with the adaxial—trichome containing—side down.
5. For the earliest stages, the first leaf pair is exposed by gently pulling apart the cotyledons before placing the cover slip.

3.1.3. Leaf Area Measurements

1. The blade area is imaged using a binocular microscope. The magnification is adjusted to the size of the leaves under examination (1.25–6.3×), making sure to photograph a ruler at the same magnification for image calibration.
2. Images are acquired for about ten leaves per genotype/treatment and per time point.
3. The leaf area is measured using an image analysis program, e.g., ImageJ.
4. First of all, the program is calibrated using the ruler image, which allows converting pixels to corresponding distances in mm.
5. The "polygon selection" tool is used to outline the leaf blade and the area is measured.
6. Because the leaves of Arabidopsis often have a curved surface, the leaf edge is sometimes folded double. To accommodate for this, it is necessary to measure the area of the folded parts of the leaf that can be easily recognized in addition to the outline of the whole leaf area.

7. The measurements are then copied into a spreadsheet program and added to determine the total leaf blade area.
8. For the kinematic analysis at least five medium-sized leaves are selected and average area and standard errors calculated.

For the earliest time points, it is not possible to determine the leaf area using a dissecting microscope. In these cases the area is drawn using a DIC microscope and drawing tube (see below).

3.1.4. Cell Area Measurements

The cells can be visualized in the cleared samples with a DIC microscope. Depending on the size of the cells a 20×, 40×, or 63× magnification is used. Preferably the abaxial epidermis of the first leaf pair is used for the analysis (see Note 3 for a discussion on cell types). As there is a gradient in cell development from leaf tip to bottom, the cells are analyzed at two positions of the leaf at about a quarter from the tip and bottom of the leaf and halfway between the leaf margin and the mid-vein. Regions directly above the vasculature are avoided because the epidermal cells can be more elongated and harder to visualize due to the optical disturbance of the underlying dense vascular strands.

Preferably, the leaf epidermal cells are outlined using a drawing tube (for an alternative see Note 4). This device consists of a tube containing a mirror, which allows visualization through the microscope ocular piece of both the sample and the drawing being made on a piece of paper next to the microscope. To gather sufficient data, we draw about 50 cells for each epidermal area examined. While drawing, it is important to avoid creating a bias in cell sizes. Keep in mind to avoid the edge of the paper and it is imperative not to draw a disproportionately high number of small cells. This problem may arise because small cells more easily fit on the paper compared with bigger ones. This can be avoided by drawing approximately the same area of your paper (about 1/3 area works well in our hands). To be able to apply the correct scaling, a drawing needs to be made of a calibration grid for each magnification used.

3.1.5. Determination of Number of Stomata

Stomatal pores are a feature that needs to be dealt with specifically. They are usually open and might erroneously be recognized by automated analyzes as cells. One solution to avoid this artifact is to completely fill in the stomatal pores. The number of stomata per drawing is counted manually and recorded.

3.1.6. Processing of the Images

When the drawings are finished, they need to be digitized.

1. Beforehand, it is important to correct potential artifacts. Cell walls that show gaps need to be closed, and cell walls at the edge of the drawing that do not delineate full cells need to be erased. Also, avoid cell walls that touch the edges of the paper.

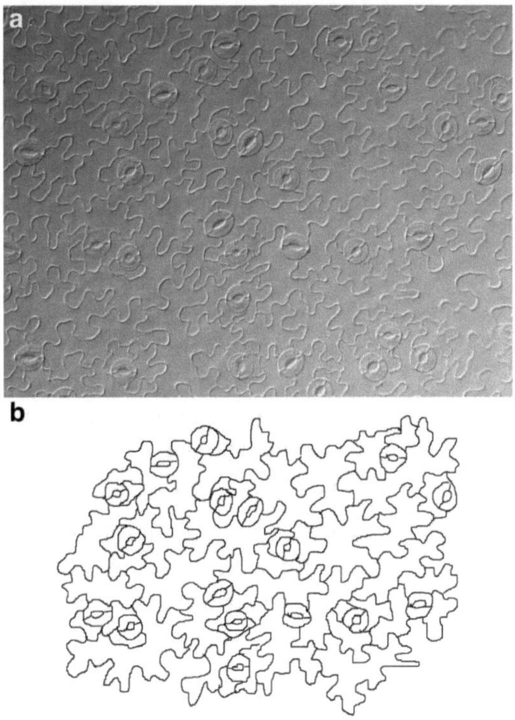

Fig. 1. Image processing of the adaxial epidermis of *Arabidopsis thaliana* leaves. (**a**) DIC image of cleared leaves. (**b**) Typical drawing of cell outlines.

Do not forget to record the origin of the drawing by annotating next to the drawn area: genotype/treatment, time point, position in the leaf (tip or bottom), and magnification used for visualization (for a typical example of the epidermis and a drawing see Fig. 1).

2. Next, the drawings are scanned. Depending on the available scanner, the settings might take some trial and error. Scanning in color at 300 dpi to a jpeg file suffices in our hands.
3. The images are processed using image analysis software. Here we outline the detailed procedure based on the freely available program ImageJ.
4. First of all, the calibration needs to be set.
   (a) Open the appropriate calibration image and use the Straight Line tool to select a length of the grid.
   (b) Use the "Analysis/Set Scale" command and fill in the known distance and the units.
   (c) Selecting the option "global" the program will apply these settings for all images that are subsequently opened.
   (d) Under "Analysis/Set Measurements" verify that "Area" and "Display label" are selected.

5. Open an image and convert it to 8-bit greyscale under "Image/Type."
6. Apply a threshold (using "Image/Adjust/Threshold") that allows the cell walls to be best visualized.
7. Apply a closing step ("Process/Binary/Close") to close small gaps in the drawing possibly created by the thresholding.
8. Use the magic wand tool and click to the left of the drawn area. This selects the outline of the entire area of the cell drawing. If the selection includes the interior of some cells not all cell walls are closed. This issue can be resolved by drawing lines (selecting a line across the opening and clicking "Ctrl-D"), as this is tedious, drawings are best checked beforehand for these artifacts.
9. Measure the drawn area by clicking "Ctrl-M."
10. Next, count all cells in the drawing. It may help to select every cell with the wand tool and click "Ctrl-F" to fill it to easily keep track of which cells were counted.
11. The obtained data can be copied from the Result Window to a spreadsheet program (e.g., MS Excel).

3.1.7. Calculations

Following the above procedures, the primary data for each leaf consist of the leaf blade area and two cellular drawings (tip and base of leaf): number of cells, total area of the drawn cells, and number of stomata. From these data we can derive the following:

1. Average leaf area can be determined as the average (and standard error) of all leaf areas per genotype/treatment at each time point (units: mm^2). Due to the exponential nature of the growth process, this parameter is best presented on a log-scale.
2. The total area of the drawn cells divided by the number of cells in it (pavement cells + guard cells (2 per stoma)), yields the average cell area in that drawing. Calculate an average (and standard error) of all the cell areas for all the leaves for each genotype and time point.
3. Calculate for each drawing the Stomatal Index (SI) by dividing the number of guard cells (number of stomata (S) multiplied by two to correct for the presence of two guard cells) by the total number of cells (pavement cells (PC) + guard cells):

$$SI = 2 * S / (PC + 2 * S).$$

4. Divide the leaf area by the average cell area from the same leaf to obtain the number of cells per leaf. Then averages (and standard error) of the number of cells per leaf for all the leaves per genotype/treatment and time point can be calculated. Due to

the exponential nature of the division process cell numbers are typically represented in log-scale with base 2.

5. The leaf expansion rate (LER) is the derivative of leaf area over time (on a log*n* scale). This derivative can be calculated using the LocPoly algorithm or using a spreadsheet such as MS Excel. An R-script for the LocPoly or an example of an Excel sheet with those calculations set up can be obtained from the corresponding author of this chapter. The basis of these methods is a local quadratic fitting of a polynomial to a small section of the data, which allows smoothing the data and calculating the local differential.
   (a) In Excel this requires setting up three columns: (1) time (in days after germination/sowing), (2) the quadrate of time, and (3) the natural log of average leaf area.
   (b) Typically a five-point quadratic fitting is used to calculate adjusted leaf area and derivative (LER) for the midpoint. Because this is not possible for the first and last two points of the series, these are calculated from the same fit as the third from the beginning and end of the series, respectively.
   (c) Calculations start from the third time point: in columns 4–7 the polynomial coefficients will be calculated using the function "Linest."
      - The first argument for this function is the *y*-values for 5 points (C2:C6).
      - The second argument is the X and X^2-values for 5 points (A2:B6).
      - The third and fourth arguments are "TRUE" for the use of a constant and "FALSE" for the output of the fitting statistics (these co-ordinates for cells are assuming that the data columns contain a row of headers).
      - This results in the following formula: = Linest(C2:C6, A2:A6, TRUE, FALSE).
      - To invoke the array-calculation after entering the data push the combination "Ctrl-Shift-Enter." This will write in the three selected cells the c, b, and a value of the quadratic fitting $ax^2 + bx + c$, respectively.
      - This array calculation can then be copied down up to the third last point.
      - Now it is possible to calculate the smoothed *y*-values ($a{*}X^2 + b{*}X + c$) and LER, the derivative ($2{*}a{*}X + b$), in two extra columns (see Note 5 for the rationale behind this smoothing approach).

– The first and last two points are calculated from their own *X* values and the coefficients of the fitting for the third points.

– When the *X* values are given in days, the obtained LER needs to be divided by 24 to obtain the rates as $mm^2/mm^2/h$.

6. Average cell division rate is the derivative of the cell number data with respect to time. The calculation is similar to the LER as explained above, using the $\log_2$ of the number of cells (cell/cell/h). Cell cycle duration can be calculated as the inverse of cell division rate. For additional information on effects on cell cycle phase duration these data can be combined with flow-cytometry measurements (see Note 6).

3.2. Analysis of Maize Leaf

The cells in the epidermis of monocotyledonous leaves are arranged in linear files. A kinematic method based on cell length profiling in function of position along the leaf was developed. This method assumes that during steady-state growth the cell length profiles are constant. It entails measurements of leaf elongation rate, the cell length profile along the axis of the leaf, and the estimation of the size of the leaf intercalary meristem. The method will be explained and discussed based on our experience with maize, but can be adapted to other monocotyledonous species. The procedures are designed to obtain all data from individual leaves, so that all parameter values can be replicated for each treatment to facilitate statistical analysis.

3.2.1. Plant Growth and Measurements of Leaf Elongation Rates

1. To perform a kinematic analysis of leaf growth in maize, a batch of at least ten plants per condition is necessary.
2. The leaf elongation rate (LER) in function of time is recorded for a first subset of at least five of these plants by measuring with a ruler the length of the leaf (soil surface to leaf tip) under study at regular time intervals, preferably daily (see Note 7 for automation and higher resolution approach).
3. This measurement involves straightening the leaf by hand and some caution is required not to break or damage as touching may influence growth rates. This procedure is carried out from leaf appearance (emergence from the sheath at the base of the shoot) until its complete extension.
4. From the recorded data, LER is calculated as the difference in leaf length on two successive time points divided by the time interval between them (in mm/h). For a typical monocotyledonous leaf, leaf elongation is generally linear during the first days after appearance, followed by a period of progressive decline depending on leaf position and treatment (environmental conditions). The first days of linear increase are considered as a situation of steady-state growth. At the cell level, it is

assumed that also cell production and cell length profiles in the growth zone are constant during the same time period (8).

5. To ensure that the plants grow at steady-state they should preferentially be grown under controlled conditions to avoid variations of leaf elongation rate. Especially fluctuations in temperature, light intensity, or humidity should be minimized, because they strongly influence growth rates (8, 13).

3.2.2. Sampling

For practical purposes the 4th or 5th leaf from maize seedlings are convenient for these analyses as they appear relatively early (2–3 weeks) after sowing and have a period of steady-state growth that lasts for several days. From a total population of at least ten plants per treatment, minimally 3–5 representative leaves are selected for kinematic analysis based on LER measurements during the first 3 days after emergence. For simplification of the kinematic calculations, the whole growth zone needs to be harvested during steady-state growth. The size of the growth zone depends on the environmental conditions, the species, the genotype, and the developmental stage examined. Therefore, it is important to make an estimate of this size to make sure that the samples encompass the full extent of the growth zone. For maize the growth zone of leaf 4 or 5 of maize B73 grown at 25°C spans about 100 mm (9), which can be used as a reference for the analysis of maize leaf growth.

1. To sample the growth zone, we need to remove the older leaves that surround the growing leaf.
2. Special care should be taken not to damage the basal part of the growing leaf because this is where the leaf basal meristem is located. Some practice beforehand to optimize the dissection technique is required.
3. Next, the growth zone is segmented into smaller pieces (for example, segments of 10 or 20 mm), because the growth zone is too large to mount as a whole on microscope slides.
4. Depending on their position along the leaf, samples are treated differently: in the meristematic zone, both observation of DAPI-stained nuclei and cell length measurements are needed and occur sequentially.
   (a) Therefore the basal 3–4 cm of each leaf are placed in 3/1 (v/v) absolute ethanol:acetic acid for fixation of cell walls and clearing of chlorophyll.
   (b) After determination of the size of the intercalary meristem, the samples used for DAPI staining are placed in lactic acid, for further clearing.
5. The samples of the remaining part of the growth zone are placed in absolute ethanol for 48 h for chlorophyll removal and fixation and afterwards they are cleared in lactic acid.

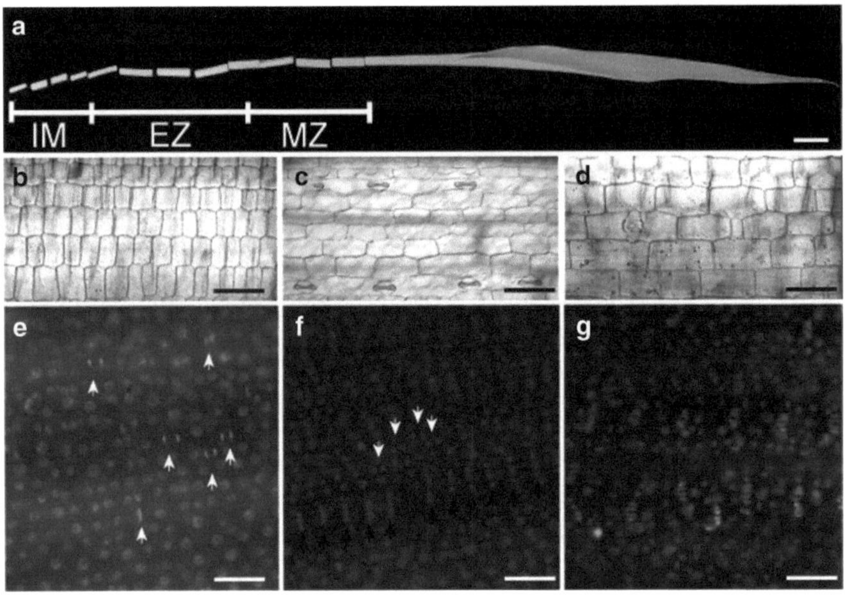

Fig. 2. Image analysis of the maize leaf. (**a**) Dissection of the growth zone of a growing leaf 4 into 4 0.5-cm and 8 1-cm pieces. (**b–d**) DIC of the leaf intercalairy meristem (IM: **b**), elongation zone (EZ: **c**) and mature zone (MZ: **d**). (**e–g**) DAPI fluorescence image of the leaf intercalairy meristem (basal IM: **e**, distal IM: **f**), early elongation zone (**g**). Scale bars indicate:1 cm (**a**) and 50 μm (**b–g**). *Arrowheads* indicate anaphase and telophase cells, identified as mitotical cells (**e**) and the result of formative divisions (**f**).

6. Once in lactic acid, the samples can be kept at 4°C from 24 h up to several weeks.

3.2.3. Estimation of the Extent of the Linear Extent of the Leaf Basal Meristem

1. On the basal sections, we first measure the size of the meristematic zone of the leaves. This involves the observation of the evolution of mitotically dividing cells in the meristematic zone. Therefore, we visualize the nuclei with DAPI.
2. The samples are rinsed with a buffer solution containing 50 mM NaCl, 5 mM EDTA, and 10 mM Tris–HCl (pH 7).
3. Next, the samples are incubated in the dark for 1–5 min in the same buffer solution containing DAPI at a concentration of 1 μg/ml (for an alternative stain, see Note 8).
4. Too intense staining should be avoided, since only staining of the epidermal cells and not of underlying cell layers is desired. Therefore, we check the samples for fluorescent signal emission after a short incubation.
5. According to this first microscopic assessment, the reaction is stopped when a satisfactory signal is achieved.
6. To stop the reaction, the DAPI is removed by rinsing in the buffer solution. Sample re-incubation is always possible if it is necessary to achieve a higher signal level.

7. The samples are mounted in a drop of the buffer solution on a microscope slide and covered with a cover slip. Take care to mount the samples with the abaxial epidermis faced up.
8. For image analysis a digital camera connected to a personal computer with image capturing software is required. If available software that facilitates measurements directly on the live image is most convenient.
9. Starting from the leaf base, recognizable mitotic cells (metaphase, anaphase, telophase) are scored per frame, preferably using a magnification, allowing to visualize around 1,000 cells per frame in order to be able to have a reasonable amount of mitotically dividing cells per frame.
10. From the base to the tip, first a steady amount of mitotically dividing cells is observed per frame (4–5% of all cells on the screen). At some point this percentage will drop, until finally no more mitotic figures are observed, which is considered as a proxy for the distal boundary of the meristem.
11. In this analysis only proliferative mitotically dividing cells are taken in account and not formative cell divisions (Fig. 2f). In the distal part of the meristem formative division occurs as part of the formation of the guard cells. They can be recognized as asymmetrical divisions, with often a division plane perpendicular to the proliferative divisions.
12. In practice, the measurement is done by counting the amount of frames from the base to the last mitotic cell dividing cell and multiplying this number by the corresponding distance along the leaf.
13. Once the measurements are completed, the samples are placed in lactic acid, for the cell length measurements.

3.2.4. Cell Length Measurements

1. For the cell length measurements the leaf (segments) are mounted on microscope slides.
2. The samples are unrolled on the object slide and the mid-vein is removed with a scalpel.
3. The samples are mounted in lactic acid, so that the abaxial epidermis, which generally contains fewer stomata compared to the adaxial side, is placed face up.
4. The specimens are analyzed under a microscope fitted with differential interference contrast (DIC) optics.
5. Find the orientation of the samples and start from the most basal part. In maize you can use the trichomes at the edge of the leaf, which point toward the leaf tip as a reference to distinguish base from tip directions.
6. Adjust the microscope to optimize the image focusing on the epidermal cell walls.

7. Moving in distal direction from the base of the leaf sample, all the cells belonging to one epidermal cell file are measured.
8. Typically, a cell file adjacent to stomatal rows on the abaxial side is chosen, since these cell files are convenient to recognize and consists of a single cell type.
9. The easiest and fastest way to handle these measurements is to directly measure the cells on the images captured "live" by a digital camera connected to a PC running image analysis software.
10. For ImageJ, plug-ins for a range of cameras are available from the program Web site.
11. Alternatively, it is possible to capture a series of overlapping pictures, which are subsequently used to make a composite image by stitching the individual images together. Several commercial photo-stitching programs are available, recent versions of Photoshop (Adobe) have a routine to do this (file/automate/photomerge) and there are also public domain ImageJ plug-ins available that can do this. When any of these routines are used make sure that there is no deformation of the images to make them fit better as this will lead to errors in the measured cell sizes.
12. To take into account the variation in cell size at different positions across the leaf, these measurements are repeated for a few equivalent cell files (e.g., cell files adjacent to stomatal rows) for each leaf.
13. The distance from the base to the midpoint of each cell is calculated in a spreadsheet. This is done by determining the cumulative length of all cells in more basal positions in the same file + 0.5x the length of the cell itself.
14. Afterwards, all data for a leaf can be combined by copying the data for the different cell files into the same position and size columns and sorting them for ascending position.
15. To allow averaging between leaves and further calculations, the next step is to determine the average cell length, $l(x)$, at specific, equally spaced positions along the leaf. To this end, we developed a local polynomial smoothing procedure (10). We have implemented this procedure as an R script that is available upon request from the corresponding author. This script gives a series of data with increasing smoothing. The amount of smoothing required is somewhat arbitrary and ideally should just remove the local noise, but not affect the overall curve. The script also calculates the local derivative of the cell length distribution. The data for $l(x)$ can be averaged (at x) between leaves and standard errors can be calculated.

3.2.5. Calculation of Overall Kinematic Parameters

Based on the measured LER, cell length profile and meristem size, the kinematic parameters for leaf elongation growth can be calculated:

1. The size of the growth zone (L_{gz}) is estimated as the distance from the leaf base to the position where the cells reach 95% of their mature length on the smoothed cell length profile.
2. The size of the elongation zone (L_{el}) is the difference between L_{gz} and L_{mer} (which was determined from the distribution of mitotic figures, see Subheading 3.2.3).
3. To determine the number of cells in each zone, the cell length data need to be transformed into a function of cell density. To this end the number of cells for each interval of the interpolated cell length data are calculated by dividing the size of the interval by the average size of the cells in it.
4. The number of cells in the meristem (N_{mer}) equals the cumulative number of cells located in the intervals corresponding to the meristem.
5. The number of cells in the elongation zone (N_{el}) equals the cumulative number of cells in all the intervals located within the elongation zone.
6. The number of cells in the whole of the growth zone (N_{gz}) equals $N_{mer} + N_{el}$.
7. Mature cell size l_{mat} can be calculated as the average cell length in the mature region.
8. Cell production in the whole of the meristem can be calculated as LER/l_{mat}.
9. The average cell division rate (D_{avg}) equals cell production rate divided by N_{mer} and the average cell cycle duration $T_c = \ln(2)/D_{avg}$.
10. In addition, the local cell elongation rate $r(x)$, which is calculated directly in roots, can be estimated in monocotyledonous leaves from the cell length profile. This parameter is calculated for all positions from derivative of cell length, given by the LocPoly algorithm, multiplied by the cell production: r(×)=P * l(×). Because in the meristem *P* is not constant, this formula can only be used for positions distal to the meristem boundary.

3.2.6. Flow Cytometry

Similar to the approach used for the *Arabidopsis* leaves, a deeper insight into the role of cell cycle progression in the leaf elongation can be obtained by complementing the kinematics with flow-cytometry. This technique allows the relative quantification of DNA nuclear content in the different regions of the growth zone and thereby it enables inferences of cell cycle progression status in the meristem and of endoreduplication in the elongation zone (9). However, bear in mind that this approach uses the entire leaf rather than the specific cell type analyzed using the microscopy, so that some caution needs to be taken into account with the interpretation.

4. Notes

1. To facilitate regular spacing of the plants, plates with a bottom grid are convenient. These provide 32 full squares on a 12 cm dish. For analysis of young leaf material up to 7 DAG (days after germination) the seeds can be sown densely with four seeds per square. For the time points 8–11 DAG two seeds are sown per square while at later time points only one seed per square can be used. This way the plants have enough space to develop without overlapping with each other.
2. Starch—Sometimes the leaf material does not clear completely after transfer to lactic acid. This is typically due to the presence of relatively high amounts of starch, which makes the imaging of the epidermal cells very difficult. To improve the clearing of the leaves, they can be transferred to Hoyer medium. The duration of clearing in Hoyer medium depends on the stage of the leaf material and on the amount of starch present. For young leaves 15–30 min are usually sufficient, while older leaves may require up to 2 h. It takes some trial and error to determine the desired clearing time. If the leaf material is kept too long in Hoyer medium, the clearing is excessive and the cell walls are no longer visible using DIC. Finally, the material needs to be transferred back to lactic acid after Hoyer treatment and mounted on object slides using lactic acid as mounting medium. The cell analysis needs to be done shortly (hours) after the Hoyer clearing, as the effect of the Hoyer clearing dissipates and clearing needs to be redone.
3. Cell types—For practical reasons we have opted to work on the abaxial epidermis of the first leaf pair as it does not have trichomes, which complicate the cell measurements. Cell divisions in this tissue are in transverse direction only. However, conceptually it is also possible to perform the calculations for the adaxial epidermis by either ignoring or taking into account the trichomes and accompanying cell complex. For the palisade parenchyma layers calculations are theoretically possible by either excluding the divisions perpendicular to the surface that increase the number of mesophyll layers or by including measurements of the number of palisade cell layers. In the spongy mesophyll individual cell layers are not easily defined, complicating the analysis probably beyond what is feasible to

work on. For addressing specific research questions it may be interesting to compare calculations on multiple cell layers.

4. Alternative to drawing tubus—In absence of a drawing tubus, images of cells can also be made using a digital camera mounted to the microscope and tracing the walls on a computer using the mouse or a drawing pad. The disadvantage of this method is that some areas of the image may not be focused optimally, resulting in some loss of accuracy.
5. Smoothing approach—Smoothing and interpolation of the primary cell length and velocity data is an important step in the calculations. Note that by using a local polynomial approach instead of fitting a predefined function, no assumptions as to the overall shape of the curve need to be made.
6. Flow-cytometry can be used to analyze the (nuclear) DNA content of the cells of the leaf using DAPI as a fluorochrome. This allows estimating the relative cell cycle phase duration during cell proliferation. Because the standard preparation involves chopping the entire leaf blade this analysis is not restricted to a single tissue as for the case of cell measurements. Nevertheless, we found that the transitions from proliferation to expansion and from expansion to maturity as defined by the kinematic analysis are closely reflected in the DNA profile (9), so that with some caution these data can effectively be combined.
7. LVDTs—A more advanced alternative to the daily manual (ruler) measurements of leaf length is using electronic systems, based on linear velocity displacement transducers (LVDTs). Using LVDTs the measurements of leaf extension are acquired at much higher time resolution (minutes or even seconds) (11, 12).
8. Alternative to DAPI/fluorescence—An alternative staining protocol for nuclei visualization is using Fuelgen staining, which allows doing the analysis without the need of an epifluorescence microscopy (12).

References

1. Poorter H, Garnier E (1996) Plant growth analysis: an evaluation of experimental design and computational methods. J Exp Bot 47(302):1343–1351
2. Goodwin RH, Stepka W (1945) Growth and differentiation in the root tip of Phleum pratense. Am J Bot 32:36–46
3. Erickson RO, Sax KB (1956) Rates of cell division and cell elongation in the growth of the primary root of Zea mays. Proc Am Philos Soc 100(5):499–514
4. Walter A, Silk WK, Schurr U (2009) Environmental effects on spatial and temporal patterns of leaf and root growth. Annu Rev Plant Biol 60:279–304
5. Fiorani F, Beemster GTS (2006) Quantitative analyses of cell division in plants. Plant Mol Biol 60(6):963–979
6. Candela H, Martínez-Laborda A, Micol JL (1999) Venation pattern formation in Arabidopsis thaliana vegetative leaves. Dev Biol 205(1):205–216

7. Donnelly PM et al (1999) Cell cycling and cell enlargement in developing leaves of Arabidopsis. Dev Biol 215:407–419
8. Ben-Haj-Salah H, Tardieu F (1995) Temperature affects expansion rate of maize leaves without change in spatial distribution of cell length. Analysis of the coordination between cell division and cell expansion. Plant Physiol 109:861–870
9. Beemster GTS et al (2005) Genome-wide analysis of gene expression profiles associated with cell cycle transitions in growing organs of Arabidopsis. Plant Physiol 138(2):734–743
10. Beemster GTS, Baskin TI (1998) Analysis of cell division and elongation underlying the developmental acceleration of root growth in Arabidopsis thaliana. Plant Physiol 116(4): 515–526
11. Basu P et al (2007) A novel image-analysis technique for kinematic study of growth and curvature. Plant Physiol 145(2):305–316
12. van der Weele CM et al (2003) A new algorithm for computational image analysis of deformable motion at high spatial and temporal resolution applied to root growth. Roughly uniform elongation in the meristem and also, after an abrupt acceleration, in the elongation zone. Plant Physiol 132(3):1138–1148
13. Granier C, Tardieu F (2009) Multi-scale phenotyping of leaf expansion in response to environmental changes: the whole is more than the sum of parts. Plant Cell Environ 32: 1175–1184

Chapter 18

Regeneration in *Arabidopsis* Tissue Culture

Kaoru Sugimoto and Elliot M. Meyerowitz

Abstract

An entire *Arabidopsis* plant can be regenerated from a small piece of tissue by two sequential hormonal treatments in tissue culture. Currently this in vitro regeneration system is a good system to study the mechanism by which plants show regenerative plasticity. Also, it is useful to test the hormone sensitivity of plants and to propagate sterile lines in *Arabidopsis*. Here we describe a standard protocol for regenerating *Arabidopsis* plants in tissue culture, and for preparing and observing samples using confocal microscopy to study cells during regeneration.

Key words: *Arabidopsis thaliana*, Plant regeneration, Tissue culture, Plant hormone, Callus formation, Shoot induction, Root induction

1. Introduction

It is well known that a small piece of plant somatic tissue (explant) can reproduce a full array of tissue types by regeneration. In 1957, Skoog and Miller reported that the relative level of plant hormones auxin and cytokinin directs the developmental fate of regenerating tobacco pith tissues (1). High cytokinin/auxin ratios induced shoots, low ratios induced roots, and when the both hormones were at high concentrations, a mass of growing cells (callus) was induced in explants. Since then, people have applied various culture conditions to a wide variety of tissue explants and have successfully observed regeneration in a large number of plant species.

The pattern of organ regeneration varies depending on the system used. Hicks (1980) classified various types of regeneration

Ive De Smet (ed.), *Plant Organogenesis: Methods and Protocols*, Methods in Molecular Biology, vol. 959,
DOI 10.1007/978-1-62703-221-6_18, © Springer Science+Business Media New York 2013

into two categories: direct or indirect organogenesis (2). In the case of direct organogenesis, shoots or roots are directly induced from tissue explants, while indirect organogenesis requires callus formation prior to shoot or root induction. The traditional scheme for *Arabidopsis* organ regeneration in culture involves indirect two-step procedures (3). Explants are first treated with auxin-rich callus-inducing medium (CIM) to form callus. Then, subsequent culture of the callus on shoot- or root-inducing medium (SIM or RIM), containing different ratios of auxin to cytokinin, induces shoot or root tissues, respectively.

In the 1980s, a number of *Arabidopsis* regeneration procedures were further optimized and developed together with *Agrobacterium*-mediated transformation experiments (4–6). Because dividing cells forming callus are susceptible to bacterial infection (6), *Agrobacterium* is inoculated in callus explants and regenerated plantlets are screened on CIM and SIM selective media containing antibiotics. Through these optimization experiments to achieve an efficient transformation system, various parameters have been investigated, and it has been found that regeneration capabilities vary depending on ecotype, tissue, and period of callus induction (3–6), although what mechanisms provide these differences have remained unknown. One of the most commonly used regeneration systems in *Arabidopsis* was developed by Valvekens et al. (4).

In 1993, Bechtold et al. reported a new transformation method in *Arabidopsis thaliana*, which requires neither tissue culture nor plant regeneration (7). Clough and Bent (1998) modified this method and developed the protocol of floral dipping *Agrobacterium*-mediated transformation (8). Thanks to this method, making transformants became much easier and more efficient in *Arabidopsis*, which was the only species to which this method was successfully applied. Since then, regeneration assays are rarely done for the purpose of transformation in *Arabidopsis*. They are now done for propagating sterile mutants, for testing sensitivity to hormones, and for studying the mechanism of plant regeneration, which is still largely unknown.

Recent technology of live imaging and transcriptome analysis has just started revealing the characteristics of regenerating cells. It is expected that further investigation will explain the phenomena previously observed during the optimization regeneration for transformation. Here we introduce a standard protocol for the formation of *Arabidopsis* callus and subsequent induction of shoots and roots to regenerate a whole plant based on previous publications (4, 6, 9, 10) and our own experience, and briefly describe how we prepare and observe samples using confocal microscopy for the study of regenerating cells.

2. Materials

2.1. Hormone Stocks

Prepare hormone stocks in the solvent indicated in parentheses. Generally, 100% ethanol or DMSO is used for dissolving auxins (NAA, IAA, IBA, and 2,4-D), and 1 N KOH or DMSO is used for dissolving cytokinins (kinetin, zeatin, BAP, and 2-iP) and ABA. Store the stock solution at −20°C.

1. 20,000× 2,4-D: 10 mg/ml (100% ethanol or DMSO).
2. 20,000× kinetin: 1 mg/ml (1 N KOH or DMSO).
3. 5,000× SIM hormone mix: 10 mg zeatin, 5 mg d-biotin, 2 mg IBA/ml (DMSO).
4. 10,000× IBA: 5 mg/ml (100% ethanol or DMSO).

2.2. Plant Growth Media and Tissue Culture Media

Media formulations introduced here are based on previous publications (4, 6, 11, 12) or slightly modified from them, in which case we have confirmed that plants are regenerated efficiently. Other conditions are also effective.

1. Murashige and Skoog (MS) medium (for 1 l): 1× MS salts (4.33 g), Sucrose (3 g).

   Adjust pH to 5.8 with 1 N KOH, make final volume up to 1 l, and add 8 g of phytoagar (Caisson Laboratories, North Logan, UT, USA). After autoclaving, cool medium to ~60°C, and then add antibiotics if necessary for selection. Pour into round Petri dishes in tissue culture hood.
2. CIM (for 1 l): 1× Gamborg B5 salts (3.1 g), Glucose (20 g), MES (2-(*N*-Morpholino)ethanesulfonic acid) (0.5 g), Gamborg Vitamins (1 ml).

   Adjust pH to 5.7 with 1 N KOH, make final volume up to 1 l, and add 8 g of phytoagar. After autoclaving, cool to ~60°C, and then add the following hormone stocks in tissue culture hood (see Notes 1 and 2): 20,000× 2,4-D (50 μl) and 20,000× kinetin (50 μl). Mix well and pour into round Petri dishes.
3. SIM: 1× MS salts (4.3 g), Sucrose (10 g), MES (2-(*N*-Morpholino) ethanesulfonic acid) (0.5 g), Gamborg Vitamins (1 ml).

   Adjust pH to 5.7 with 1 N KOH, make final volume up to 1 l, and add 3 g of phytogel. After autoclaving, cool medium to ~60°C, and then add SIM hormone mix in tissue culture hood (see Note 3): 5,000× SIM hormone mix (200 μl). Mix well and pour into round Petri dishes.
4. Root-inducing medium (RIM): The base medium for SIM and RIM is the same. Instead of SIM hormone mix, add 10,000× IBA stock solution (100 μl) to the above base medium.

   Mix well and pour into round Petri dishes, plastic boxes, or test tubes (see Subheading 2.3).

2.3. Containers for Pouring Media and Tools for Making Explants

1. Round Petri dishes 150 × 15 mm (for MS).
2. Round Petri dishes 100 × 15 mm (for CIM and SIM).
3. Sterile Plastic boxes (Durphy Packaging Company, Ivyland, PA, USA), or round Petri dishes 100 × 15 mm, or sterile test tubes.(for RIM).

   If plastic boxes and test tubes are not sterile, sterilize by autoclaving (if autoclavable) or by immersing in 95% ethanol for at least 30 min and then drying out in a tissue culture hood.
4. Gas-permeable filter tape (Carolina Biological Supply Company, Burlington, NC, USA).
5. Fine-point forceps (and a sterile razor blade if you make explants from cotyledons, leaves, and stems).
6. Alcohol lamp.

2.4. Dye Solutions for Counterstaining Tissues

Dissolve dyes in distilled water or DMSO and store in dark wrapped in aluminum foil or in amber tubes at 4°C.

1. 100× Propidium iodide (PI) stock solution

   5 mg/ml.
2. 10× FM4-64 and FM1-43 stock solution (Molecular Probes, Eugene, OR, USA)

   100 μg/ml.

3. Methods

3.1. Sterilization of Seeds and Growth of Seedlings

A vapor-phase method for sterilizing seeds is introduced here based on in ref. 8 (a brief liquid sterilization method is written in Note 4). Keep everything sterile as much as possible from the step of plating seeds.

1. Carry out this procedure in fume hood. Place seeds in open tubes in plastic sealed box (Tupperware). Add 3 ml of HCl carefully into a 150–250 ml beaker containing 100 ml bleach, and then place the beaker in the box. After sealing the box, let it stand for 3–5 h (Fig. 1). The box will be filled with chlorine fumes (see Note 5).
2. Open the box just a fraction in the fume hood and leave it 5–10 min to let chlorine fumes dissipate. Bring the seed tubes to the tissue culture hood (see Note 6).
3. Sprinkle the sterilized seeds over MS plates in the tissue culture hood by gently tapping the tube (an alternative method is given in Note 7). Seal the plate by wrapping gas-permeable filter tape around it twice (see Note 8).

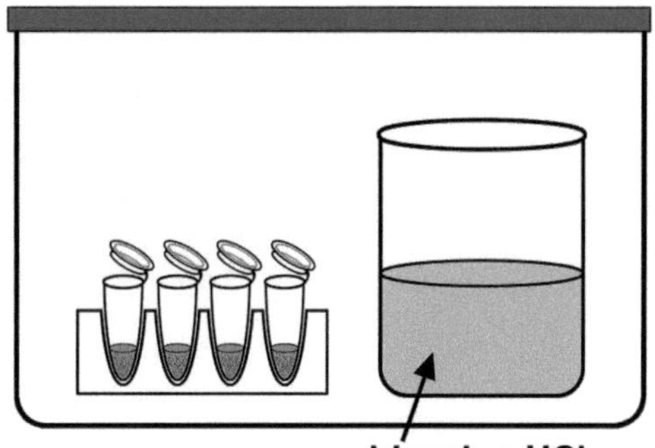

Fig. 1. A vapor-phase method for sterilizing seeds. Seeds tubes are kept open in plastic sealed box. Just prior to sealing the box, 3 ml of HCl is added into the bleach. The box with chlorine fumes should remain sealed in a fume hood 3–5 h.

4. Keep the plate in 4°C for 2–5 days to stratify the seeds so that they germinate simultaneously (see Note 9).
5. Transfer the plates to plant growth room with constant light at 22°C (see Note 10).
6. Allow the seeds to grow for 5–12 days (about 3–10 days after germination) (see Note 11).

3.2. Making Explants and Callus Induction

When handling explants, carry out the procedures in a tissue culture hood with sterile technique. Root is introduced as a tissue source for explants here, similar methods can be used for other tissues, though (see Note 12).

1. Dip forceps in the jar containing 95% ethanol, and then evaporate alcohol by flaming using alcohol lamp.
2. Pick up a seedling using the forceps from MS plate and put it on CIM plate (see Note 13). Hold the seedling with forceps and sever the root at 1 cm from the tip using another forceps (see Notes 14 and 15) (Fig. 2). Discard the above region including aerial tissues. Make sure entire explant contacts with the medium surface.
3. Seal plates by wrapping with gas-permeable filter tape twice (see Note 8).
4. Incubate the explants for 5–10 days under continuous light at 22°C (see Notes 10 and 16). Callus formation should be clearly visible by 5 days after the start of incubation.

3.3. Shoot and Root Induction

The shoot regeneration efficiency depends upon ecotype. It is reported that WS and C24 are efficient, while Col-0 and L-er are not (6, 13). According to our recent experiments, Col-0 also yields a reasonable number of shoots from root explants in the following

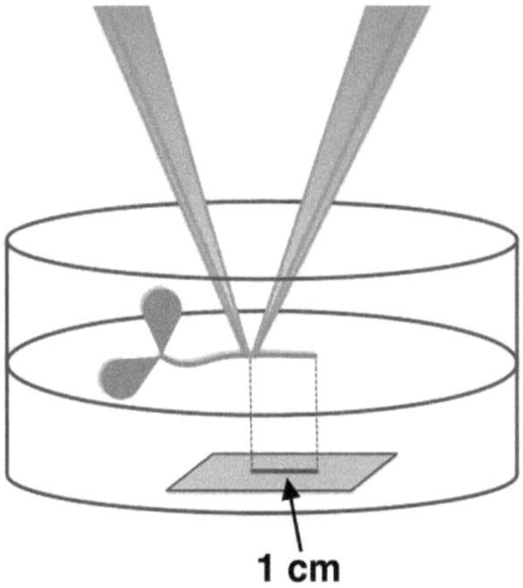

Fig. 2. Making 1 cm of root explants. CIM plate is placed on the tape with a 1 cm line. The line is visible through the agar. Seedlings are placed on CIM, and the position of the plate is adjusted so that the root of the seedling overlaps the line. The explant is excised at 1 cm from the root tip using two sets of sterile forceps.

method. This protocol is based on the *Arabidopsis* laboratory manual by Weigel and Glazebrook (9, 10).

1. Transfer explants onto SIM plates and seal the plates with filter tape (see Notes 8 and 17).
2. Culture explants under continuous light at 22°C (see Note 10). The explants will turn green after about 5 days. Shoots should appear within 2 weeks (especially at the edge part of the explants).
3. If the shoots are big (>4 mm in length), remove callus with forceps and transfer shoots directly to RIM plates or boxes (see Notes 18 and 19) (Fig. 3). If the shoots are still small (<4 mm in length), excise the shoots with callus (2–3 mm in diameter) from the explant and place on another fresh SIM plate. For the non-shoot forming region, break large callus into small pieces (2–3 mm in diameter) and place on fresh SIM plates. Calli go on to produce shoots when divided into small pieces. Incubate the small shoots or calli on SIM another 1–2 weeks until appropriately sized shoots are formed on the explants, and then transfer onto RIM. Seal the plates or boxes with filter tape.
4. Culture shoots on RIM plates under continuous light at 22°C (see Note 10). Roots will appear within 1–3 weeks. Carefully remove the plantlets from the medium and place in damp soil (see Note 20). Cover the soil pot with plastic wrap to maintain humidity.
5. After a few days, remove the plastic cover when the stems straighten up. Allow the plants to grow to set seed.

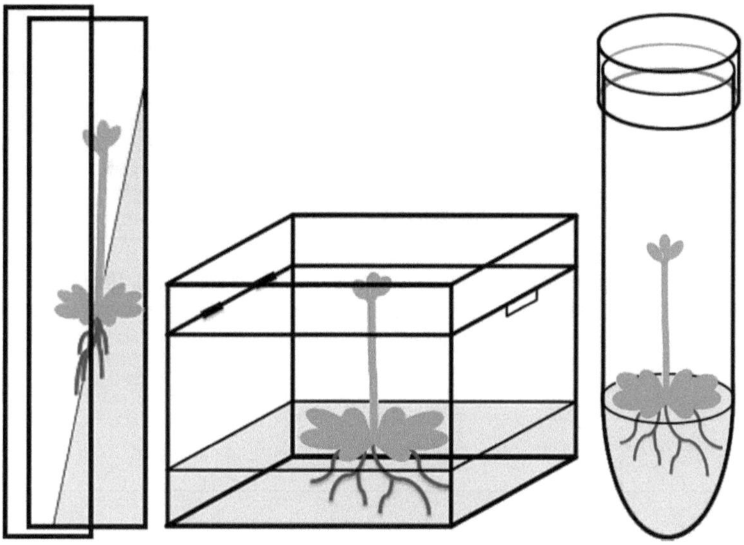

Fig. 3. Regenerating plantlets on RIM. RIM containers should be tall enough for the plantlets to grow. In the case of round Petri dishes, medium is poured at a slant and the dishes are placed vertically when regenerating plantlets are cultured.

3.4. Confocal Microscope Observation of Fluorescent Signals in Explants

Propidium iodide (PI) stains cell walls of a wide range of tissue types and nuclei of dead cells. FM4-64 and FM1-43 dyes nicely stain the plasma membrane of cells in shoot-derived tissues, but not in root and callus tissues (see Note 21). An alternative way is visualization of autofluorescence of chlorophyll.

1. Dilute dye-stock solutions with distilled water (see Note 22).
2. Drop the staining solution on explants. Remove air bubbles if there are any, using forceps or pipette. Leave the sample for more than 2–3 min at room temperature for PI staining and for at least 30 min at 4°C for FM staining (see Notes 23–25).
3. To stabilize the sample, overlay molten 1–1.5% agarose on the sample and medium surface in a thin layer (Fig. 4). Pour distilled water into the plate so as to cover the explant, and place on the upright microscope stage.
4. Use either 488 nm, 514 nm, or 543 nm laser to excite PI, and either a >585 nm long pass filter or 581–620 nm band pass filter for emission, depending on what other fluorescence sources are present in the sample. For example, if the subject tissue contains a lot of chlorophyll (excited by any laser, collected by a >650 nm long pass filter), choose a 581–620 nm band pass filter to cut off the autofluorescence of the chlorophyll. For FM4-64, use either a 488 nm, 514 nm, or 543 nm laser to excite, and >650 nm long pass filter for emission. For FM1-43, use either 477 nm or 488 nm laser to excite, and either >560 nm long pass filter or 530–615 nm band pass filter for emission.

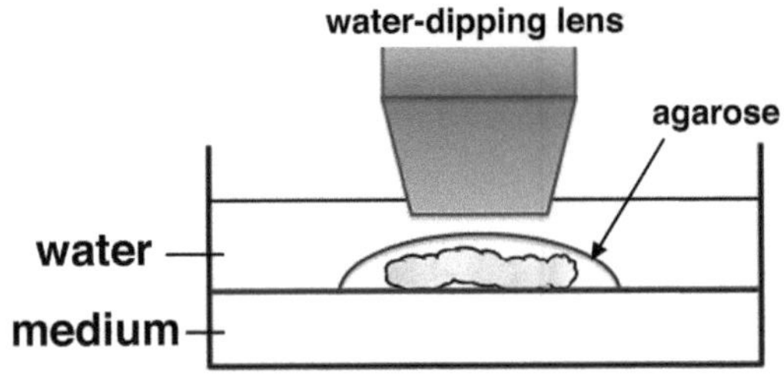

Fig. 4. Observation of explants using water-dipping lens. The sample is stabilized with a thin layer of molten 1–1.5% agarose. The plate is filled with water and placed on the upright microscope stage.

4. Notes

1. The sensitivity to auxin varies depending on tissue types. To induce callus from root, hypocotyl, cotyledon, pedicel, carpel explants, use standard CIM described above (in Subheading 2). To induce callus from leaf, stem, and petal explants, increase the concentration of 2,4-D to four times as much as the standard CIM.
2. If CIM contains transportable auxin such as NAA, instead of the non-transportable auxin 2,4-D, root explants form big lateral root-like structures, which elongate more than the normal 2,4-D-induced callus. When the concentration of NAA is raised higher (10 μM), apparently similar type of callus is induced as is in 2,4-D CIM, which does not elongate (14). But we find that this high concentration of NAA-induced callus rarely generates shoots, but roots, even when it is transferred to SIM. Therefore, 2,4-D is more suitable than NAA to regenerate shoots from root explants.
3. A common basal medium can be also used for all of the tissue culture media. Akama et al. (1995) reported very simple media formulations, where MS basal media (1× MS, 20 g/l glucose, 5 g/l gellan gum) supplemented with different hormone stock solutions were used for CIM, SIM, and RIM (13).
4. For liquid sterilization, soak seeds in 95% ethanol for 30–60 s, then 50% (v/v) bleach with 0.05% (v/v) Triton X-100 or Tween-20 for 5 min. Wash the seeds with sterile water three times. Spread the seeds on filter paper and dry in a tissue culture hood, or suspend in 0.1% MS top agar (see Note 7).
5. Use the permanent alcohol/water proof pen when you label the tubes, otherwise all the labels are bleached by the chlorine fumes.

6. Make sure that chlorine fumes gets out into the air only in a fume hood. Leave the bleach beaker in a fume hood overnight and discard once chlorine fumes have completely dissipated.
7. Melt down sterile MS top agar (1× MS basal medium with 0.5% phytoagar) using a microwave oven and cool the agar to 60–50°C (MS top agar should be autoclaved when it is first made. Melt the agar with microwave before use from a solidified, previously autoclaved stock). Suspend the sterilized seeds in MS top agar (approximately 5 ml per Petri dish). Spread the seed suspension over the plates, and let them dry a little, so that the seeds do not move when the plates are moved.
8. Wrap tape around the plate at least twice to avoid contamination from outside.
9. We find that leaving the plate in 4°C too long causes the seeds to rapidly form long thin roots with many lateral roots. The callus growth from these roots is also abnormal compared to the case where the cold treatment of the seeds is only a few days.
10. Make sure not to heat the plate from the bottom due to lights of a shelf below, otherwise the medium evaporates and seeds or explants are damaged.
11. The competency for shoot regeneration varies depending on the stage of the original tissues. For root explants, 3–10 days after germination is the optimal stage to make explants for efficiently yielding shoots. For cotyledon explants, younger stages (3–6 days after germination) of the tissues generate shoots.
12. For making cotyledon and leaf explants, excise cotyledon and leaf tissue from the plant and place on a sterile Petri dish. Cut into small pieces using a sterile scalpel or razor blade and place on CIM with the abaxial side in contact with the medium. Callus will be formed at the cut surface and along the vasculature. For petal explants, use a whole petal as an explant and place on CIM with the abaxial side down. The stalk region forms callus well. For stem and pedicel, cut into small pieces on a sterile Petri dish and place on CIM.
13. If there are many water drops on a lid of medium plate, remove them all by tapping or air dry the lid in a tissue culture hood. Wet plates are more likely to get contaminated by fungi.
14. The explants derived from apical side (the tip part) of the roots form callus, and later shoots, more efficiently than those derived from basal (hypocotyl) side of the root.
15. To make root explants, of 1cm length prepare a piece of tape, on which a 1 cm line is drawn with alcohol-proof pen. Put the tape in a tissue culture hood and place a CIM plate on it, and observe the line through the CIM agar. Put the seedling on

the CIM plate and adjust the position of plate so that the root of seedling overlaps the line. Cut the root at 1 cm from the tip (Fig. 2).

16. Competence for shoot regeneration varies depending on the stage of the original tissue and the pre-incubation time on CIM (see also Note 11). Shoot regeneration is severely perturbed when explants are made from old seedlings or pre-incubated on CIM for long times (>15 days).
17. The calli expand and shoots and leaves are formed on explants on SIM plates. Make enough room between explants so that the shoots derived from different explants will not get tangled.
18. Choose the shoots that contain apical meristem surrounded by leaves with regular phyllotaxy. Avoid glassy and bright-colored shoots.
19. In the case of round Petri dishes, medium is poured with the dish at a slant and placed vertically when regenerating plantlets are cultured (Fig. 3) (9, 10).
20. Plants from tissue culture are sometimes sterile, although flowers often form on the plantlets. It is suggested that the key step is root generation and soil transfer (10). Feldmann and Marks (3) reported that short exposure to RIM (10 days) followed by transfer to hormone-free medium (1–2 weeks) yields more fertile plants with a normal appearance. Also, lower concentration of IBA in RIM may promote root growth more greatly (IBA 0.3 mg/l in the case of Kakimoto T. (Osaka University, Japan), 0.02 mg/l in the case of Akama K. (Shimane University, Japan), personal communication) (6).
21. FM4-64 is useful to observe shoot primordium formation upon SIM as FM4-64 strongly stains shoot progenitor cells, but not the peripheral zone (15).
22. Resuspend dye-stock solution precipitate and mix well every time before diluting.
23. For time-lapse imaging, sterilize objective lens using ethanol and lens paper every time before use.
24. FM dye initially clearly stains plasma membrane and subsequently it becomes incorporated into endocytic vesicle membranes (16). To suppress this membrane dynamic activity, stain samples in 4°C and observe immediately after they are taken out of the cold.
25. To stain untreated leaf or petal explants, immerse in staining solution containing 0.1% Triton-X. If the tissue contains air bubbles between cells (such as young stage of cotyledon explants), mount the sample on a glass slide with 0.1% Triton-X and push down from the top using a coverslip.

Acknowledgments

We thank Kazuhito Akama, Toshiro Ito, and Tatsuo Kakimoto for valuable information about tissue culture media formulations, Cory Tobin and Paul Tarr for information and technical suggestions about FM dye staining, and Carolyn Ohno for information about seeds sterilization. This work was supported by the US National Science Foundation (Grant IOS-0846192 to E.M.M.) and the Japan Society for the Promotion of Science (to K.S.).

References

1. Skoog F, Miller CO (1957) Chemical regulation of growth and organ formation in plant tissues cultured in vitro. Symp Soc Exp Biol 54:118–130
2. Hicks GS (1980) Patterns of organ development in plant tissue culture and the problem of organ determination. Bot Rev 46:1–23
3. Feldmann KA, Marks MD (1986) Rapid and efficient regeneration of plants from explants of Arabidopsis thaliana. Plant Sci 47:63–69
4. Valvekens D, Montagu MV, Lijsebettens MV (1988) Agrobacterium tumefaciens-mediated transformation of Arabidopsis thaliana root explants by using kanamycin selection. Proc Natl Acad Sci USA 85:5536–5540
5. Schmidt R, Willmitzer L (1988) High efficiency Agrobacterium tumefaciens-mediated transformation of Arabidopsis thaliana leaf and cotyledon explants. Plant Cell Rep 7:583–586
6. Akama K, Shiraishi H, Ohta S, Nakagawa K, Okada K, Shimura Y (1992) Efficient transformation using hypocotyl explants of Arabidopsis thaliana: comparison of the efficiencies with various organs, plant ecotypes and Agrobacterium strains. Plant Cell Rep 12:7–11
7. Bechtold N, Ellis J, Pelletier G (1993) In planta Agrobacterium- mediated gene transfer by infiltration of adult Arabidopsis thaliana plants. C R Acad Sci Paris, Life Sci 316: 1194–1199
8. Clough SJ, Bent AF (1998) Floral dip: a simplified method for Agrobacterium-mediated transformation of Arabidopsis thaliana. Plant J 16:735–743
9. Weigel D, Glazebrook J (2002) Arabidopsis: a laboratory manual. Cold Spring Harbor Laboratory Press, New York
10. Huang H, Ma H (1992) An improved procedure for transforming Arabidopsis thaliana (Landsberg erecta) root explants. Plant Mol Biol Rep 10:372–383
11. Higuchi M, Pischke MS, Mahonen AP, Miyawaki K, Hashimoto Y, Seki M, Kobayashi M, Shinozaki K, Kato T, Tabata S, Helariutta Y, Sussman MR, Kakimoto T (2004) In planta functions of the Arabidopsis cytokinin receptor family. Proc Natl Acad Sci USA 101:8821–8826
12. Ozawa S, Yasutani I, Fukuda H, Komamine A, Sugiyama M (1998) Organogenic responses in tissue culture of srd mutants of Arabidopsis thaliana. Development 125:135–142
13. Akama K, Puchta H, Hohn B (1995) Efficient Agrobacterium-mediated transformation of Arabidopsis thaliana using the bar gene as selectable marker. Pant Cell Rep 14:450–454
14. Sugimoto K, Jiao Y, Meyerowitz EM (2010) Arabidopsis regeneration from multiple tissues occurs via a root development pathway. Dev Cell 18:463–471
15. Gordon SP, Heisler MG, Reddy GV, Ohno C, Das P, Meyerowitz EM (2007) Pattern formation during de novo assembly of the Arabidopsis shoot meristem. Development 134:3539–3548
16. Jelinkova A, Malinska K, Simon S, Kleinevehn J, Parezova M, Pejchar P, Kubes M, Martinec J, Friml J, Zazimalova E, Petrasek J (2010) Probing plant membranes with FM dyes: tracking, dragging or blocking? Plant J 61:883–92

Chapter 19

Isolation and Analysis of mRNAs from Specific Cell Types of Plants by Ribosome Immunopurification

Angelika Mustroph, M. Eugenia Zanetti, Thomas Girke, and Julia Bailey-Serres

Abstract

Multiple ribosomes assemble onto an individual mRNA to form a polyribosome (polysome) complex. The epitope tagging of specific ribosomal proteins can enable the immunopurification of polysomes from crude cell extracts derived from cryopreserved tissue samples. Through expression of the epitope-tagged ribosomal protein in cell-type and regional specific domains of *Arabidopsis thaliana* and other organisms it is feasible to quantitatively assess the mRNAs that are associated with ribosomes with cell-specific resolution. Here we present detailed methods for development of transgenics that express a FLAG-tagged version of ribosomal protein L18 (RPL18) under the direction of individual promoters with specific domains of expression, the immunopurification of ribosomes, and bioinformatic analyses of the resultant datasets obtained by microarray profiling. This methodology provides researchers with the opportunity to assess rapid changes at the organ, tissue, regional or cell-type specific level of mRNAs that are associated with ribosomes and therefore engaged in translation.

Key words: Ribosome immunopurification, Polysomes, Translatome, Cell-specific gene expression, Microarray, Differential expression analysis

1. Introduction

Each multicellular organ of a higher plant contains a number of specific cell types, which exhibit particular functions at structural, morphological, developmental, and biochemical levels. For example, the *Arabidopsis thaliana* root consists of an estimated 15 cell types (1). These are organized in radial symmetry from the epidermis to the central stele: epidermis (nutrient uptake, outer barrier), cortex (storage), endodermis (horizontal transport and inner barrier), pericycle (origin of lateral root meristems), phloem (assimilate transport), and xylem (water and mineral transport).

Ive De Smet (ed.), *Plant Organogenesis: Methods and Protocols*, Methods in Molecular Biology, vol. 959,
DOI 10.1007/978-1-62703-221-6_19, © Springer Science+Business Media New York 2013

Other specific root cell types include the lateral root cap, columella, and quiescent center. Individual cell types have distinguishing features. For example, cortex cells are generally equipped with a thin cell wall and endodermal cells are characterized by the Casparian strip that forms from radial deposition of suberin and/or lignin to limit diffusion of water to the central cylinder. Even cells of individual layers can have distinct identities, such as hair forming (trichoblast) and non-hair forming (atrichoblast) cells of the root epidermis. Cells of a specific type can also be distinguished by position, such as along the longitudinal axis of a root divided into the stem cell niche, division, elongation, and maturation zones.

During organ development, cells emanating from the meristem progressively establish specific identities. Cell differentiation is mediated by a number of processes involving hormonal signaling networks and other factors (i.e., (2, 3)). Key players in cell differentiation are transcription factors of both cell autonomous and non-cell autonomous function (4). For example, the differentiation between root cortex and endodermis in *Arabidopsis* is mediated by two transcription factors, SCARECROW (SCR) and SHORTROOT (SHR) (5, 6). Another example is the differentiation of root epidermal hair-cells and non-hair-cells via a complex regulatory network involving transcriptional regulators including TRANSPARENT TESTA GLABRA (TTG1), GLABRA3 (GL3), ENHANCER OF GLABRA3 (EGL3), WEREWOLF (WER), CAPRICE (CPC), GLABRA2 (GL2), and several others (7, 8). The processes of differentiation, determination, and function of individual cell types remain to be further explored. Questions that can be asked of developmental processes include: (1) how is the differentiation of specific cell types initiated; (2) which genes mediate the cell differentiation processes; (3) what is the temporal and spatial regulation of genes within specific cell types; and (4) which genes function to provide specific cellular identities? Subsequent to cell differentiation, questions can be asked about the common and exceptional processes of differentiated cell types. For example, (1) how does position within an organ influence gene regulation; (2) how do environmental parameters including the diurnal cycle, nutrient availability, abiotic, and biotic stress influence activities of specific cells?

Until now most of the genes involved in cell differentiation, identity, metabolism, and response mechanisms have been discovered through the characterization of mutants in *Arabidopsis* and other plants. However, forward and reverse mutational approaches cannot reveal all key regulators and their associated networks involved in cell-specific processes because these regulators are frequently encoded by functionally redundant genes. Another complication is that mutants affecting essential cell-specific regulatory processes can be lethal. An added challenge is that the mRNAs of individual cell types are not easily isolated from a multicellular organ. Further detailed evaluation of genes and regulatory networks of individual cell types demands approaches that allow investigators to quantitatively evaluate the mRNA of targeted cells.

1.1. Deciphering Gene Expression in Subpopulations of Cells Within Organs

The gene transcripts of individual cell types and populations can be isolated by at least three different methods: (1) mechanical dissection of cells from organs, for example, by Laser Capture Microdissection (LCM) (i.e., (9–12)); (2) fluorescence-assisted cell or nuclei sorting (FACS) of a subpopulation of cells or nuclei expressing GFP (i.e., (1, 13–17)); and (3) immunopurification of cell-specific-tagged ribosomes (18, 19) or nuclei (20).

The advantage of the first method is that it can be used with all kinds of plants, regardless of whether or not they can be genetically transformed. The second and third methods require the use of transgenic plants and have only been reported for the model plant *Arabidopsis thaliana* (see above). However, LCM is readily performed with tissues of relatively large size, but is extremely challenging for thin cell layers or low abundance cell types because of contamination by non-target cell types. Furthermore, this approach requires specific equipment and preparation of the tissue, precluding its use to evaluate transient changes in gene regulation.

The use of FACS to isolate a subpopulation of GFP-tagged cells from a multicellular organ can be used for all cell types of the root, including the quiescent center (21). This approach has been adapted for the shoot apical meristem (22). However, until now this approach has not yet been used for the analysis of cell types within photosynthetic organs. This is most likely because of technical difficulties associated with fluorescence of chlorophyll or difficulty in efficiently protoplasting all leaf cells equally well, including the epidermal cell layer. One experimental consideration with this technique is that it takes 1–2 h to generate protoplasts. Even when performed in the presence of an inhibitor of transcription, transcriptional changes may occur. This may be especially relevant if the goal is to capture transient changes in gene regulation. Nevertheless, the sorting of GFP-tagged cell types has been elegantly used to reveal a wealth of cell-type specific data from *Arabidopsis* seedlings (1, 13–15, 21, 22) and is a rich source for the selection of putative target genes involved in development. A slight modification of this technique was established by GFP-tagging of a core histone in specific cell types, which results in GFP-labeled nuclei, which can be obtained from gently disrupted tissues without the requirement for protoplasting (16). However, this method would only allow the isolation of nuclear RNA from specific cell types.

The third technique involves the immunopurification of ribosomes associated with mRNAs, either as monoribosomes (monosomes) or polyribosomes (polysomes). The method of immunopurification of mRNAs associated with ribonucleoprotein complexes (mRNPs), including the ribosome, is generally referred to as m*R*NP *i*mmuno*p*urification (RIP) (23), but has also been called *t*ranslating *r*ibosome *a*ffinity *p*urification (TRAP) (19). RIP of plant ribosomes is effectively accomplished by use of an amino-terminally FLAG-epitope-tagged version of ribosomal protein L18 (RPL18) (24). This method was established using the near-constitutive Cauliflower mosaic virus 35S promoter to drive FLAG-RPL18 in transgenic *Arabidopsis*. Subsequently, the method was tailored to capture mRNAs from specific cell types by driving FLAG-RPL18 expression with promoters expressed at a regional or cell-type-specific level (18, 19). This method has several advantages over LCM or FACS of GFP-marked protoplasts. First, RIP can be used with frozen material, allowing samples to be harvested with minimal manipulation and no fixation. Second, it can be applied to the seed, vegetative, and reproductive organs, as well as pollen. Third, the technique analyzes mRNAs associated with ribosomes, providing information on gene activity that better approximates protein production than steady-state transcript levels. We refer to the ribosome-associated mRNAs of the cell as the translatome, in contrast to the total mRNA pool or transcriptome.

Several studies have demonstrated quantitative differences between the transcriptome and translatome of plants, especially during stress conditions (25–31). These distinctions arise from the competition between mRNAs for translation coupled with active regulation of mRNA sequestration and degradation (32–34). In addition, the translation of a subset of cellular mRNAs in plants may be regulated by miRNAs (35) as well as targeting to subcellular locales (34). Because translation is primarily controlled during the recruitment of the 43S initiation complex to the mRNA, it is presumed that most mRNAs associated with polysomes are actively translated. By tagging RPL18, a protein of the large ribosome subunit (60S), the mRNAs isolated by RIP includes only those associated with a 80S ribosome or polyribosome complexes. In the following section, we describe the use of this technique and give points for consideration in the planning and execution of this methodology.

Regardless of the technique chosen for the isolation of cell-specific populations of mRNAs, several things need to be considered. First, all three methods yield small amounts of RNA for analysis. The yield depends on the number of cells per cell type in an organ and on the developmental stage (Table 1). Therefore, it may be necessary to amplify the mRNAs obtained before hybridization to gene or tiling microarrays (RIP-chip) or high-throughput RNA sequencing (RIP-RNAseq). Second, the mRNA analysis requires comparison to other mRNA populations, such as the total mRNA of an organ, or RNA of non-overlapping cell types. In the following sections, we provide guidance for the management of these challenges.

Table 1
Yields of immunopurified polysomal RNA from seedlings using different cell-type specific promoters[a]

		Expected RNA yield (ng/mL of tissue)	
Promoter	**Target cell type**	**Shoot**	**Root**
p35S	Near constitutive	1,000–1,500	200–600
pRBCS	Shoot photosynthetic	300–800	n.d.
pRPL11C	Root proliferating cells	n.d.	200–600
pSUC2	Root and shoot phloem companion cells	15–50	10–30
pSULTR2;2	Root phloem companion cells, shoot bundle sheath	15–50	10–30
pGL2	Root atrichoblast epidermis, shoot trichomes, but is also expressed at low levels in root phloem cells	40–150	20–60
pCO_2	Root cortex meristematic zone	n.d.	10–30
pPEP	Root cortex elongation and maturation zone	n.d.	20–60
pSCR	Root endodermis, quiescent center	n.d.	50–150
pSHR	Root vasculature	n.d.	50–150
pWOL	Root vasculature	n.d.	50–150
pCER5	Cotyledon and leaf epidermis	30–100	n.d.
pKAT1	Cotyledon and leaf guard cells	5–10	n.d.

[a]These yields are from 7-day-old *Arabidopsis thaliana* (Col-0) seedlings of specific transgenic genotypes grown on 0.43% (g/v) MS medium additionally containing 1% (g/v) sucrose (18). Yields from other starting material will vary depending upon expression of promoters in the cell types of the selected sample. *n.d.* not determined

1.2. Selection of Suitable Promoters

The RIP technique works as follows: a protein subunit of the large ribosomal subunit of the cytoplasmic ribosome was engineered with an amino His_6-FLAG or carboxyl FLAG-His_6 and overexpressed in *Arabidopsis* plants (24). Of several ribosomal proteins tested, RPL18 appeared to be the most suitable target based on a number of criteria. It should be noted that transgenics that produce FLAG-RPL18 encode three endogenous *RPL18* genes. Since ribosomes possess a single copy of RPL18, only a subset of the ribosomes in a cell that produces FLAG-RPL18 contains the tagged protein. However, since multiple ribosomes associate with mRNAs to form polysomes, RIP can be used to efficiently isolate all of the polysomes that possess at least one FLAG-tagged ribosome. The actual immunopurification is accomplished by generation of a cell extract with a membrane and cytoskeleton-disrupting buffer that maintains polysome complex integrity, followed by ribosome capture with anti-FLAG-agarose beads. After the release of the ribosomes from the agarose beads by use of excess 3×-FLAG peptide, the ribosome-associated mRNA can be purified and used for subsequent analyses. The complexes obtained include 60S ribosomal subunits, monosomes, and small to large polysomes, in proportions to the amounts of ribosome complexes present in crude cell extracts (24).

To isolate mRNAs from a specific cell type, the FLAG-RPL18 construct can be expressed under the control of a cell-type specific promoter. We have done this using stable transformation of *Arabidopsis* plants, but transient transformation of plant cells may also be possible (i.e., transformation of *Medicago truncatula* hairy root cultures by the use of *Agrobacterium rhizogenes* or viral-mediated transient transformation of *Nicotiana benthamiana*). To date, we have used several promoters active in the root (*pGL2* for epidermis (36), *pCO2* and *pPEP* for cortex (37, 38), *pSCR* for endodermis (39), *pSHR* and *pWOL* for stele (40, 41), *pSUC2* and *pSULTR2;2* for phloem companion cells (42, 43), and *pRPL11* for meristem (44)) and shoot cell types (*pGL2* for trichomes (45), *pCER5* for epidermis (46), *pKAT1* for stomata (47); *pSUC2* and *pSULT2;2* for phloem companion cells (42, 43), and *pRBCS* for leaf mesophyll cells (48)). *pLAT52* was used to drive *FLAG-RPL18* in developing microspores and pollen (49) (J.B.S. unpublished). Others have used *APETALA1* (*pAP1*), *APETALA3* (*pAP3*), and *AGAMOUS* (*pAG*) to drive FLAG-RPL18 expression in specific domains during early flower development (19). The use of a promoter with yeast upstream activation sequence (UAS) elements and GAL4 expressed in specific cell types (50, 51), or any other effective promoter-driver system, might also be used to regulate the expression of FLAG-RPL18 in specific cell types.

The selection of a promoter for cell-specific RIP is crucial: (1) the promoter needs to be highly specific, without background activity in other cells of the same organ; (2) the promoter needs to be active during ribosome biogenesis, in this case the FLAG-RPL18 protein will be incorporated into ribosomes; and (3) the promoter needs to be highly active in the desired cell type so that the tagged RPL18 is abundant.

During the establishment of the method, we made Gateway-compatible constructs for recombinational introduction of specific promoters upstream of the epitope-tagged RPL18 in a binary T-DNA plasmid. The use of the Gateway system (Invitrogen, Carlsbad, CA, USA) is described elsewhere (52). We suggest the use of two different gene constructs to establish the method for a new promoter: pGATA:HF-RPL18 for RIP, and pGATA:HF-GFP-RPL18 to test the promoter specificity, strength, and sensitivity. Both are available in a binary T-DNA plasmid. Although it would be convenient to use the pGATA:HF-GFP-RPL18 construct for both RIP and promoter characterization, we do not know if the additional GFP fusion would interfere with ribosome function or polysome formation. An alternative would be to use only pGATA:HF-RPL18, and test the promoter specificity by immunolocalization of the FLAG peptide, but in our hands this approach was insufficiently sensitive.

An important consideration in the selection of the promoter is whether or not it will be expressed at a time when ribosomes are synthesized or assembled. The temporal regulation of ribosome biogenesis is not well characterized in plants. However, based on patterns of transcription of rDNA and accumulation of ribosomal protein mRNAs, the process occurs in growing tissues, particularly in zones of active cell division such as in apical and lateral meristematic regions (53). Most likely, ribosome biogenesis continues during development but becomes limited in mature tissues. Nevertheless, we have successfully immunopurified polysomal mRNAs from mature organs.

With the pGATA:HF-GFP-RPL18 construct, one should produce transgenic plants and observe the GFP signal by conventional and confocal fluorescence microscopy. We recommend evaluation of 20–40 independent transgenic T1 plants. The goal of the GFP assessment is to determine (1) if the promoter is reliably expressed in the target cell type; (2) if the promoter is strong enough for GFP to be detected within the nucleolus and cytoplasm to be visibly in the targeted cell type at the required developmental stage, and (3) if fluorescence is detectable in other cell types of the same organ (this is not desirable). It is especially important to also observe older cells that have a large vacuole and little visible cytoplasm. During our pilot project we found one promoter, *pGL2*, had the anticipated expression in non-hair epidermal cells, but was also expressed at low levels in phloem companion cells. This led to challenges when analyzing the microarray data generated from RIP with roots from the *pGL2:FLAG-RPL18* line (18).

2. Materials

2.1. General Remarks

- This method is a slightly modified version of Mustroph et al. (54).
- All solutions and equipment used in this protocol need to be free of RNAse. Glassware, Miracloth, pipette tips, tubes, and solutions must be sterilized by autoclaving for 15 min.
  All steps are carried out on ice or at 4°C.
- Unless otherwise stated, all solutions are prepared with sterile deionized water.
- Harvest plant material directly into liquid nitrogen, and grind it into a fine powder using sufficient liquid nitrogen to maintain a

frozen state. Pulverization can be accomplished with a porcelain mortar and pestle. Store the pulverized tissue at –80°C until use.

2.2. Equipment

- This technique requires the use of transgenic *Arabidopsis* or other plant expressing a FLAG-tagged ribosomal protein (18, 19).
- Preparative centrifuge with fixed angle or swinging bucket rotor accommodating 30 mL tubes (i.e., Beckman J2-21 high-speed centrifuge and JA-20 rotor, fitted with rubber inserts to accommodate 15 mL or 30 mL Corex tubes).
- Low-speed benchtop centrifuge with swinging buckets for 15 mL or 50 mL Falcon tubes (required speed 8,200 × *g*).
- Rocking shaker, capable of shaking at about 60 rpm/min.
- Eppendorf or other benchtop microcentrifuge capable of centrifugation at 16,000 × *g*.

2.3. Solutions and Chemicals

- Kanamycin.
- Heparin (see Note 1).
- EZview-α-FLAG agarose beads (Sigma, product number F2426).
- $FLAG_3$ peptide (Sigma, product number F4799).
- RNAse inhibitor (40 U/μL) (see Note 1).
- Qiagen RNeasy kit (Catalog # 74904).
- 8 M Guanidine-HCl, autoclaved.
- 99% (v/v) ethanol.

The following stock solutions are autoclaved and stored at room temperature for up to 1 month:

2 M Trizma base (Sigma T1503) adjust to pH 9.0 with HCl.

2 M KCl.

0.5 M Ethylene glycol-bis(2-aminoethylether)-*N*,*N*,*N'*,*N'*-tetraacetic acid (EGTA), adjust to pH 8.0 with 10 M NaOH (see Note 2).

1 M $MgCl_2$.

20% (v/v) Polyoxyethylene 10 tridecyl ether (PTE) (see Note 3).

10% Sodium Deoxycholate (DOC) (see Note 4).

20% Detergent mix (see Note 5): 20% (w/v) polyoxyethylene(23)lauryl ether (Brij-35), 20% (v/v) Triton X-100, 20% (v/v) Octylphenyl-polyethylene glycol (Igepal CA 630), 20% (v/v) polyoxyethylene sorbitan monolaurate 20 (Tween 20).

Solutions NOT to be autoclaved, stored at –20°C in aliquots:

0.5 M Dithiothreitol (DTT).

50 mg/mL Cycloheximide, dissolved in ethanol.

50 mg/mL Chloramphenicol, dissolved in ethanol.

0.5 M Phenylmethylsulfonyl fluoride (PMSF), dissolved in isopropanol.

2.4. Buffers

Buffers should be prepared on the day of each experiment and kept on ice. Amount of stock solution for 50 mL solution at indicated final concentration is mentioned between brackets.

- Polysome Extraction buffer (PEB): 0.2 M Tris-HCl, pH 9.0 (5 mL), 0.2 M KCl (5 mL), 0.025 M EGTA (2.5 mL), 0.035 M $MgCl_2$ (1.75 mL), 1% Detergent mix (see Notes 6 and 7) (2.5 mL), 1% DOC (see Note 8) (5 mL), 1% PTE (see Note 8) (2.5 mL), 5 mM DTT (0.5 mL), 1 mM PMSF (0.1 mL), 50 μg/mL Cycloheximide (50 μL), 50 μg/mL Chloramphenicol (50 μL), 0.5 mg/mL Heparin (see Note 1).
- Wash buffer: 0.2 M Tris-HCl, pH 9.0 (5 mL), 0.2 M KCl (5 mL), 0.025 M EGTA (2.5 mL), 0.035 M $MgCl_2$ (1.75 mL), 5 mM DTT (0.5 mL), 1 mM PMSF (0.1 mL), 50 μg/mL Cycloheximide (50 μL), 50 μg/mL Chloramphenicol (50 μL), 20 U/mL RNAse inhibitor (see Note 1) (25 μL).

3. Methods

3.1. Establishing Stable Arabidopsis Transformants

For the isolation of cell type-specific mRNA populations by RIP, stable transgenic lines can be produced by routine transformation methods. A suggested procedure for establishment of lines expressing FLAG-RPL18 or FLAG-GFP-RPL18 is given here. Usually, RIP is performed on more than 2 g of plant tissue, ideally with several biological replicate samples, therefore one should have sufficient genetic material for the desired experiment. This protocol can also be adapted for other plant species, but species-specific characteristics may need to be considered.

The following steps should be carried out to obtain stable *Arabidopsis* transgenic lines:

1. Transform *Arabidopsis* by the floral dip method (55). The T1 seeds obtained from the transformed *Agrobacterium* infected inflorescence are selected on solid 0.5× MS agar medium additionally containing 50 mg/L Kanamycin, since the vector pGATA:HF-RPL18 carries the *NPTII* gene for transformant selection.
2. For *FLAG-RPL18* lines: From leaves of T1 plants, extract DNA and verify the presence of the construct by the use of PCR with vector-specific primers. From the transgene positive plants, confirm expression of the FLAG-tagged protein by western-blot analysis of leaf extracts with a FLAG antibody

(for example, Sigma F1804). If the promoter is specific for root cell types, perform the analysis on a portion of T2 seedlings (see below). Be aware that cell types with a small percentage of the whole organ will yield only a small amount of protein. We typically select the lines with the average level of FLAG-tagged RPL18 accumulation in a group of transgenics. If the promoter is expressed in a very small proportion of cell types, it may be necessary to perform a RIP to detect the tagged protein (see below).

3. For *FLAG-GFP-RPL18* lines: Confirm expression of the GFP-tagged protein by fluorescence microscopy. This generation can be used to characterize the promoter; however, single copy insertion homozygotes can also be established as described for *FLAG-RPL18* lines.
4. Collect T2 seeds from positively tested plants.
5. Select T2 seedlings again on Kanamycin-containing MS plates (usually from 12 to 20 independent T1 plants) and chose those lines that show a 3:1 (Kanamycin-resistant: Kanamycin-sensitive) segregation. Lines with this segregation ratio most likely contain one T-DNA insertion. Plant 12–24 plants per line.
6. Collect T3 seeds from the T2 plants.
7. Select T3 seedlings on Kanamycin-containing MS plates. Choose those that show 100% resistance to Kanamycin, since those should be homozygous for the T-DNA insertion carrying the *NPTII* and *FLAG-RPL18* transgenes. To date, we have not observed abnormal growth phenotypes associated with expression of FLAG-RPL18 in diverse cell types.
8. Collect tissue samples (i.e., leaves) of the T3 plants (these can be pooled from multiple plants) and perform additional experiments to characterize the plants. These can include: (a) Southern Blot analysis to confirm single insertion events; (b) TAIL-PCR to detect insertion site of the T-DNA; and (c) a pilot IP experiment to confirm the yield of the IP procedure using the appropriate organ (see Subheading 3.2–3.5).
9. Collect T4 seeds from the T3 plants.
10. At this stage, or at the stage of T5 seeds, large-scale RIP experiments can be carried out for microarray or RNAseq analyses.

It takes about 10–12 months to get to the T4 generation for *Arabidopsis* plants. When carefully followed, the resulting transgenic line should be stable in subsequent generations. We recommend the selection of seedlings from subsequent generations on Kanamycin, isolation of inflorescences during flowering to limit cross-pollination, as well as the harvest of seeds from individual plants to avoid contamination of lines.

The RIP method is also useful when carried out in mutant background or in different *Arabidopsis* accessions. In the first case, a known mutant or another transgenic (i.e., an overexpression line) can be crossed with an established line expressing FLAG-RPL18 under the control of a near-constitutive (i.e., 35S) or cell-specific promoter. Alternatively, the mutant can be directly transformed with the promoter:FLAG-RPL18 construct according to the protocol described above. In the second case, different accessions could be transformed, but there might be variations in the transformation efficiency of different genotypes. Furthermore, it would be essential to check for the specificity of cell-type specific promoters in different *Arabidopsis* accessions using a FLAG-GFP-RPL18 construct driven by the promoter.

3.2. Tissue Extraction

- Estimate volume of pulverized tissue powder, and add two times the volume of freshly prepared PEB (see Subheading 2.4). For RIP to be followed by quantitative mRNA analyses (microarray or RNAseq), use at least 2.5 mL of packed leaf tissue and 5 mL of PEB. RIP from seedlings requires more tissue than for leaves. Note, the yield is rRNA from the isolated ribosomes and the associated mRNA.
- Let the mixture thaw on ice.
- Carefully homogenize the mixture by use of a glass homogenizer.
- Leave the mixture on ice for 10 min (or until all samples are prepared).
- Centrifuge the samples at 4°C, 16,000×*g*, for 15 min in a microcentrifuge.
- Filter the supernatant through Miracloth into another sterile tube. Repeat the centrifugation step to ensure removal of material that pellets at 16,000×*g*.
- If desired, save 10% of the clarified extract to isolate total RNA. Note, if steady-state mRNA is to be evaluated use the same extraction buffer as for the RIP.

3.3. Preparation of the α-FLAG M2 Agarose Beads

- Resuspend the α-FLAG M2 agarose gel in the reagent vial by gentle shaking to make a uniform suspension. Transfer 150 μL of the agarose gel to a new 1.5 mL tube. Use a cut pipette tip for easier transfer.
- Centrifuge at 8,200×g at 4°C for 60 s.
- Remove the supernatant with a Pasteur pipette, add 1.5 mL of wash buffer (see Subheading 2.4), and resuspend beads.
- Centrifuge at 8,200×g at 4°C for 60 s.
- Remove the supernatant with a pipette and wash one more time with 1.5 mL of wash buffer before continuing with the ribosome immunoprecipitation.

3.4. Ribosome Immunoprecipitation

- Mix 5 mL of the clarified extract (see Subheading 3.2) with 150 μL of washed α-FLAG M2 agarose beads (see Subheading 3.3) in a 15 mL plastic Falcon tube. Amount of α-FLAG M2 agarose beads can be increased when more extract is used (see Note 9).
- To bind the FLAG-tagged ribosomes to the agarose beads, incubate for 2 h at 4°C with gentle back-and-forth shaking on a rocking platform.
- Centrifuge for 60 s at 8,200 × *g* at 4°C.
- Transfer the supernatant to a new tube. This is the supernatant of the immunoprecipitation or unbound fraction (see Note 10).
- To the beads that have bound the ribosomes, add 6 mL of PEB, mix by gently inverting the tube, incubate at 4°C for 5 min with gentle shaking on a rocking platform, and centrifuge for 60 s at 8,200 × *g* at 4°C (First wash).
- Remove the supernatant with pipette and add 6 mL of Wash buffer to the beads (see Subheading 2.4). Incubate at 4°C for 5 min with gentle shaking (Second wash).
  Centrifuge for 60 s at 8,200 × *g* at 4°C.
- Remove the supernatant with pipette and add 6 mL of Wash buffer to the beads. Incubate at 4°C for 5 min with shaking (Third wash).
- Repeat wash again, for a total of four washes (it is not necessary to save the wash buffer).
  Centrifuge for 60 s at 8,200 × *g* at 4°C.
- Remove the supernatant. To elute the affinity purified ribosomes, use a fine-tipped pipette to remove as much of the supernatant as possible. Add to the beads 300 μL of wash buffer containing 200 ng/μL of $FLAG_3$ peptide, and 20 U/mL RNAse inhibitor (see Note 1). Incubate for 30 min at 4°C with shaking on a rocking platform.
- Centrifuge for 60 s at 8,200 × *g* at 4°C. Transfer the supernatant to a new tube. If the supernatant still contains the beads (white or red particles), centrifuge again at 13,000 × *g* for 2 min at 4°C, and transfer to a new tube. It is extremely important to remove all of the beads.
- The resulting solution is the eluate of the RIP that contains released FLAG-tagged ribosomes including the associated proteins and RNAs. This can be used to isolate RNA (see Subheading 3.5), analyze proteins by SDS-PAGE or mass spectrometry, or for further fractionation on sucrose gradients to assess the size distribution of the purified ribosomal subunits, monosomes and polysomes (24).

3.5. RNA Extraction

- For extraction of RNA from the eluate, use the Qiagen RNeasy kit (Catalog #74904).
- Add 2 volumes of 8 M guanidine-HCl to the eluate of the immunoprecipitation and vortex for 1 min.
- Add 3 volumes of 99% ethanol and vortex for 1 min.
- Precipitate the RNA at −20°C overnight.
- Centrifuge at 16,000 × *g* at 4°C for 45 min.
- Remove supernatant and let the pellet dry for 20 min.
- Prepare extraction buffer adding 10 μL of β-mercaptoethanol to 1 mL of Qiagen RLT buffer (provided with the RNeasy kit, contains guanidine thiocyanate).
- Resuspend the pellet in 450 μL of RLT buffer and vortex for 1 min.
- Add 250 μL of 99% ethanol and mix by inverting the tube. Do not vortex.
- Apply the sample into an RNeasy mini spin column. Incubate for 3 min.
- Centrifuge for 15 s at 16,000 × *g*.
- Add 700 μL of RW1 buffer (provided with the RNeasy kit, contains guanidine thiocyanate) and centrifuge for 15 s at 9,000 × *g*. Discard the flow through.
- Add 500 μL of Qiagen RPE buffer (provided with the RNeasy kit, four volumes of ethanol is added to RPE buffer before usage according to the manual) and centrifuge for 15 s at 9,000 × *g*. Discard the flow through.
- Add 500 μL of RPE buffer to the column and centrifuge for 2 min at 9,000 × *g*.
- Transfer the column to a new 2 mL microtube. Centrifuge for 1 min at 16,000 × *g* to remove remaining ethanol.
- Transfer the column to a new 1.5 mL microfuge tube and add 30–50 μL of RNAse-free water. Incubate for 5 min.
- Elute RNA by centrifuging for 1 min at 16,000 × *g*.
- RNA can now be used for further analysis (i.e., cDNA synthesis).

3.6. Methods for Analyses of mRNA Populations

The RNA (rRNA and mRNA) obtained by the RIP procedure described above can be used in subsequent experiments such as real-time PCR, microarray analysis, or RNAseq using procedures described elsewhere. Before doing so the quantity and quality of the RNA should be analyzed. Since the yield of the RNA is usually very low, the use of a sensitive method is recommended. A photometer capable of monitoring nanogram quantities in a submicroliter volume is needed to estimate the RNA concentration and purity. The quality of the RNA can be analyzed by capillary electrophoresis, for example, on an Agilent 2100 Bioanalyzer by use of RNA 6000 Nano or Pico Assay reagent kits (Agilent Technology). This evaluation is strongly recommended because it will reveal if there is degradation of the rRNA.

For real-time PCR, it is sufficient to use the purified RNA directly as a target for cDNA synthesis. However, the common microarray technology usually requires a much higher amount of RNA (1–15 μg) than the amount that is obtained by RIP or other cell-type isolation methods (10–500 ng, see Table 1). Therefore, a two-cycle eukaryotic target-labeling assay is recommended to produce Biotin-labeled cRNA. For the Affymetrix platform, the GeneChip IVT Labeling Kit (Affymetrix) can be used successfully for this purpose (18). For downstream experiments, we recommend using the same quantity of RNA from all RIP samples.

Deep sequencing of the translatome (RNAseq) can be performed on the purified RNA. The advantage of this method is that one can determine variations in mRNAs that are usually not detected by gene microarray hybridization (although possibly by tiling arrays) (e.g., splicing variants, putative non-coding mRNAs, and possibly miRNAs). RNAseq with the Illumina platform was used to evaluate RIP isolated translatomes of floral meristem cell types (19).

3.7. Bioinformatic Analysis of Microarray Data for Transcriptomes and Translatomes

3.7.1. Normalization of Data

The basics of computational analysis of DNA microarray hybridization data for steady-state mRNAs are described elsewhere (56, 57). Much of what applies for steady-state mRNA comparisons also applies for the analysis of mRNA purified from cell types by RIP. Generally, such analyses include: (1) the normalization of the microarray raw data; (2) quality controls; (3) identification of differentially expressed genes (DEGs) by pairwise comparisons; (4) cluster analyses of gene expression profiles; and (5) functional categorization of gene sets. The freely available statistical environment R along with Bioconductor packages (56) can be used for the analysis. A typical R script for the normalization of standard Affymetrix GeneChips is given below. As an example, we suggest to use selected microarray raw data (six .CEL files) from the dataset GSE14502 (18) specified below.

```
############################################
## 1. Load and Normalize microarray data
############################################

## Load libraries (R functions for microarrays)
library(affy)
library(limma)
```

```
## Create tabular experiment definition file
## ("affytarget.txt") of this format:
Name    FileName        Target
SC1     GSM362215.CEL   shoot_control_total_rep1
SC2     GSM362216.CEL   shoot_control_total_rep2
SC3     GSM362217.CEL   shoot_control_total_rep3
SH1     GSM362218.CEL   shoot_hypoxia_total_rep1
SH2     GSM362219.CEL   shoot_hypoxia_total_rep2
SH3     GSM362220.CEL   shoot_hypoxia_total_rep3

## Load experiment definition file
targets <- readTargets("affytarget.txt")

## Read in microarray data from .CEL files, defined in
FileName column
data <- ReadAffy(filenames=targets$FileName)

## Perform RMA normalization and write results to file
eset_rma <- rma(data)
write.table(exprs(eset_rma),"rma.txt", sep = "\t",
col.name = NA , quote = F)

## Present call analysis provided by MAS5 algorithm
eset_PMA <- mas5calls(data)
write.table(exprs(eset_PMA),"PMA.txt", sep = "\t",
col.name = NA , quote = F)

###################################
## 2. Quality control
###################################

## Generate quality report with arrayQualityMetrics
(58)
library("arrayQualityMetrics")
arrayQualityMetrics(expressionset = eset_rma,
outdir = "Report", force = TRUE)

## Calculate correlation coefficients between arrays
mydata <- 2^exprs(eset_rma)
colnames(mydata) <- c(targets(,1))
correlation <- cor(mydata, method="spearman")
write.table(correlation, "corr.txt", sep = "\t",
col.name = NA , quote = F)

## Clustering of samples (replicates should be
nearest neighbors)
hc <- hclust(dist(1-correlation))
```

```
# Draw sample dendrogram (tree) and store in pdf format
pdf("hc1.pdf")
plot(hc)
dev.off()

###########################################
```

The script describes the normalization of the hybridization signals by the RMA normalization method (59). Alternatively, other normalization methods can be used here, such as MAS5, gcRMA, dCHIP, VSN, or PLIER. All of these methods have different strengths and weaknesses, as described elsewhere (i.e., (60, 61)). Usually, weakly expressed genes show a higher degree of disagreement among these methods than stronger expressed genes.

Due to the low amount of RNA isolated by RIP it is important to perform the appropriate quality controls for the arrays (see above). Furthermore, the level of gene expression as compared to the background signal on the microarray needs to be considered for each gene. To estimate the amount of expressed mRNA, the present call information of the non-parametric Wilcoxon signed rank test (PMA-values (62)) can be computed in R (63), as described above. Probe sets with absent calls in most of the samples should be ignored in subsequent analyses, since it is likely that their expression level falls below the detection limit of the technology.

3.7.2. Determination of the Proportion of Individual mRNAs in Ribosome Complexes

If the transcriptomes and translatomes of an entire organism, organ, or tissue are compared, then it is also possible to determine the relative proportion of mRNA associated with polysomes (25, 26, 30). The ribosome loading (RL, Eq. 1) of any gene can be estimated from transcripts scored as P in the steady-state mRNA population (transcriptome) as well as the polysomal mRNA population (translatome) obtained by RIP with the *p35S:FLAG-RPL18* line or from a polysome pellet obtained by differential centrifugation (54).

$$\mathrm{RL} = \frac{\text{Signal of an mRNA in polysomal RNA}}{\text{Signal of an mRNA in total cellular RNA}}. \quad (1)$$

If the goal is to compare RL between genes under different biological conditions or in distinct genotypes that have different global levels of protein synthesis, then a normalization step that takes into account differences in total cellular polysome levels is required. This requires the determination of the relative proportion of rRNA in polysome complexes from the different samples (treatment, genotype, etc.) by using a specialized spectrophotometer as described in detail by Mustroph et al. (54). Typically, polysome gradient analysis is performed on three biological replicates, and the A254 absorbance in region of the gradient containing polysome complexes is divided by the A254 absorbance of the entire sucrose gradient. The values from three biological replicates are averaged to obtain the relative polysome content. The hybridization signal data obtained after the RMA normalization method (see Subheading 3.7.1) are further adjusted based on the relative polysome content of the corresponding biological sample. Although it is possible to analyze the immunopurified polysomes from individual cell types by sucrose gradient fractionation, it is not possible to obtain the total cellular mRNAs from the exact same cell population. Therefore RL values can only be measured at the organ or region-specific level using organs, portions of organs, or ribosomes immunopurified with the *35S:FLAG-RPL18* line.

3.7.3. Determination of Differentially Expressed Genes

After normalization, one approach would be to determine the genes with significant differences in mRNA abundance between samples by pairwise comparisons. Here, we provide a simple R script for the analysis of differentially expressed genes by pairwise comparisons of RMA-normalized expression values. This script uses the LIMMA package (64). It can be coupled with other microarray normalization methods such as MAS5, dCHIP, or GC-RMA. However, the expression data from MAS5 or dCHIP need to be log-transformed to perform the analysis. Important for such an approach is the selection of the reference sample (see Subheading 3.7.6).

```
##################################
## 3. DEG Analysis
##################################

## Load library
library(limma)

## Define replicated samples
design <- model.matrix(~ -1+factor(c(1,1,1,2,2,2)))
colnames(design) <- c("control","hypoxia")

## Fit normalized data to design matrix
fit_rma <- lmFit(eset_rma, design)

## Define which contrasts (comparisons) to make
contrast.matrix<-makeContrasts(hypoxia-control,
levels=design)

## Perform DEG analysis and store data
fit2_rma <- contrasts.fit(fit_rma, contrast.matrix)
fit2_rma <- eBayes(fit2_rma)
write.fit(fit2_rma,  results=NULL,  file="DEG_rma.
txt", digits=9, adjust="fdr", sep="\t")

##########################################
```

The outcome of the analysis will be a table containing the DEG results for all the probe sets (gene identifiers) represented on the GeneChip. The most important values for the user are the log2-fold changes (log base 2 of intensity ratios) and the *P*-values of the moderated *t*-statistic adjusted for multiplicity testing. In order to select significantly regulated genes, the user needs to define a confidence threshold for the log2-fold changes (usually set to >|1|, corresponding to a twofold change or higher) and the adjusted *P*-value (often set to <0.01). The latter is corrected for multiple testing to determine false discovery rates (FDRs), for example, by the Benjamini and Hochberg method (65).

3.7.4. Cluster Analysis of Expression Data

Besides a simple pairwise comparison, it is also suggested to perform cluster analysis of gene expression profiles. Clustering of microarray data by genes will assign them to groups according to the similarity of their expression patterns across different cell types. Usually, this is only possible if an experiment set contains sufficient samples to compute robust similarity measures among the genes. Over 50 clustering methods have been applied to gene expression data. Here we only describe three of the most commonly used methods, including (a) hierarchical clustering, (b) *k*-means, and (c) fuzzy *k*-means clustering. An important user decision for most clustering routines is the choice of the distance measure. Among the many distance measures available, the most commonly used ones in the microarray field are the Euclidean distance or distance-transformed correlation coefficients. Pearson's correlation coefficient is one of the most popular methods, but it is very sensitive to outliers (extreme expression values). This limitation can be overcome by choosing Spearman's correlation coefficient instead, where a rank transformation step minimizes the influence of outliers. To convert correlation values (similarity metric) into a distance measure, they can be subtracted from 1. If one wants to treat negative correlations the same way as positive ones, one can use their absolute values. Clustering as described here can be performed in various software environments, such as R or the freely available GUI (graphical user interface) program TIGR-MeV (66).

If RMA-normalized expression values are used, then they should be un-logged for the cluster analysis. Scaling of expression values of individual genes/probe sets across all samples is recommended when using Euclidean distances for clustering, because this measure is not a scale invariant. Scaling will transform the expression profiles for each gene to comparable expression ranges with a mean value close to zero and a standard deviation of one. This transformation is not required when clustering is performed on log2-fold changes (see above). Because clustering can be computationally very time consuming on large data sets, it is usually recommended to apply it only to a subset of the genes represented on a microarray, such as the set of DEGs identified above. If clustering more than 20,000 genes, then the use of a server with large memory capacities (e.g., >16 GB RAM) may be necessary.

For hierarchical clustering, one can provide a distance matrix as input. If the software takes care of the computation of the distance matrix internally, then the expression table can be used instead. Most hierarchical clustering algorithms join the most to least similar items (here genes or samples) in an agglomerative manner using as cluster joining rule either single, average, complete, or other linkage methods. Average or complete linkages are often preferred. The output of hierarchical clustering is a similarity tree (dendrogram) that can be used to sort the rows of the underlying expression table according to the gene order defined by the dendrogram. For visualization purposes, one can present the dendrogram next to a heat map where the expression values are represented by a color scheme.

Both *k*-means and fuzzy *k*-means clustering partition the data into a user-defined number of *k* clusters. Because choosing an optimal *k* value can be difficult, it is recommended to try out several *k* values or to perform permutation tests where one can monitor the enrichment of Gene Ontology (GO) terms (see below) and then choose the settings resulting in the highest enrichment. The fuzzy *k*-means clustering is a variant of the *k*-means method that enables a gene to be member of several clusters, which might have advantages in large data sets. The degree of fuzziness is another user-definable parameter that needs to be optimized. Below we provide an R script to perform a basic clustering analysis. A more detailed script of fuzzy *k*-means clustering is given elsewhere (67).

```
########################################
## 4.0 loading of libraries and data
########################################
library(cluster)
library(e1071)

## Import expression table (rows are genes and col-
umns are samples)
mydataS<-read.delim("selected_data.txt",header=T,
row.names=1, sep="\t")
mydataS <- as.matrix(mydataS)

## Scale data when using Euclidean distances
mydataSc <- t(scale(t(mydataS)))

##################################
## 4.1 hierarchical clustering
##################################

## Perform hierarchical clustering with
```

```
## correlation-based distances (here Pearson)
hr <- hclust(as.dist(1-cor(t(mydataS), method=
"pearson")), method="complete", members=NULL)

## Cut tree in order to get clusters ##
(cut height needs to be optimized)
mycl <- cutree(hr, h=max(hr$height)/2)
write.table(mycl, file="cut2.txt", sep="\t")

## Perform hierarchical clustering using the
Euclidean distance measure
mydataSc <- t(scale(t(mydataS)))
hr <- hclust(dist(mydataSc, method = "euclidean"),
method="complete", members=NULL)

## If RMA data are used, un-log them before usage

##############################
## 4.1 k-means clustering
##############################

## Fast clustering with varying outcome for each run
## (use scaled values, only Euclidean distance), 10
clusters.
## This can be used to run analysis with different
## K (= centers) = numbers of clusters for ##
selection of "best" outcome.
kl_fast <- kmeans(mydataSc, centers = 10, iter.max
= 10000, nstart = 100)
write.table(kl_fast$cluster, file="kmean_10.txt",
sep="\t")

## Slow, but reproducible clustering due ##
to determination of medoids.
## This should be used for the final analysis,
## since it gives reproducible (optimal) results.
kl_slow <- pam(dist(mydataSc, method = "euclid-
ean"), k=10, diss=TRUE)
write.table(kl_slow$clustering, file="pam-10-Eucl.
txt", sep="\t")

## The slow, accurate method also allows ##
the use of Pearson distances.
## When using Pearson distance, one does not need ##
to scale data.
kl_slow_Pea <- pam(as.dist(1-cor(t(mydataS),
method="pearson")), k=10, diss=TRUE)
write.table(kl_slow_Pea$clustering, file="pam-
10-Pear.txt", sep="\t")
```

```
#################################
## 4.2 Fuzzy k-means clustering
#################################

## Fast clustering with different outcome for each run
## (use scaled values, only Euclidean distance), 10
clusters.
## This can be used to run analysis with different
## K = numbers of clusters for selection of "best"
outcome.
## The degree of fuzzyness is determined by m,
## when m is higher (2.0), results become very fuzzy.
cl<-cmeans(mydataSc,10,5000,verbose=TRUE,dist=
"euclidean",method="cmeans",m=1.2)
write.table(cl$membership,  file="cmean_10.txt",
sep="\t")
write.table(cl$cluster, file="cmean_10C.txt", sep="\t")

## Slow, but more robust clustering.
## This should be used for the final analysis,
## since it gives reproducible (optimal) results.
## The degree of fuzzyness is determined by memb.exp,
## when m is higher (2.0), results become very fuzzy.
fanny_Euc <- fanny(dist(mydataSc, method = "euclid-
ean"), 10, diss=TRUE, memb.exp = 1.2, maxit = 10000)
write.table(fanny_Euc$membership, file="Fanny-10-Euc.
txt", sep="\t")

## The slow, accurate method also allows the ##
use of Pearson distances.
## When using Pearson distance, one does not need ##
to scale data.
fanny_Pea   <-   fanny(as.dist(1-cor(t(mydataS),
method="pearson")), 10, diss=TRUE, memb.exp = 1.2,
maxit = 10000)
write.table(fanny_Pea$membership,   file="Fanny-
10-Pea.txt", sep="\t")

###########################################
```

3.7.5. Functional Categorization of Gene Sets

The gene lists obtained from DEG and cluster analysis can be functionally annotated by enrichment tests. These methods allow the researcher to pinpoint annotation terms that are more abundant in a gene list than expected by chance. For instance, one can test the enrichment of Gene Ontology (GO) terms, Pfam domains, or KEGG pathways. Various online or standalone software tools can be used to perform these analyses (see www.geneontology.org for overview). In R these analyses can be performed with the GOstats package from Falcon and Gentleman (68). An alternative is to use the MapMan and PageMan programs developed for *Arabidopsis* and other plant species (69) that perform similar enrichment tests.

3.7.6. Selection of a Reference Sample for Comparative mRNA Analyses

When doing a quantitative mRNA analysis, a reference sample is required that can be used for the basis of comparison. At least three controls are possible, with one chosen by the investigator according to the specific requirements: (1) the total mRNA of the respective organ (i.e., total organ transcriptome); (2) the RIP mRNA obtained by use of the p35S promoter or another ubiquitously expressed promoter (i.e., p35S:FLAG-RPL18 translatome); or (3) the RIP mRNA obtained from non-overlapping cell types in that same organ. Based on the goal of each experiment, the researcher needs to carefully consider advantages and disadvantages of different reference samples as described below.

The use of total RNA from the same extract used for RIP (see Subheading 3.2) has obvious advantages. However, there will be discrepancies between the association of mRNAs with ribosomes (the translatome) and the total amount of individual mRNA (the transcriptome), especially under situations where the plant is stressed or when conditions are changing rapidly (25–31). Therefore, if the experimenter aims to evaluate environmental responses, total mRNA should also be evaluated from stressed and non-stressed tissue samples. Further analysis of the proportion of individual mRNAs associated with ribosomes in an organ can be estimated by sucrose density centrifugation (30, 54) (see Subheading 3.7.2).

The comparison of RIP mRNA from specific cell types to that of a RIP mRNA from a line in which FLAG-RPL18 is under the control of a ubiquitous promoter, such as p35S, is a second possibility. This is most suitable in tissues where the p35S promoter is strongly and evenly active. However, in our study with *Arabidopsis* seedlings (18), it was found that the p35S promoter is not evenly active in root cell types based on imaging 35S:FLAG-GFP-RPL18 transgenics and analysis of the immunopurified mRNAs. The mRNAs isolated with the p35S:FLAG-RPL18 line were enriched in transcripts found in the stele. This could be a consequence of either stronger 35S promoter activity in the stele or a higher proportion of ribosomes in stele cell types relative to other cell types of the root. As an alternative to p35S, the promoter of RPL18 was used in the study of floral meristem translatomes defined by the *AGAMOUS*, *APETELA1*, and *APETELA3* promoter driving HF-RPL18 (19).

As alternative to the use of the p35S or pRPL18 as a control, one might use a promoter that is active in a non-overlapping cell type in the same organ. In our study (18) we performed RIP for a large number of different cell types, allowing comparisons to be made between translatomes. Similarly, different cell types were also compared in the studies using cell-specific GFP expression, protoplasting, and FACS (1). When a large number of samples are compared, mRNAs enriched in specific cell types can be recognized with greater confidence.

4. Notes

1. Heparin and RNAse inhibitor are only required for tissues with high RNAse content. We generally do not use these unless there is evidence of rRNA or mRNA degradation in the samples. If so, the addition of heparin to the extraction buffer usually alleviates the problem. RNA degradation can result if samples are thawed prior to addition of the extraction buffer or the extraction is performed above 4°C.
2. EGTA dissolves only after adjusting the pH.
3. Shake bottle before pipetting the solution.
4. Use lung protection while weighing DOC.
5. Dissolve while heating to about 60°C.
6. To isolate soluble ribosomes only, omit the detergents. The yield of ribosomes will be much lower.
7. Warm solution at 42°C before use; pipette with a 1,000 μL tip enlarged by cutting 0.5 cm from the end.
8. This detergent is included to disrupt ribosome-cytoskeleton association.
9. For large-scale experiments, adjust the amount of α-FLAG agarose gel according to the amount of extract. Since this gel is the most costly reagent, when using lines with cell-type specific promoters expressed in a limited number of cells of an organ/tissue, use two to three times more tissue for the same amount of beads.
10. The unbound fraction can be used to estimate the efficiency of binding. Mix the extract with SDS gel loading buffer, run a SDS-acrylamide gel, generate a protein blot, and incubate with α-FLAG antibody (for example, Sigma F1804). If FLAG-protein is still present in unbound fraction, increase amount of beads in the RIP.

Acknowledgments

We thank all of the individuals who have contributed to the development of polysome isolation and analyses methods in our group over the years. This work was supported by the U.S. National Science Foundation (DBI 0211857, IBN-0420152, and IOS-0750811 to J. B.-S. and 2010-0820842 and ABI-0957099 to T. G.).

References

1. Brady SM, Orlando DA, Lee JY et al (2007) A high-resolution root spatiotemporal map reveals dominant expression patterns. Science 318:801–806
2. Benfey PN, Scheres B (2000) Root development. Curr Biol 10:R813–R815
3. Van Norman JM, Breakfield NW, Benfey PN (2011) Intercellular communication during plant development. Plant Cell 23:855–864
4. Iyer-Pascuzzi AS, Benfey PN (2009) Transcriptional networks in root cell fate specification. Biochim Biophys Acta 1789: 315–325
5. Nakajima K, Benfey PN (2002) Signaling in and out: control of cell division and differentiation in the shoot and root. Plant Cell 14(Suppl):S265–S276
6. Petricka JJ, Benfey PN (2008) Root layers: complex regulation of developmental patterning. Curr Opin Genet Dev 18:354–361
7. Ishida T, Kurata T, Okada K et al (2008) A genetic regulatory network in the development of trichomes and root hairs. Annu Rev Plant Biol 59:365–386
8. Schiefelbein J, Kwak SH, Wieckowski Y et al (2009) The gene regulatory network for root epidermal cell-type pattern formation in Arabidopsis. J Exp Bot 60:1515–1521
9. Spencer MW, Casson SA, Lindsey K (2007) Transcriptional profiling of the Arabidopsis embryo. Plant Physiol 143:924–940
10. Jiao Y, Tausta SL, Gandotra N et al (2009) A transcriptome atlas of rice cell types uncovers cellular, functional and developmental hierarchies. Nat Genet 41:258–263
11. Klink VP, Hosseini P, Matsye P et al (2009) A gene expression analysis of syncytia laser microdissected from the roots of the *Glycine max* (soybean) genotype PI 548402 (Peking) undergoing a resistant reaction after infection by *Heterodera glycines* (soybean cyst nematode). Plant Mol Biol 71: 525–567
12. Matas AJ, Agustí J, Tadeo FR et al (2010) Tissue-specific transcriptome profiling of the citrus fruit epidermis and subepidermis using laser capture microdissection. J Exp Bot 61:3321–3330
13. Birnbaum K, Shasha DE, Wang JY et al (2003) A gene expression map of the Arabidopsis root. Science 302:1956–1960
14. Dinneny JR, Long TA, Wang JY et al (2008) Cell identity mediates the response of Arabidopsis roots to abiotic stress. Science 320:942–945
15. Gifford ML, Dean A, Gutierrez RA et al (2008) Cell-specific nitrogen responses mediate developmental plasticity. Proc Natl Acad Sci USA 105:803–808
16. Zhang C, Barthelson RA, Lambert GM et al (2008) Global characterization of cell-specific gene expression through fluorescence-activated sorting of nuclei. Plant Physiol 147:30–40
17. Galbraith DW, Janda J, Lambert GM (2011) Multiparametric analysis, sorting, and transcriptional profiling of plant protoplasts and nuclei according to cell type. Methods Mol Biol 699:407–429
18. Mustroph A, Zanetti ME, Jang CJ et al (2009) Profiling translatomes of discrete cell populations resolves altered cellular priorities during hypoxia in Arabidopsis. Proc Natl Acad Sci USA 106:18843–18848
19. Jiao Y, Meyerowitz EM (2010) Cell-type specific analysis of translating RNAs in developing flowers reveals new levels of control. Mol Syst Biol 6:419
20. Deal RB, Henikoff S (2010) A simple method for gene expression and chromatin profiling of individual cell types within a tissue. Dev Cell 18:1030–1040
21. Nawy T, Lee JY, Colinas J et al (2005) Transcriptional profile of the Arabidopsis root quiescent center. Plant Cell 17:1908–1925
22. Yadav RK, Girke T, Pasala S et al (2009) Gene expression map of the Arabidopsis shoot apical meristem stem cell niche. Proc Natl Acad Sci USA 106:4941–4946
23. Jain R, Devine T, George AD et al (2011) RIP-Chip analysis: RNA-binding protein immunoprecipitation-microarray (chip) profiling. Methods Mol Biol 703:247–263
24. Zanetti ME, Chang IF, Gong F et al (2005) Immunopurification of polyribosomal complexes of Arabidopsis for global analysis of gene expression. Plant Physiol 138:624–635
25. Kawaguchi R, Girke T, Bray EA et al (2004) Differential mRNA translation contributes to gene regulation under non-stress and dehydration stress conditions in *Arabidopsis thaliana*. Plant J 38:823–839
26. Kawaguchi R, Bailey-Serres J (2005) mRNA sequence features that contribute to translational regulation in Arabidopsis. Nucleic Acids Res 33:955–965
27. Branco-Price C, Kawaguchi R, Ferreira RB et al (2005) Genome-wide analysis of transcript abundance and translation in Arabidopsis seedlings subjected to oxygen deprivation. Ann Bot 96:647–660

28. de Jong M, van Breukelen B, Wittink FR et al (2006) Membrane-associated transcripts in Arabidopsis; their isolation and characterization by DNA microarray analysis and bioinformatics. Plant J 46:708–721
29. Nicolaï M, Roncato MA, Canoy AS et al (2006) Large-scale analysis of mRNA translation states during sucrose starvation in Arabidopsis cells identifies cell proliferation and chromatin structure as targets of translational control. Plant Physiol 141:663–673
30. Branco-Price C, Kaiser KA, Jang CJ et al (2008) Selective mRNA translation coordinates energetic and metabolic adjustments to cellular oxygen deprivation and reoxygenation in *Arabidopsis thaliana*. Plant J 56:743–755
31. Piques M, Schulze WX, Höhne M et al (2009) Ribosome and transcript copy numbers, polysome occupancy and enzyme dynamics in Arabidopsis. Mol Syst Biol 5:314
32. Bailey-Serres J (1999) Selective translation of cytoplasmic mRNAs in plants. Trends Plant Sci 4:142–148
33. Kawaguchi R, Bailey-Serres J (2002) Regulation of translational initiation in plants. Curr Opin Plant Biol 5:460–465
34. Bailey-Serres J, Sorenson R, Juntawong P (2009) Getting the message across: cytoplasmic ribonucleoprotein complexes. Trends Plant Sci 14:443–453
35. Brodersen P, Sakvarelidze-Achard L, Bruun-Rasmussen M et al (2008) Widespread translational inhibition by plant miRNAs and siRNAs. Science 320:1185–1190
36. Masucci JD, Rerie WG, Foreman DR et al (1996) The homeobox gene GLABRA2 is required for position-dependent cell differentiation in the root epidermis of *Arabidopsis thaliana*. Development 122:1253–1260
37. Heidstra R, Welch D, Scheres B (2004) Mosaic analyses using marked activation and deletion clones dissect Arabidopsis SCARECROW action in asymmetric cell division. Genes Dev 18:1964–1969
38. Mace DL, Lee J-Y, Twigg RW et al (2006) Quantification of transcription factor expression from Arabidopsis images. Bioinformatics 22:e323–e331
39. Wysocka-Diller JW, Helariutta Y, Fukaki H et al (2000) Molecular analysis of SCARECROW function reveals a radial patterning mechanism common to root and shoot. Development 127:595–603
40. Helariutta Y, Fukaki H, Wysocka-Diller J et al (2000) The SHORT-ROOT gene controls radial patterning of the Arabidopsis root through radial signaling. Cell 101:555–567
41. Estelle M (2001) Cytokinin receptor: just another histidine kinase. Curr Biol 11: 271–273
42. Imlau A, Truernit E, Sauer N (1999) Cell-to-cell and long-distance trafficking of the green fluorescent protein in the phloem and symplastic unloading of the protein into sink tissues. Plant Cell 11:309–322
43. Takahashi H, Watanabe-Takahashi A, Smith FW et al (2000) The roles of three functional sulphate transporters involved in uptake and translocation of sulphate in *Arabidopsis thaliana*. Plant J 23:171–182
44. Williams ME, Sussex IM (1995) Developmental regulation of ribosomal protein L16 genes in Arabidopsis thaliana. Plant J 8:65–76
45. Szymanski DB, Jilk RA, Pollock SM, Marks MD (1998) Control of GL2 expression in Arabidopsis leaves and trichomes. Development 125:1161–1171
46. Pighin JA, Zheng H, Balakshin LJ et al (2004) Plant cuticular lipid export requires an ABC transporter. Science 306:702–704
47. Nakamura RL, McKendree WL Jr, Hirsch RE et al (1995) Expression of an Arabidopsis potassium channel gene in guard cells. Plant Physiol 109:371–374
48. Donald RG, Cashmore AR (1990) Mutation of either G box or I box sequences profoundly affects expression from the Arabidopsis rbcS-1A promoter. EMBO J 9:1717–1726
49. Twell D, Yamaguchi J, McCormick S (1990) Pollen-specific gene expression in transgenic plants: coordinate regulation of two different tomato gene promoters during microsporogenesis. Development 109:705–713
50. Haseloff J (1999) GFP variants for multispectral imaging of living cells. Methods Cell Biol 58:139–151
51. Kiegle E, Moore CA, Haseloff J et al (2000) Cell-type-specific calcium responses to drought, salt and cold in the Arabidopsis root. Plant J 23:267–278
52. Hartley JL, Temple GF, Brasch MA (2000) DNA cloning using in vitro site-specific recombination. Genome Res 10:1788–1795
53. McIntosh KB, Bonham-Smith PC (2005) The two ribosomal protein L23A genes are differentially transcribed in *Arabidopsis thaliana*. Genome 48:443–454
54. Mustroph A, Juntawong P, Bailey-Serres J (2009) Isolation of plant polysomal mRNA by differential centrifugation and ribosome immunopurification methods. Methods Mol Biol 553:109–126

55. Clough SJ, Bent AF (1998) Floral dip: a simplified method for Agrobacterium-mediated transformation of *Arabidopsis thaliana*. Plant J 16:735–743
56. Gentleman RC, Carey VJ, Bates DM et al (2004) Bioconductor: open software development for computational biology and bioinformatics. Genome Biol 5:R80
57. Horan K, Jang C, Bailey-Serres J et al (2008) Annotating genes of known and unknown function by large-scale coexpression analysis. Plant Physiol 147:41–57
58. Kauffmann A, Gentleman R, Huber W (2009) arrayQualityMetrics – a bioconductor package for quality assessment of microarray data. Bioinformatics 25:415–416
59. Irizarry RA, Hobbs B, Collin F et al (2003) Exploration, normalization, and summaries of high density oligonucleotide array probe level data. Biostatistics 4:249–264
60. Millenaar FF, Okyere J, May ST et al (2006) How to decide? Different methods of calculating gene expression from short oligonucleotide array data will give different results. BMC Bioinformatics 7:137
61. Binder H, Preibisch S, Berger H (2010) Calibration of microarray gene-expression data. Methods Mol Biol 576:375–407
62. Liu WM, Mei R, Di X et al (2002) Analysis of high density expression microarrays with signed-rank call algorithms. Bioinformatics 18:1593–1599
63. Gautier L, Cope L, Bolstad BM et al (2004) Affy-analysis of Affymetrix GeneChip data at the probe level. Bioinformatics 20:307–315
64. Smyth GK (2004) Linear models and empirical bayes methods for assessing differential expression in microarray experiments. Stat Appl Genet Mol Biol 3:Article3
65. Benjamini Y, Hochberg Y (1995) Controlling the false discovery rate: a practical and powerful approach to multiple testing. J R Stat Soc B 57:289–300
66. Saeed AI, Bhagabati NK, Braisted JC et al (2006) TM4 microarray software suite. Methods Enzymol 411:134–193
67. Orlando DA, Brady SM, Koch JD et al (2010) Manipulating large-scale Arabidopsis microarray expression data: identifying dominant expression patterns and biological process enrichment. Methods Mol Biol 553:57–77
68. Falcon S, Gentleman R (2007) Using GOstats to test gene lists for GO term association. Bioinformatics 23:257–258
69. Thimm O, Blaesing O, Gibon Y et al (2004) MAPMAN: a user-driven tool to display genomics data sets onto diagrams of metabolic pathways and other biological processes. Plant J 37:914–939

Chapter 20

Analyzing Small and Long RNAs in Plant Development Using Non-radioactive *In Situ* Hybridization

Pilar Bustos-Sanmamed, Carole Laffont, Florian Frugier, Christine Lelandais-Brière, and Martin Crespi

Abstract

In the past decade, hundreds of non-coding RNAs (small and long RNAs) have been identified as crucial elements in developmental processes and stress response in plants. Among small RNAs, the microRNAs or miRNAs control levels of specific mRNA by inhibiting translation or reducing the stability of their mRNA targets through integration into different ribonucleoproteins (RNP). Spatio-temporal expression of small and long RNAs, using reporter genes or *in situ* hybridization, is essential to understand their functions. We are interested in understanding the role of various non-coding RNAs (including miRNAs) in the regulation of root and nodule development in legumes, which are agriculturally important crops. Here, we present the protocol we are currently using for detection of small and long RNA in model legume plants and tissues, like nodules and roots. The probe selection, as well as the fixation and permeabilization steps allowing to preserve tissues and cell integrity and to increase accessibility to RNA targets, will be specifically discussed.

Key words: *Medicago truncatula*, *Lotus japonicus*, Legume, Root, Nodule, Flower, Shoot apical meristem, Locked nucleic acid (LNA), Non-coding RNA

1. Introduction

Non-coding RNAs (ncRNAs) are specific regulators of gene expression. Long and small ncRNA may regulate diverse processes such as: dosage compensation, genomic imprinting, developmental patterning, cell differentiation, and/or stress responses (1). Spatial detection of ncRNAs and location of their specific putative targets by *in situ* hybridization (2–5) could help to better understand their regulatory functions.

Ive De Smet (ed.), *Plant Organogenesis: Methods and Protocols*, Methods in Molecular Biology, vol. 959,
DOI 10.1007/978-1-62703-221-6_20, © Springer Science+Business Media New York 2013

Although other methods have been described to detect small and long RNA, such as miRNA_PCR (5) or Northern blot (6), they do not reveal any information about spatial expression profiles and cellular localizations (3). In addition, for these techniques, a total RNA extraction from bulked tissues is required, which could cause a dilution of targets/miRNAs of interest that may be abundant only in specific cell types (7).

We are interested in understanding the role of various non-coding RNAs (including miRNAs) in the regulation of root and nodule development in legume crops. We previously developed protocols (adapted from ref. 8) to determine in the model legume *Medicago truncatula* the spatial expression pattern of mature miRNA and of their targets using *in situ* hybridization on root and nodule sections, as described in Boualem et al. (9) and Lelandais-Brière et al. (10). The protocol was additionally successfully applied on flowers and shoot apical meristems as well as in the other model legume *Lotus japonicus* (11).

1.1. Critical Points for In Situ Hybridization

1.1.1. Tissue Fixation and Permeabilization

The first crucial step for *in situ* hybridization is the fixation of the material (4), which must preserve tissue and cell integrity. As cryofixation requires a high input in specific equipment for subsequent sectioning, chemical fixation is generally used. This alternative is very effective for proteins and nucleic acids and involves treatments with aldehydes that penetrate into the cells and form covalent cross-linking between intracellular components. The most commonly used agents are glutaraldehyde, formaldehyde, or paraformaldehyde. Glutaraldehyde is very efficient in preserving cellular structures but its molecular weight limits its ability to diffuse into thick samples and, in addition, it may induce autofluorescence. Cross-linking with formaldehyde and paraformaldehyde (PFA) is not as effective as glutaraldehyde, but tissue penetration is better (12). Although PFA is toxic and must thus be handled in the fumehood, this compound is recommended as it gives a stronger staining and a better preservation of tissue and nucleic acid integrity (13). With some samples (thick or covered by hydrophobic lipidic compounds like cutin or suberin), the fixation step can be improved by a short and mild vacuum infiltration step.

For *in situ* hybridization, another difficulty is that the access of the probe to the target can be masked by associated proteins. Additional treatments can thus be added during or after the fixation step to enhance cell permeability and thus improve resolution and stringency. Classically, in our protocols, we add non-ionic detergents (mainly Triton ×-100 and Tween 20) to the fixative solution, which help to permeabilize the membranes by extracting lipids. Moreover, removal of proteins from the target RNA is performed by a protease treatment before hybridization. Note that tissues fixed with PFA need a shorter protease treatment than formaldehyde (12), thus limiting the risk of injuring tissue and cellular structures during this step.

1.1.2. Choice and Labeling of Probe

The choice of the probe (length, DNA or RNA, labeling method) is also essential to carry out successful *in situ* hybridization (7).

As long ncRNA are generally between 200 and 1,000 nucleotides in length, the size of the probe has first to be determined. It must be taken into account that with too long probes the signal may be weak due to poor penetration into the cell. On the other hand, if the probe is too small, the melting temperature of target/probe hybrid is low, thus enhancing the risk of unspecific and/or weak binding (2). To overcome these problems, we generally use RNA probes, for long ncRNAs, that are antisense from a 100 to 250 nt region of the transcript, does not share more than 70% identity with other genes, and that presents a high GC content (as G/C pairs are more stable than A/U). Although the probe itself is sensitive to RNAse degradation, RNA probes are preferred to DNA ones as RNA/RNA duplexes are more stable and allow a further RNase treatment to eliminate non-hybridized RNA, thus reducing background.

To detect plant small RNAs from 20 to 24 nt, oligonucleotidic probes are used that correspond to the complementary sequence of the targeted small RNA. These short oligonucleotidic probes display low melting temperature, making it hard to detect small RNAs with sufficient sensitivity and specificity. This issue is generally solved by modifying some of the nucleotides using the locked nucleic acid (LNA) modified oligonucleotide technology (Exiqon, Danemark; http://www.exiqon.com). By LNA modification, the altered nucleotide becomes more rigid in 3D space (5) which enhances target affinity and specificity (6, 14). LNA-modified oligonucleotides show an increase in melting temperature of target–probe hybrid by ≈2–10°C per nucleotide modified. At least, 25–30% LNA-modification of the probe is recommended (2, 4, 7).

Probes can be labeled with radioactive and non-radioactive methods, including biotin, digoxigenin (DIG), and fluorescent dyes like fluorescein, Texas Red, or ALEXAs. Radioactive labels are generally more sensitive than non-radioactive ones, but need longer exposure time and involve safety and disposal problems associated with radioactivity. Fluorescent labels allow direct detection of one or several probes at the same time but require a fluorescent microscope. Thus, biotin and digoxigenin are the more common non-radioactive labeling used for in situ hybridization probes. However, as biotin is present endogenously in some plant tissues false-positive results may be obtained and digoxigenin is the most appropriate for this (12). DIG-labeled probe detection requires the use of a specific anti-DIG antibody generally conjugated either to a fluorescent dye or to alkaline phosphatase, whose activity is detected by production of a purple precipitate when incubating in presence of a NBT/BCIP substrate. The sensitivity of DIG-labeled probe detection is high but controlling the probe concentration and time of revelation by using sense probes or sequences not present in plant RNAs (that do not detect any existing transcript in the tissues) is crucial to avoid background and artifacts (4).

2. Materials

The protocol described here is adapted from Ref. 8.

2.1. Equipment

1. Rotary microtome (e.g., Leica RM 2155 from Leica Microsystems, Wetzlar, Germany).
2. Hybridization oven (60°C).
3. Light microscope with camera (e.g., Leica DMI6000 B provided with Leica DFC 350 FX camera).
4. Rotary shaker.
5. Hotplate (42 and 65°C).
6. UV crosslinker.
7. Positive charged Poly-l-lysine-coated slides and coverslips (see Note 1).
8. Truncated (22 mm square top, tapered to 12 mm bottom) embedding molds Peel-Away.

2.2. Chemicals

1. Ethanol 100%.
2. Paraformaldehyde (see Note 2).
3. Eosin Y disodium salt for microscopy.
4. Histo-Clear II.
5. Paraplast X-Tra.
6. pGEM-T Easy vector (Promega, Madison, Wisconsin, USA).
7. miRCURY LNA-modified oligonucleotides (Exiqon, Danemark; http://www.exiqon.com).
8. DIG RNA Labeling (SP6/T7) kit, for long RNAs, or DIG Oligonucleotide 3 -End Labeling kit for small RNAs (Roche, La Rochelle, France).
9. Positively charged nylon membrane.
10. Blocking Reagent (for nucleic acid hybridization and detection, Roche).
11. Antidigoxigenin-AP Fab Fragment from sheep.
12. Nitro Blue Tetrazolium (NBT) 100 mg/ml in 70% dimethylformamide and 5-bromo-4-chloro-3-indolyl-phosphate (BCIP) 50 mg/ml in 100% dimethylformamide.
13. Pronase from *Streptomyces griseus*.
14. Acetic anhydride.

15. Deionized formamide (Ambion, Austin, Texas, USA).
16. RNAse A (10 mg/ml).
17. Albumin from bovine serum fraction V (BSA).
18. Acetic acid.
19. Alcian Blue 8GX.
20. DPX mountant for histology.

2.3. Buffer and Solutions Composition

Use sterile distilled water to prepare all buffers and solutions.

1. Fixative solution (see Note 3): 4% (p/v) paraformaldehyde diluted in microtubules stabilizing buffer preheated at 60°C: 50 mM 1,4-Piperazinediethanesulfonic acid (PIPES), 5 mM Ethylene glycol-bis(β-aminoethyl ether)-*N*,*N*,*N*′,*N*′-tetraacetic acid (EGTA), 5 mM $MgSO_4$. Adjust pH of MTSB at 6.9–7 with solid KOH (MTSB, or see Note 4). After adding paraformaldehyde to MTSB, readjust fixative solution pH at 6.9–7 with concentrate H_2SO_4. Just before using, add 0.1% (v/v) Triton X-100 and 0.1% (v/v) Tween 20.
2. 10× Saline solution: 8.5% (w/v) NaCl in sterile water.
3. 10× Tris buffered saline (TBS): 1 M Tris–HCl, 1.5 M NaCl, pH 7.5.
4. 10× Substrate buffer: 1 M Tris–HCl, 1 M NaCl, pH 9.8.
5. 20× Pronase buffer: 1 M Tris–HCl, 0.1 M Ethylene-diaminetetraacetic acid (EDTA), pH 7.5.
6. Pronase solution: 0.13 mg/ml Pronase in 1× Pronase buffer.
7. 10× Phosphate buffered saline (PBS): 1.3 M NaCl, 70 mM Na_2HPO_4, 30 mM NaH_2PO_4 (pH 6.5–7).
8. Acetic anhydride solution: 1.25% (v/v) triethanolamine, 0.5% (v/v) HCl, 0.5% (v/v) acetic anhydride.
9. 10× Salt solution: 3 M NaCl, 0.1 M Tris–HCl, 0.1 M $NaPO_4$, 50 mM EDTA.
10. Prehybridization buffer: 1.25% (v/v) 10× salt solution, 50% (v/v) formamide, 2.5% (v/v) dextran sulfate sodium salt 50%, 0.25% (v/v) 50× Denhardt's solution. Mix well.
11. Hybridization buffer (see Note 5): 1.25% 10× Salt solution, 50% formamide, 2.5% dextran sulfate sodium salt 50%, 0.13% (v/v) tRNA 100 mg/ml, 0.25% 50× Denhardt's solution.
12. 10× Sodium chloride-Sodium Citrate (SSC): 3 M NaCl, 0.3 M sodium citrate.
13. 10× NaCl–Tris–EDTA buffer (NTE): 5 M NaCl, 100 mM Tris–HCl pH 7.5, 10 mM EDTA.
14. Antibody buffer: 1% (p/v) BSA, 0.3% (v/v) Triton X100 in 1x TBS.

15. Staining solution (10 ml): 9.7 ml 1× Substrate buffer (0.1 M Tris–HCl, 0.1 M NaCl, pH 9.8.) with 250 μl $MgCl_2$ 2 M, 15 μl NBT 100 mg/ml, and 33 μl BCIP 50 mg/ml.

3. Methods

3.1. Sample Fixation and Dehydration

When detecting RNA by *in situ* hybridization, it is important to consider that RNA can easily degrade. Although fixation and dehydration of the tissue before embedding allow a fast inhibition of RNase activity, it is critical to fix tissues quickly and avoid sample contamination with RNAses till the hybridization stage by working with sterile RNAse-free material.

1. Collect plant samples and place them immediately in cold-ice Fixative solution.
2. Fix the samples at 4°C a few hours to overnight depending on their thickness on a rotary shaker with gentle agitation. A vacuum pump can be used for larger samples (e.g., nodules or flowers).
3. Take out the Fixative solution and rinse with 1× Saline solution at 4°C for 30 min with gentle agitation.
4. To embed the samples in paraffin, tissues must be completely dehydrated to avoid water and wax to emulsify. Dehydrate samples at 4°C with gentle shaking by passing through graded ethanol solutions as follows: 50% EtOH:1× Saline (v/v) for 3 h, 70% EtOH:1× Saline for 3 h, 85% EtOH:1× Saline overnight, 95% EtOH:distilled water, and 100% EtOH for 4 h each.
5. Repeat once with 100% EtOH overnight.
6. Samples can be kept at 4°C for a week after either 70% EtOH:1× Saline or 100% EtOH.

3.2. Embedding

Before sectioning, samples must be solidified into paraffin. As dehydration can enhance the transparency of the tissues, samples can be stained with eosin before embedding to orient the tissue for later sectioning.

1. For eosin staining, perform the following incubations at room temperature with gentle shaking before starting embedding process: 100% EtOH for 2 h; 1.6 μl eosin Y (Sigma) in 100 ml of EtOH:Histo-Clear II 1:1 (v/v) for 1 h; 100% Histo-Clear II for 1 h. Repeat this last step three times.
2. From this stage, embedding should be carried out inside an oven at 60°C to prevent paraffin solidification (see Note 6).

Put the samples into Petri dishes or glass vials containing Histo-Clear II:Paraplast X-Tra 1:1 (v/v) and incubate overnight.

3. During a day, add every hour half of the volume of Histo-Clear II of melted Paraplast X-Tra. Add Paraplast X-Tra slowly for preventing bubbles formation.
4. Renew, twice a day for 3 days, half of the Paraplast X-Tra for freshly melted Paraplast X-Tra.
5. For block preparation, slowly move Petri dishes or plastic molds to a hotplate pre-heated at 65°C. Fill the mold with freshly melted Paraplast X-Tra.
6. Slowly place samples inside the mold, for preventing bubbles formation. Dispose samples in parallel rows and at the bottom of the mold. Remove the mold from the hotplate and reorient the samples if necessary before the Paraplast X-Tra hardens.
7. Let the Paraplast X-Tra containing the samples solidify overnight at room temperature. Keep the molds at 4°C until use (they can be stored for several months).

3.3. Sectioning

1. Remove the block from the mold and place it on a foil in ice.
2. Shape the block into a trapezium with a sterile scalpel (trapezoid shaped blocks helps the identification of a single section). Make sure the face that will be cut with the microtome is flat.
3. Next, cool the block in ice for 30 min (put the side to be sliced in touch with the foil).
4. Serial sections of 10–15 μm thickness are generated with a retracting rotatory microtome.
5. Pre-heat slides on a glass plate placed on a hotplate at 42°C.
6. Put down a layer of distilled water along each slide.
7. Lay the sections on a clean black paper and then transfer them into the pre-heated water over the slide (two lines of about seven stretched sections on each slide). Bright face of the section should be in contact with the slide.
8. Place vertically the slide over a tissue paper to drain away the excess of water. Avoid air or water bubbles formation on the sections.
9. Dry the slides up at 42°C overnight.
10. Keep the slides in a slide rack closed with parafilm at 4°C for 1 week maximum or in a box containing dessicant for long-term storage.

3.4. Probe Preparation and Non-radioactive Labeling

1. For long RNA detection, the cDNA corresponding to the probe is cloned into the pGEM-T Easy vector (Promega, see Note 7). This vector (as many others) contains T7 and SP6 promoters on each of the side of the multicloning site in oppo-

site orientation, so that RNA probes can be synthesized in vitro either in antisense or in sense orientation for negative control (see Note 8). Before RNA probe synthesis, the recombinant plasmid has to be linearized by digestion with an appropriate restriction enzyme (see Note 9), in order to avoid in vitro transcription to go beyond the expected region. The linearized plasmid purified from agarose gel is used as template for labeling. For small RNA, LNA-modified oligonucleotides (Exiqon), antisense from the target, are used as probes (see Note 10).

2. Labeling with Digoxigenin is performed using the DIG RNA Labeling Kit (SP6/T7) (Roche) or the DIG Oligonucleotide 3′-End Labeling kit (Roche; France, see Note 11) for long or small RNA detection, respectively, according to the manufacturer's instructions.
3. For checking probe labeling efficiency, add 0.5 μl of the probe next to 0.5 μl of the control provided by the kit onto a Nylon positive charged membrane (Roche). Let the membrane dry for a few minutes and then, UV cross-link for 60 s.
4. Next, incubate the membrane with gentle agitation as follows: 1× TBS for 1 min; 30 min in 0.5% Blocking Reagent (Roche, France) in 1× TBS; 30 min in a dilution 1/5,000 of Antidigoxigenin-AP Fab Fragment (Roche, France) in 1× TBS; wash twice in 1× TBS for 5 min each; and, finally, 5 min in 1× Substrate buffer.
5. Reveal in 1× Substrate buffer containing NBT and BCIP (7.5 μl NBT 100 mg/ml and 16 μl BCIP 50 mg/ml to 5 ml Substrate buffer) for 30 min to 1 h. Stop the reaction washing the membrane with water. Signal strength must be similar to the control provided by the kit (see Note 12).

3.5. Hybridization and Washing

A prehybridization treatment is essential to reduce background staining. The hybridization step allows binding of the probe to the target and, subsequent washes aim removing unbound probe or non-selectively and partially annealed probe. The probe concentration, time of hybridization, and the stringency of the washes are crucial steps during the *in situ* hybridization process. All steps before hybridization are carried out at room temperature. Unless otherwise noted, hybridization and later washes are carried out at 50°C (see Note 13).

1. First, paraffin must be removed from sections (see Note 14). Deparaffinization procedure is made by immersing the tissue into Histo-Clear II twice for 10 min each.
2. Get rid of loose Histo-Clear II flushing with 100% EtOH twice.

3. Then, rehydrate sections by 5 min washes with decreasing ethanol concentrations (90, 80, 60, 50, and 30% in 1× Saline solution) and a final incubation in 1× Saline solution for 10 min.
4. Incubate the slides for 5 min in 1× Pronase buffer followed by a treatment in Pronase solution for 10 min (see Note 15).
5. Stop the reaction with two incubations in 1× PBS for 2 min each.
6. For increasing tissue preservation, slides are incubated in 4% formaldehyde in 1× PBS for 30 min (see Note 16).
7. Wash three times the sections with 1× PBS for 2 min each.
8. Incubate the samples in Acetic anhydride solution during 10 min (see Note 17).
9. Next, perform two washes with 1× PBS followed by another wash with 1× Saline solution, for 2 min each.
10. Back dehydrate the sections through graded ethanol concentrations (30, 50, 80, 95%) in saline 1× buffer and 100% EtOH (see Note 18), for 2 min each.
11. Before starting hybridization, place a glass plate inside a glass bowl horizontally and approximately 3 cm above the bottom to obtain a "humidity chamber."
12. Trim pieces of parafilm at the size of the slides.
13. Pre-warm all buffers and solutions at the needed temperature according to the protocol.
14. Wash the sections in 1× Salt solution for 10 min.
15. Pre-incubate each slide 30 min in 250 μl Prehybridization buffer.
16. Heat the probe 5 min at 80°C and then place it rapidly into ice.
17. Add 5 μl of the probe (see Note 19) to 220 μl Hybridization buffer, mix well, and spin briefly.
18. Take one slide, dry the back side with a tissue paper.
19. Put a waterdrop onto the glass plate and place the slide over (waterdrop makes the slide to be fixed on the plate).
20. Drop 220 μl of the "Hybridization buffer-probe" mix on the slide.
21. Spread the "Hybridization buffer-probe" mix over the slide with a pipette tip avoiding to touch the sections.
22. Place one of the piece of parafilm over the Hybridization buffer. Be careful not to form bubbles.
23. Repeat with every slide individually.
24. Pour formamide:1× SSC 1:1 (v/v), prewarmed at 50°C, in the bottom of the bowl.
25. Seal the bowl with cling film and hybridize for 14 h at 50°C.

26. Put the slides covered by the parafilm in a rack and incubate them at 50°C in formamide: 1× SSC 1:1 (v/v) three times for 30, 60, and 60 min, respectively.
27. Remove carefully the parafilm covering the slides after the first incubation.
28. Immerse the sections with 1× NTE buffer three times for 5 min each at 37°C.
29. Digest with 20 μg/ml RNase A in 1× NTE 30 min at 37°C to remove the excess of probe.
30. Rinse twice in 1× NTE for 5 min each at room temperature.
31. Incubate the slides in formamide: 1× SSC 1:1 (v/v) for 1 h at 50°C.
32. Dip the sections with 1× SSC for 5 min. Then, wash three times with TBS for 5 min each at room temperature.

3.6. Signal Detection

Perform the next steps at room temperature.

1. Incubate the slides for 30 min in 0.5% Blocking reagent (Roche) diluted in 1× TBS.
2. Next, immerse the slides in the Antibody buffer for 30 min.
3. Place a glass plate inside a glass bowl, as for the hybridization process, and pour water in the bottom of the bowl.
4. Put the slides on the plate, pour 500 μl per slide of 0.5 μl/ml "Anti-Digoxigenin-AP fab fragments" (Roche) in Antibody buffer, and carefully spread over the sections.
5. Seal the bowl with cling film and protect sections from the light.
6. Incubate the slides 90 min (see Note 20) at room temperature.
7. Place the slides in a rack and stop the reaction by immersing the rack five times in 1× TBS for 5 min each.
8. Equilibrate the sections by washing in 1× Substrate buffer for 5 min.
9. Prepare again the plate in the bowl and add water at the bottom.
10. Dry the back of the slide, put a waterdrop on the plate and place the slide over.
11. Pour 800 μl of Revealing solution and spread over the entire surface of the slide.
12. Cover the bowl with cling film and foil as previously done.
13. Incubate in the Revealing solution for 3 h to overnight (see Note 21).

14. Stop the reaction at the desired signal intensity by rinsing the slides in water for 2 min.
15. Wash the slides, for 2 min each, in EtOH (40, 70% in water, and 100%) subsequently.
16. Repeat the washes in the reverse order.
17. For counterstaining, dip the slides in 0.25% Alcian Blue 8GX in 3% acetic acid.
18. Next, rinse two or three times in water.
19. Place again the slides onto a glass plate and let them dry (10–15 min at room temperature).
20. Cover them with a foil in order to protect them from light.
21. Cover the sections with a coverslip, pouring 150 μl of DPX per slide.
22. Let dry at least 1 day at room temperature in darkness before observation using a standard light microscope.

4. Notes

1. Positive charged slides have a permanent electrostatic charge that attracts sections without using special adhesives or protein coatings. This is particularly important for *in situ* hybridization procedure, where the sections undergo several lengthy and sometimes harsh treatments and thus tend to be detached from the slide during the procedure.
2. Paraformaldehyde is one of the best fixatives as it favors a good balance in preserving tissue, keeping RNA within the tissue and allowing a proper recognition of the target transcript.
3. Fixative solution may be aliquoted and preserved at −20°C for several months. When thaw, if precipitated, warm at 37°C until complete dissolution.
4. PBS could be used instead of MTSB.
5. Composition of the Hybridization solution is one critical point for the efficiency of the *in situ* hybridization. Temperature of hybridization, pH, monovalent cation concentration, and organic solvents influence probe--target annealing. Dextran sulfate hydrates reduce the amount of water needed for dissolving the nucleotides and therefore increase probe concentration resulting in higher hybridization efficiency. Formamide allows to decrease the hybridization temperature by lowering the thermal stability of the bonds. Denhardt's solution and tRNA reduce the risk of unspecific binding of the probe.

6. Remember to pre-heat at 60°C all material needed for embedding before use.
7. Alternatively, probes can be made directly by PCR amplification. In order to synthesize antisense probe, amplify by using a forward primer and a reverse one, which contains T7 promotor sequence in 3' (forward + T7-reverse). For sense probe, amplify with a reverse primer and with a forward one with the T7 promotor sequence in 5′ (T7-forward + reverse).
8. To test that the observed labeling is specific to the target, several controls should be used. Negative controls serve to establish the non-specific background level and positive controls allow to test the complete procedure. Probes known to work on their targets in other tissues or plants but corresponding to genes not expressed in the tissue or cell type analyzed could be good negative controls (e.g., probes corresponding to species-specific genes). The use of sense RNA can also be used in case no other probe is available, although the presence of antisense RNAs for a particular locus can give a signal. For small RNA, an oligonucleotide probe in sense orientation (same sequence as the target) or an oligonucleotide antisense from a small RNA absent in that species can be used as negative controls. For non-coding RNAs, probes that are species-specific are much easy to find as these transcripts diverge rapidly between plants and even ecotypes. Positive controls could be probes that recognize small RNA expressed profusely in all cell types or have a specific pattern in the tissues analyzed.
9. pGEM-T Easy vector contains a large set of unique restriction sites flanking its multicloning site that allow linearization on the opposite side of the promoter to allow gene transcription. However, the use of SP6 or T7 primers for amplification allows the obtention of a linearized delimited template for in vitro transcription by PCR (see Note 7).
10. LNA-oligonucleotides makes possible to generate probes as short as 20 nucleotides because LNA-modified nucleotides have an increased affinity for target transcripts. LNA-oligonucleotides show an increase in melting temperature of target–probe hybrid by ≈2–10°C per nucleotide modified. At least, 25–30% LNA-modification of the probe is recommended (2, 4, 14).
11. Alternatively, LNA-modified probes could be 5′-end labeled using the DIG Oligonucleotide 5′-End Labeling kit (Roche, France). Occasionally, increasing signal intensity can be achieved by both 5′- and 3′-end labels. However, this double labeling might change the specificity of the probe because of extra nucleotides.

12. As a high probe signal could generate a strong background and artifacts, it is interesting to test several probe dilutions, to determine the concentration that maximizes the signal relatively to the background.
13. Optimal temperature for prehybridization, hybridization, and washing may depend on the length and sequence of the probe. The optimal temperature can be calculated using the melting temperature (Tm) of each probe.
14. Incomplete removal of paraffin causes poor staining of the sections.
15. Fixation can induce RNA-protein cross-linking and may thus affect probe binding. In addition, RNA are usually associated to RNP proteins which may hinder their accessibility. To expose the RNA out of protein complexes, a protease treatment is used, such as pronase, a mixture of proteases isolated from *Streptomyces griseus*. However, a too long incubation may severely affect cell integrity in the section. Again, several dilutions and time of treatment need to be tested depending on the plant material to find the optimal ratio of pronase treatment allowing a maximal signal but retaining cellular integrity.
16. Slides can also be incubated within 4% paraformaldehyde in PBS at 4°C overnight.
17. The acetylation treatment reduces background by removing nonspecifically bound probes. It also appears to inactivate RNAses (12).
18. Slides can be left in 100% ethanol for several hours.
19. Optimal probe concentration: 100–500 ng/ml (2). Signal intensity is limited by the abundance of the target, whereas unspecific background appears to be caused by high probe concentration.
20. Alternatively, it is possible to incubate overnight at 4°C.
21. The reaction should be stopped when background becomes barely evident in the negative control. For better contrast, immersion may be carried out for short time and, for better morphologic interpretation, longer times. If revealing overnight, add enough Revealing solution to prevent slides from getting dry.

References

1. Brosnan CA, Voinnet O (2009) The long and the short of non coding RNAs. Curr Opin Cell Biol 21:416–425
2. Nouvo GJ (2008) In situ detection of precursor and mature microRNA in paraffin embedded, formalin fixed tissues and cell preparations. Methods 4:39–46
3. Song R, Ro S, Yan W (2010) In situ hybridization detection of microRNAs. Methods Mol Biol 629:287–294
4. Planell-Saguer M, Rodicio MC, Mourelatos Z (2010) Rapid in situ codetection of noncoding RNAs and proteins in cells and formalin-fixed

paraffin-embedded tissue sections without protease treatment. Nat Protoc 5:1061–1073

5. Nuovo GJ, Elton TS, Nana-Sinkam P, Volinia S, Croce CM, Schmittgen TD (2009) A methodology for the combined in situ analyses of the precursor and mature forms of micro RNAs and correlation with their putative targets. Nat Protoc 4:107–115
6. Válóczi A, Hornyik C, Varga N, Burgyán J, Kauppinen S, Havelda Z (2004) Sensitive and specific detection of microRNAs by northern blot analysis using LNA-modified oligonucleotide probes. Nucleic Acid Res 32:e175. doi:10.1093/nar/gnh171
7. Yang H, Wanner IB, Roper SD, Chaudhari N (1999) An optimized method for in situ hybridization with signal amplification that allows the detection of rare mRNAs. J Histochem Cytochem 47:431–445
8. Válóczi A, Várallyay E, Kauppinen S, Burgyán J, Havelda Z (2006) Spatio-temporal accumulation of microRNAs is highly coordinated in developing plant tissues. Plant J 47:140–151
9. Boualem A, Laporte P, Jovanovic M, Laffont C, Plet J, Combier JP, Niebel A, Crespi M, Frugier F (2008) MicroRNA166 controls root and nodule development in *Medicago truncatula*. Plant J 54:876–887
10. Lelandais-Brière C, Naya L, Sallet E, Calenge F, Frugier F, Hartmann C, Gouzy J, Crespi M (2009) Genome-wide *Medicago truncatula* small RNA analysis revealed novel microRNAs and isoforms differentially regulated in roots and nodules. Plant Cell 21:2780–2796
11. Bustos-Sanmamed P, Tovar-Méndez A, Crespi M, Sato S, Tabata S, Becana M (2011) Regulation of nonsymbiotic and truncated hemoglobin genes of *Lotus japonicus* in plant organs and in response to nitric oxide and hormones. New Phytol 189:765–776
12. Braissant O, Wahli W (1998) A simplified *in situ* hybridization protocol using non-radioactively labelled probes to detect abundant and rare mRNAs on tissue sections. Biochemica 1:10–16
13. Martínez A, Miller MJ, Quinn K, Unsworth EJ, Ebina M, Cuttitta F (1995) Non-radioactive localization of nucleid acids by direct *in situ* PCR and *in situ* RT-PCR in paraffin-embedded sections. J Histochem Cytochem 43:739–747
14. Obernosterer G, Martinew J, Alenius M (2007) Locked nucleic acid-based *in situ* detection of microRNAs in mouse tissue sections. Nat Protoc 2:1508–1514

Chapter 21

Analyzing Protein Distribution in Plant Tissues Using "Whole-Mount" Immunolocalization

Pilar Bustos-Sanmamed, Carole Laffont, Florian Frugier, Christine Lelandais-Brière, and Martin Crespi

Abstract

Proteins are distributed in different cellular compartments. Our group studies the role of non-coding RNAs and associated RNPs in the development and stress response in legumes. Ribonucleoproteins (RNPs) are RNA-protein complexes that play different roles in many cellular processes. Long and small non-coding RNAs determine the specificity of action of several RNPs as the RNA Induced Silencing Complex (RISC), or affect mRNA translation, splicing and stability by interacting with other RNPs such as P-bodies, spliceosome or polysomes. Together with small and long RNAs (Chapter 20), the precise localization of the associated RNPs or the translational products regulated by small RNAs (ie target proteins regulated by miRNAs, or translationally-regulated products) by immunocytochemistry could bring novel insights into these regulatory processes. The protocol described is currently used for detection of RNP associated proteins in nodules and roots of *Medicago truncatula* but could be extended to any other protein. The critical points, as the choice of the antibody and the fixation and permeabilization steps, that allow preservation of tissue and cell integrity and increase the accessibility to epitopes, will be discussed.

Key words: *Medicago truncatula*, Legume, Root, Nodule, Ribonucleoproteins

1. Introduction

Ribonucleoproteins (RNPs) are proteins associated with transcribed RNA. They define the lifetime, cellular localization, processing, and the rate at which a specific mRNA is translated. Different RNPs, as DICER, RNA-dependent RNA polymerase or ARGONAUTE proteins, are also involved in the biogenesis and function of small non-coding RNAs (ncRNAs) which are specific regulators of gene. Although, ncRNAs expression studies can give a lot of information about genetic interactions and networks, spatial detection of ncRNAs-associated proteins by immunolocalization could help to

Ive De Smet (ed.), *Plant Organogenesis: Methods and Protocols*, Methods in Molecular Biology, vol. 959,
DOI 10.1007/978-1-62703-221-6_21, © Springer Science+Business Media New York 2013

better understand their regulatory functions (1, 2). In addition, as small RNAs control the level of specific proteins by post-transcriptional regulation (through endonuclease cleavage of the mRNA target or by blocking its translation), immunolocalization would also let us assess the localization of proteins encoded by different miRNA targets.

Although, Western blots (3) and recombinant proteins (4) are methods already described to detect proteins, Western blots do not reveal any information about their spatial expression profile and cellular localization (5) and, recombinant proteins could cause misexpression, mistargeting, altered stability, or toxic effects in cells (4).

We are interested in understanding the role of RNPs in the regulation of root and nodule development in legumes. For this purpose, we have developed a fast and simple immunolocalization protocol (adapted from ref. 6) on thick vibratome ("whole-mount") sections successfully applied in *Medicago truncatula* roots and nodules (7).

1.1. Critical Points Immuno-localization

1.1.1. Tissue Fixation and Permeabilization

For immunolocalization as well as for in situ hybridization, the first crucial step is the fixation of the material (1, 8). In this way, the general aspects to taking into account to fix the plant material for immunolocalization are the same as for in situ hybridization (see Subheading 1.1.1 in Chapter 20).

1.1.2. Choice and Labeling of the Antibody or Probe

The choice of the antibody is also essential to carry out successful immunolocalization (6). In general, polyclonal antibodies are relatively quick, easy, and cheap to produce. Moreover, these antibodies may recognize epitopes even after the antigen has been denaturated by fixation or other procedures. In some cases, however, the specificity and sensitivity of the signal obtained with polyclonal antibodies will not be high or specific enough (e.g., for highly similar proteins isoforms of one family). To overcome the problem of sensitivity, two successive immunoreactions can be performed: first, the antigen is detected by the specific antibody, and this primary antibody is itself recognized by a second antibody linked to a visual marker. Nevertheless, when the antigen of interest shares a common epitope with an apparently unrelated protein or when the antigen used for immunization was not purified enough, the primary antibody may show some irrelevant cross-reactivity. In such case, monoclonal antibodies or antibodies raised against specific peptides from the protein or to tags added to the protein of interest (e.g., HA or myc) can be used.

Antibodies, the same as probes for in situ hybridization, can be labeled with radioactive and non-radioactive methods including biotin, digoxigenin, and fluorescent dyes like fluorescein, Texas Red, or ALEXAs (see Subheading 1.1.2 in Chapter 20 for general aspects).

2. Materials

The protocol described here is adapted from Ref. 6

2.1. Equipment

1. Vacuum pump or Speed-Vac.
2. Horizontal rotary shaker.
3. Vibratome (e.g., Leica, VT 1200S).
4. Epi-fluorescence microscope (e.g., Leica DM6000 B) or confocal microscope (e.g., Leica TCS SP2).
5. Positive charged Poly-L-lysine-coated slides and coverslips (see Note 1).

2.2. Chemicals

1. Paraformaldehyde.
2. Albumin from bovine serum fraction V (BSA).
3. 4′,6-Diamidino-2-phenylindole dihydrochloride (DAPI).
4. Anti-fading reagent (e.g., Citifluor Glycerol Medium, Ted Pella).
5. 100% Ethanol.
6. Driselase.
7. DMSO.
8. Nonidet P40.

2.3. Buffers

1. Fixative solution: 3% paraformaldehyde, 0.1% Triton X-100 in PBS.
2. Microtubule stabilization buffer (MTBS): 50 mM PIPES, 5 mM EGTA, 5 mM $MgSO_4$ (pH 6.9–7.0 with KOH; do not add Na to the buffer).

3. Methods

3.1. Fixation

If not otherwise indicated, steps are performed at room temperature.

1. Immerse plant samples (ideally about 1–2 cm in length) into the Fixative solution (3% PFA, 0.1% Triton X-100 in MTSB).
2. Perform mild vacuum infiltration if required (thick samples, e.g., nodules) using a vacuum pump (see Note 2).
3. Fix the material for 2 h with gentle agitation.
4. Wash twice with MTSB 0.1% Triton X-100 (see Note 2) for 15 min each, and then twice with water 0.1% Triton X-100 for 15 min each.

5. Embed in hot (microwave) agarose 3% (for gel electrophoresis). When the agarose is cooling, slowly orient the samples according to the section (e.g., transversal, longitudinal) you need, avoiding making bubbles.
6. Make 80–100 μm thick sections using a vibratome and perform immunolabeling the same day.

3.2. Immunolabeling and Counter Staining

1. Transfer sections to slides and incubate 15 min in MTSB.
2. To permeabilize the sections, incubate first 20–30 min in 2% Driselase in MTSB (see Note 3).
3. Wash three times with MTSB, 10 min each
4. Incubate then 1 h with DMSO 10% and Nonidet P40 3% in MTSB.
5. Wash three times with MTSB, 10 min each
6. Incubate with 3% BSA in MTBS for 1 h to block unspecific binding sites.
7. Incubate with the primary antibody (see Note 4) diluted in 3% BSA in MTBS at 4°C overnight (see Note 5).
8. Wash three times with MTBS for 10 min.
9. When an antibody conjugated with fluorescent dye is used (see Note 6), protect the slides from light to prevent fading of the dye. Incubate at room temperature for 2 h with the secondary antibody (see Note 7), diluted at 1/300–1/600 in 3% BSA-MTSB (see Note 8).
10. Wash three times with MTBS 10 min each.
11. Counterstain with DAPI (1 μg/ml) in MTBS for 10 min to stain DNA (see Note 9).
12. Wash twice with water, 10 min each.
13. Mount sections in an Anti-fading reagent.
14. Use nail-varnish to seal the coverslip on the slide.
15. Samples are ready for microscopical observation (see Note 10), and can be stored for months at –20°C in dark.

4. Notes

1. Positive charged slides have a permanent electrostatic charge that attracts sections without using special adhesives or protein coatings. This is particularly important for in situ hybridization and immunolocalization procedures, because sections undergo several lengthy and sometimes harsh treatments and can be detached during the procedure.

2. Due to Triton X-100, the Fixative solution starts to foam under vacuum. If a lot of foam rises, stop the vacuum and aerate the samples. Then start a second (if necessary a third) time with vacuum. Stop vacuum infiltration when no more air bubbles are coming out of the samples. Alternatively, you can use a Fixative solution without Triton X-100 and use vacuum for 1–3 h depending on the sample size (when appropriately fixed, samples should be located near the surface of the liquid).
3. Concentration and time of treatment with Driselase should be determined for each type of tissue, and also may vary depending on the Driselase stock and storage (take care that no rehydratation occurs during storage at −20°C).
4. The simplest way to demonstrate antibody specificity is one- or two-dimension immunoblotting (Western blot). Additionally, immunoblots can be used to find the best dilution for the respective antibody. Any way, it should be taken into account that sometimes clearly detectable bands occurring in immunoblots are not reflected by strong signals in immunocytochemistry and vice versa. Usually, we start immunolocalization experiments with a two fold higher concentration of antibody than that used for Western blot.
5. Alternatively, incubation in presence of the primary antibody can be carried out for 1–2 h at room temperature or 37°C. Use 200 μl solution per slide and incubate without coverslip and within a humid chamber.
6. Choosing the secondary antibody conjugated to a fluorophore depends on the excitation and emission spectra available by the microscope. Alexa dyes are more fluorescent and more photostable than their used analogues like amino-methyl-coumarin-acetate, lucifer yellow, fluorescein, rhodamine 6G, tetramethylrhodamine, Cyanine-3, or texas red (9).
7. Secondary antibody dilution may need to be optimized. Secondary antibody could bind unspecifically to the section and the comparison with control labeling is needed. A sample without primary antibody is required to determine the background signal produced by the secondary antibody. Negative controls can also be done by incubation with the pre-immune serum or by the use of tissues from mutants lacking the protein (1, 7).
8. Slides can also be incubated in the presence of secondary antibody for 1 h at room temperature in a humid chamber.
9. To reduce autofluorescence in some plant tissues, alternatively, counterstain with 0.01% toluidine blue in MTBS for 10 min (10).
10. For vibratome thick sections (80–100 μm), use of a confocal laser scanning microscope is recommended to avoid blurring

from the out-of-focus signals. The multi-wavelength mode is useful to determine the emission spectrum of the signal analyzed and prevent false-positive signals due to the autofluorescence of plant tissues. In addition, multiple immunostaining can be done by combining different dyes with non-overlapping emission spectra.

References

1. Hause B, Frugier F, Crespi M (2007). Immunolocalization. In: Mathesius U, Journet EP, Summer LV, (eds) The *medicago truncatula* handbook. ISBN 0-9754303-1-9 . Available in: http://www.noble.org/MedicagoHandbook/
2. Sauer M, Paciorek T, Benkova E, Friml J (2006) Immunocytochemical techniques for whole-mount *in situ* protein localization in plants. Nat Protoc 1:98–103
3. Válóczi A, Hornyik C, Varga N, Burgyán J, Kauppinen S, Havelda Z (2004) Sensitive and specific detection of microRNAs by northern blot analysis using LNA-modified oligonucleotide probes. Nucleic Acid Res 32:e175. doi:10.1093/nar/gnh171
4. Sauer M, Friml J (2010) Immunolocalization of proteins in plants. In: Henning L, Köhler C (eds) Plant developmental biology: methods and protocols, 1st edn. Springer, New York, pp 253–263
5. Song R, Ro S, Yan W (2010) In situ hybridization detection of microRNAs. Methods Mol Biol 629:287–294
6. Friml J, Benková E, Mayer U, Palme K, Muster G (2003) Automated whole mount localisation techniques for plant seedlings. Plant J 34:115–124
7. Plet J, Wasson A, Ariel F, Le Signor C, Baker D, Mathesius U, Crespi M, Frugier F (2011) MtCRE1-dependent cytokinin signaling integrates bacterial and plant cues to coordinate symbiotic nodule organogenesis in *Medicago truncatula*. Plant J 65:622–633
8. Walker RP, Chen ZH, Johnson KE, Famiani F, Tecsi L, Leegood RC (2001) Using immunohistochemistry to study plant metabolism: the examples of its use in the localization of amino acids in plant tissues, and of phosphoenolpyruvate carboxykinase and its possible role in pH regulation. J Exp Bot 52:565–576
9. Panchuk-Voloshina N, Haugland RP, Bishop-Stewart J, Bhalgat MK, Millard PJ, Mao F, Leung WY, Haugland RP (1999) Alexa dyes, a series of new fluorescent dyes that yield exceptionally bright, photostable conjugates. J Histochem Cytochem 47:1179–1188
10. Ruzin SE, Sylvester AW (1994) Light microscopy. II. Observation, photomicrography and image analysis. In: Freeling M, Walbot V (eds) The maize handbook, 1st edn. Springer, New York, pp 95–107

Chapter 22

Culture Methods and Mutant Generation in the Filamentous Brown Algae *Ectocarpus siliculosus*

Aude Le Bail and Bénédicte Charrier

Abstract

Ectocarpus siliculosus is a small filamentous alga that has recently emerged as the new model for fundamental research on brown algae. Here, we describe the basic culture protocols for propagating and collecting *E. siliculosus* material that can then be used in all types of molecular biology, biochemistry and cell biology techniques. In addition, procedures for carrying out genetic experiments (generation of mutants and genetic segregation analyses) on *E. siliculosus* are described.

Key words: Brown algae, *Ectocarpus siliculosus*, Tissue culture, Genetics, Mutant

1. Introduction

Ectocarpus siliculosus has emerged as the new biological model of brown algae for genetic and genomic analyses (1, 2). The emergence of this model has been promoted by the advent of various technologies (genetics, transcriptomics, proteomics, etc.). Proper handling and control of *E. siliculosus* cultures is therefore a prerequisite for any study on this alga. In this chapter, we describe how to obtain different types of *E. siliculosus* cultures, depending on the experiment to be carried out.

The sexual life cycle of *E. siliculosus* has two generations: the haploid gametophyte and the diploid sporophyte (Fig. 1). Male and female gametophytes produce gametes in plurilocular (PL) gametangia. The fusion of a male gamete with a female gamete generates a diploid zygote which develops into a diploid sporophyte. Meiosis

Ive De Smet (ed.), *Plant Organogenesis: Methods and Protocols*, Methods in Molecular Biology, vol. 959,
DOI 10.1007/978-1-62703-221-6_22, © Springer Science+Business Media New York 2013

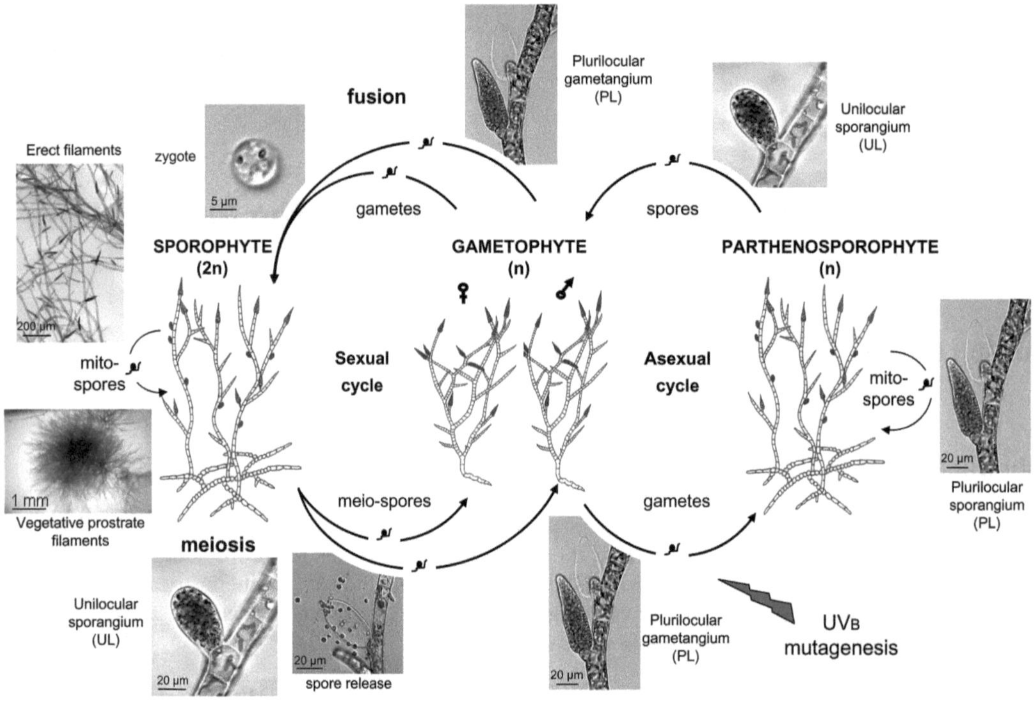

Fig. 1. Life cycle of *Ectocarpus siliculosus*. Both the sexual (*left*) and the asexual (*right*) cycles are displayed. Sporophytes, gametophytes, and parthenosporophytes are drawn. Sporophytes (diploid, *left*) are composed of prostrate filaments, carrying erect filaments differentiating reproductive organs (both unilocular and plurilocular sporangia). Gametophytes (haploid, *center*) are lacking the prostrate basis and contain only erect filaments and plurilocular gametangia. Diploid sporophytes produce meiospores (by meiosis) in unilocular sporangia (UL, *bottom left*). After release, meiospores swim, settle down, and start to divide and grow into male or female gametophytes (dioecy). Gametophytes produce gametes in plurilocular gametangia (PL, *top*). Fusion of gametes (isogamy) produces a zygote that grows into a diploid sporophyte, completing the sexual cycle. Zygotes contain two eye spots (*little black dots* inside the zygote), which are temporary remnants of each of the two gamete flagellar apparatus. Unfused gametes (*bottom right*) may grow parthenogenetically and form a parthenosporophyte (haploid, *left*), which is morphologically indistinguishable from the diploid sporophyte. Both sporophytes and parthenosporophytes can reproduce themselves asexually by the production of mitospores in plurilocular sporangia (*far left* and *far right*). The photos illustrate the different parts of the *E. siliculosus* organism: filaments and reproductive organs (sporangia and gametangia) are displayed either full (immature) or empty after spores or gametes being released. Note the difference of shapes between UL (*round*) and PL (*pointed*), as well as the presence of multiple stacked locules in the PL. Mutagenesis is performed on the gametes.

takes place in unilocular (UL) sporangia present on the erect filaments of sporophytes. Released meiospores swim, settle, and germinate into male or female gametophytes. *E. siliculosus* also reproduces asexually: unfertilized gametes germinate and give rise to haploid sporophytes, called parthenosporophytes. Parthenosporophytes generate gametophytes from their UL sporangia. Both diploid sporophytes and parthenosporophytes also differentiate plurilocular sporangia that produce mitospores by mitosis, giving, respectively, new sporophytes and parthenosporophytes.

2. Materials

2.1. Culture of E. siliculosus

1. Strains: All *E. siliculosus* strains are cultivated in the same way. These strains can be obtained from the Culture Collection of Algae and Protozoa (CCAP). The strain whose genome has been sequenced is CCAP 1310/4 (also named Ec32).
2. Artificial seawater (ASW): 450 mM NaCl, 10 mM KCl, 9 mM $CaCl_2$, 30 mM $MgCl_2$, 16 mM $MgSO_4$, pH 7.8.
3. Natural seawater (NSW): NSW is collected 1–2 nautical miles away from the coast, filtered (0.7 μm), and autoclaved. It is kept up to 3 months.
4. Provasoli medium (3):

   Prepare each of the five following solutions separately; autoclave, and conserve them at 4°C (solution 2 must be stored in the dark).

   (a) Solution 1 (10×): 30 mM H_3BO_3, 0.3 mM $FeCl_3$, 1.6 mM $MnSO_4$, 0.13 mM $ZnSO_4$, 0.028 mM $CoSO_4$, 5.7 mM EDTA.

   (b) Solution 2 (100×): 0.0067 g/L vitamin B12 (cyanocobalamin), 0.33 g/L thiamin (vitamin B1), 0.0033 g/L biotin, 333 g/L Tris.

   (c) Solution 3 (10×): 3 mM $(NH_4)_2Fe(SO_4)_2$, 3.4 mM EDTA pH 8.

   (d) Solution 4 (10×): 270 mM $NaNO_3$.

   (e) Solution 5 (10×): 15.4 mM $C_3H_7Na_2O_6P$.

   Make up each solution (i.e., 1× solutions), adjust the pH to 7.8 with HCl, complete with milliQ H_2O, aliquot in small bottles and autoclave. Provasoli medium should be stored in the dark at 4°C and used at 10 mL/L seawater (ASWp or NSWp).
5. Antibiotic mix: Add 1 g of penicillin G and 0.5 g of streptomycin to 100 mL milliQ H_2O, add 0.1 g of chloramphenicol to 10 mL of ethanol, mix and filter through a 0.22 μm membrane, aliquot and store at −20°C, use 5 μL of this antibiotic mix/liter culture.
6. Controlled environment cabinet at 13°C with a 14:10 light:dark cycle (light intensity 29 μmol photon/m^2 s).
7. Sterile laminar flow hood.
8. Sterile Petri dishes.
9. Sterile 10 L bottles with a whole cap for air bubbling.
10. Fine forceps.
11. Parafilm.
12. Sterile glass coverslips.
13. Pasteur pipettes.
14. Pipettors.

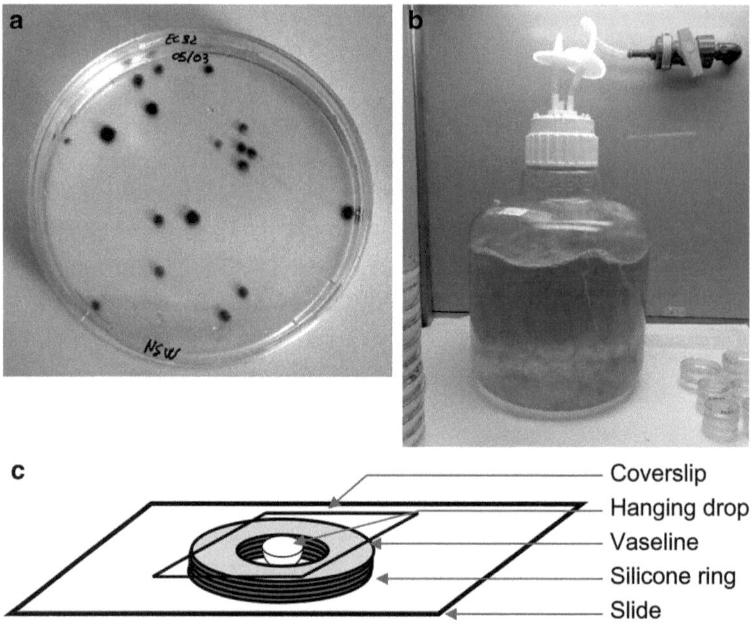

Fig. 2. Culture and crosses of *E. siliculosus*. (**a**) Petri dish containing several *E. siliculosus* organisms (looking like individual small tufts) growing in either Natural Sea Water (NSWp) or Artificial Sea Water (ASWp) supplemented with Provasoli medium (Subheading 2.1). (**b**) *E. siliculosus* culture growing in a 10 L bottle oxygenized with air bubbling. Water is renewed every 2–3 weeks. (**c**) Slide device to observe the gamete fusion and the growth of the embryo (Subheading 3.5).

2.2. Mutagenesis and Crosses

1. UV-B lamp.
2. Glass slides and coverslips.
3. Silicone ring.
4. Vaseline.

3. Methods

All experiments are performed in sterile conditions.

3.1. Routine Cultures and Sample Collection

All experiments should be carried out under a laminar flow hood with sterile equipment. The following procedures are suitable for both gametophytes and sporophytes. However, routine cultures are performed on parthenosporophytes, and diploid sporophytes are cultured essentially when crosses are necessary.
For culture:

1. Grow the cultures in Petri dishes containing ASWp and seal the dishes with parafilm (see Note 1, Fig. 2a).
2. Change the medium every 2–3 weeks (see Note 2).

For collection:

3. Scrape the cultures from the bottom of the Petri dish and filter the media through a sterile 50 μm nylon filter.
4. Blot the algal material on absorbent paper to eliminate excess water.
5. Place the material in a piece of foil, freeze the sample in liquid nitrogen, and store at −80°C.

3.2. Production of Gametophytes

The gametophytes grow from germinating meiospores, which are released from the ULs of sporophytes and parthenosporophytes (Fig. 1). To obtain cultures composed of gametophytes only, contamination of meiospores by mitospores (which are also produced by sporophytes and parthenosporophytes, see Fig. 1) must be prevented by physically isolating ULs from the sporophytes. Given their characteristic round shape, ULs can be easily distinguished from PLs (Fig. 1).

1. Prepare small Petri dishes with a sterile glass coverslip and place a drop of ASWp on the coverslip (see Note 3).
2. Cut the end of a Pasteur pipette and flame-sterilize it.
3. Under a dissecting microscope, cut a section of a sporophyte filament bearing a UL with the sharp broken end of the Pasteur pipette and suck it into the pipette.
4. Release the excised filament section into the drop of ASWp. Several filament sections can be placed in the same drop.
5. Seal the Petri dish with parafilm and place it in standard culture conditions.
6. After 1 or 2 days, check whether spores have been released from the UL. Released spores can be seen deposited around the UL membrane where they begin to germinate (the empty UL is transparent).
7. Remove the filament section with fine forceps. This step is necessary because the filament bearing the now-empty UL can survive and grow, leading to a mix of sporophytes and gametophytes in the dish.
8. Fill the dish with additional seawater.
9. Once small filaments have grown, the gametophytes can be transferred into a bigger dish if fertile gametophytes are required within a few weeks. Otherwise, they can be stored at 13°C under dim light for up to 6 months.

3.3. Gamete Release

Using this protocol, a large number of gametes can be produced for making synchronized and uniform cultures of parthenosporophytes (see Note 4).

1. Once the gametophytes are fertile (with dark brown PLs, Fig. 1), remove the filaments with forceps and place them in a pile in a new dish.
2. Remove excess water with a pipette (the filaments must be damp, but without any standing water) and place the dish overnight in the dark at 10–13°C.
3. To induce synchronized release, add 0.5–1 mL of warm (13°C or room temperature) ASWp/NSWp and wait 10–15 min.
4. Check for the presence of gametes under a microscope equipped with a mirror (gametes look like little bright spots swimming in the water around the pile of gametophytes).
5. Separate the gametes from the gametophytes by pipetting them. Place them immediately in one or several new dishes filled with seawater (see Note 5) (gametes stored in an Eppendorf tube or a Petri dish will quickly sediment and adhere to the wall).
6. Cover, seal, and place the dishes in standard culture conditions.
7. Several releases can be performed from the same gametophyte filaments. To do so, the filaments are transferred to a new dish filled with fresh NSWp. They will become fertile again after 1 week (filaments can be reused up to five times). To promote growth (gently) tease apart the pile of filaments with two forceps.

3.4. Production of Mutants

Because parthenosporophytes are haploid, recessive mutations can be directly identified. Mutagenesis is thus performed on gametes (Fig. 1) using UV-B irradiation as the mutagenic agent.

1. Release gametes as described in Subheading 3.3.
2. Put the gametes in a drop in the center of a small dish (3 cm in diameter).
3. Let them stick to the bottom of the dish for 4 h.
4. Fill the dish with seawater and seal it with parafilm.
5. Place the dish (without its lid) under a UV-B lamp (316 nm) positioned at a distance of 7.5 cm for a previously determined period of time (see Note 6).
6. After about 5 weeks, isolate the individuals (parthenosporophyte) showing morphogenetic alterations using forceps under a dissecting microscope and place them in individual dishes.
7. Let them grow until they are fertile and allow the subsequent parthenogenetic generation to develop from the germination of mitospores.
8. If the phenotype is maintained in this parthenosporophyte population, isolate one parthenosporophyte and repeat step 7 for atleast five parthenogenetic generations. This eliminates

any variants that initially showed a morphogenetic alteration compared to the non-irradiated population, but for which the mutations are not transmitted to the offspring.

3.5. Crosses

Determining the recessive/dominant nature of a mutation requires generating a heterozygous F1 population. Crosses between mutant and wild-type (WT) organisms are performed as follows.

1. Prepare gamete release for each of the two strains to be crossed separately, as described in Subheading 3.3.
2. Place a silicone ring on a slide using vaseline and also add vaseline to the top of the ring (Fig. 2c).
3. Pipette 5 μL of female gametes and place them in a single drop (~20 μL) of NSWp on a sterile coverslip.
4. Pipette 5 μL of male gametes and add them to the drop containing the female gametes.
5. Carefully invert the coverslip and put it on the silicone ring placed on the slide (the drop is now upside down, hanging in the space between the slide and the coverslip).
6. Place the slide under standard culture conditions.
7. After 1 day, observe the drop under the microscope (×1,000, with immersion oil), and determine the position of the zygotes, which can be distinguished from unfertilized gametes by the presence of two stigmata (see Note 7).
8. Every 2 days for 2 weeks, remove 5 μL from the drop, and add an equivalent volume of fresh media (do not disturb the position of the zygote in the drop). Check the development of the diploid sporophyte (see Note 8).
9. Circle the position of the drop with a marker and then place the coverslip containing the germlings right side up and transfer it to a Petri dish (see Note 9).
10. Very slowly and carefully add NSWp containing the antibiotic mix to the dish (see Note 10).
11. Localize the position of the diploid sporophytes in the Petri dish with an inverted microscope (see Note 11).
12. To prevent parthenosporophytes from growing over the diploid sporophyte, peel parthenosporophytes off from the coverslip with a sterile needle and carefully change the medium.
13. Once the diploid sporophyte is composed of more than 20 cells, remove it by pipetting and place it in a new Petri dish with fresh media (see Note 12).

3.6. Analysis of the Progeny

Determining whether the mutant phenotype is due to a single mutated locus requires analyzing the progeny of the cross.

1. Once the diploid sporophyte develops ULs (Fig. 1), isolate at least 20 ULs in individual Petri dishes as described in Subheading 3.2.
2. When the spores have been released, remove the filament section bearing the empty UL and fill in the dish with seawater.
3. After about 1 week, isolate from 20 to 30 small gametophytes from each UL with forceps (see Note 13), place them in 24-well plates filled with seawater and let them grow for about 2 weeks.
4. Transfer each gametophyte into individual Petri dishes and wait for the development of PLs (Fig. 1).
5. When the gametophytes are fertile, proceed with an individual gamete release for each gametophyte, as described in Subheading 3.3.
6. Separate the gametes from the gametophytes by pipetting them and transfer them immediately to a new dish filled with seawater.
7. Keep the gametophytes in case another gamete release is necessary.
8. Analyze the phenotype of the parthenosporophytes growing from the germination of the gametes (in any given plate, all the parthenosporophytes should display the same phenotype).
9. Determine the percentage of parthenosporophytes displaying either the mutant phenotype or the wild-type (WT) phenotype. If the phenotype is due to a single-locus mutation, then about 50% of each should be observed for each and every sampled UL.
10. A Chi-square test is used to determine the statistical significance of this distribution. If it cannot be concluded that the phenotype is due to a single-locus mutation, then genetic segregation experiments (i.e., by crossing the mutant with a WT partner) must be performed to separate the multiple mutated loci.

4. Notes

1. If a large amount of material is required (e.g., protein extraction, several simultaneous experiments), cultures can also be grown in 10 L plastic flasks with bubbling filtered (0.22 μm) compressed air to avoid CO_2 depletion (Fig. 2b). If high-magnification microscopic observations are required or specific treatments require many culture media changes (immunolocalization, in situ hybridization), fixed sporophyte cultures can be grown on sterile coverslips.

2. *E. siliculosus* cultures can be maintained for up to 1 or 2 years (depending on the strain) in 5 mL tubes with ASWp at 4°C and under reduced light intensity.
3. The glass coverslips prevent the drop from moving in the dish. To avoid disturbing the coverslip in the dish, put a drop of sterile water in the dish before placing the coverslip. To prevent the drop from spreading on the coverslip, coverslips must be dried and autoclaved in an oven at 180°C for 2 h.
4. New parthenosporophytes can also be obtained from mitospores that develop in the PLs of parthenosporophytes (Fig. 1). Spore release is performed using the same method as for gamete release. To obtain parthenosporophytes only, it is necessary to ensure that no ULs are present on the sporophytes.
5. The volume depends on the density of gametes and on the quantity of material required (usually 10–20 μL). If a high concentration of gametes is required, release should be performed in a 300 μL drop and the gametes are left to settle and adhere on the bottom of the Petri dish for 3 h before the final volume of ASWp or NSWp is added.
6. The conditions described here have been determined to lead to 50% gamete mortality (LD_{50}). Live gametes are counted immediately after mutagenesis (1, 3, 5, 7, 10, 15, 20, 25, and 30 min irradiation), and after 3 days. The mortality rate can then be calculated from the difference between the two values. In our experimental conditions (i.e., UV lamp positioned 7.5 cm above the Petri dish), the LD_{50} corresponds to 10 min irradiation for the Ec32 strain and 20 min for the Ec410 strain. In different conditions, or if other strains are to be used, a mortality curve must be determined prior to mutant selection.
7. This step is very important because unfertilized gametes will also grow in the drop and will have morphology similar to the diploid sporophyte generated from fertilization. The position of the zygote can be determined by counting the number of visual fields from the edges of the drop. Alternatively, a gridded coverslip can be used.
8. Be careful to not dislodge the drop to prevent losing track of the zygote's position. Keep a record of the relative position of the zygote as it grows, by drawing a sketch of the zygote surrounded by the gametes.
9. Before placing the coverslip into a Petri dish filled with seawater, carefully clean the external surface of the coverslip with alcohol to remove the immersion oil and to sterilize it.

10. Neither the vaseline in contact with the coverslip nor the external side of the coverslip is sterile. To avoid jeopardizing the development of the zygote, add an antibiotic mix (see Subheading 2.1, item 5).
11. Because the coverslip has been inverted, the relative positions of the diploid sporophyte and the parthenosporophytes are now symmetrically inverted.
12. To ensure that the selected individuals are diploid and do not originate from unfertilized gametes, it is strongly advised to check the heterozygous state of the selected individual by PCR amplification using molecular markers that are polymorphic for the two parental strains (e.g., as described in (4)).
13. At this stage, gametophytes have about 20 cells and are not yet floating; they can thus be easily separated and collected under a dissecting microscope.

References

1. Charrier B, Coelho S, Le Bail A et al (2008) Development and physiology of the brown alga *Ectocarpus siliculosus*: two centuries of research. New Phytol 177(2):319–332
2. Cock JM, Sterck L, Rouzé P et al (2010) The *Ectocarpus* genome and the independent evolution of multicellularity in the brown algae. Nature 465(7298):617–621
3. Provasoli L, Carlucci AF (1974) Vitamins and growth regulators. In: Stewart WDP (ed) Algal physiology and biochemistry. Blackwell, Oxford, pp 741–787
4. Le Bail A, Billoud B, Le Panse S et al (2011) *ETOILE* regulates developmental patterning in the filamentous brown alga *Ectocarpus siliculosus*. Plant Cell 23(4):1666–1678. doi:10.1105/tpc.110.081919

Chapter 23

Building Simulation Models of Developing Plant Organs Using *VirtualLeaf*

Roeland M.H. Merks and Michael A. Guravage

Abstract

Cell-based computational modeling and simulation are becoming invaluable tools in analyzing plant development. In a cell-based simulation model, the inputs are behaviors and dynamics of individual cells and the rules describe responses to signals from adjacent cells. The outputs are the growing tissues, shapes and cell-differentiation patterns that emerge from the local, chemical and biomechanical cell-cell interactions. Here, we present a step-by-step, practical tutorial for building cell-based simulations of plant development with *VirtualLeaf*, a freely available, open-source software framework for modeling plant development. We show how to build a model of a growing tissue, a reaction-diffusion system on a growing domain, and an auxin transport model. The aim of VirtualLeaf is to make computational modeling better accessible to experimental plant biologists with relatively little computational background.

Key words: Plant development, Organ growth, Cell division, Cell growth, Mathematical modeling, Cell-based modeling, Systems biology, Computational modeling, Reaction-diffusion, Biomechanics, Auxin

1. Introduction

Computational modeling is becoming a key tool in the study of biological development (1–5). Instead of focusing on the phenotypes of single gene knock-out lines or over-expression lines, computational models help the researcher to unravel the function of the gene in the context of the dynamics of the regulatory gene networks. In such a *systems biology* approach, the researcher starts with a hypothesis of the biological mechanism that he/she studies. The researcher then implements this hypothetical mechanism in a computational model and studies to what extent the computational model reproduces the biological observations, and where the comparison fails. These discrepancies offer the most valuable information, because they

Ive De Smet (ed.), *Plant Organogenesis: Methods and Protocols*, Methods in Molecular Biology, vol. 959,
DOI 10.1007/978-1-62703-221-6_23, © Springer Science+Business Media New York 2013

identify components missing from the model, incorrect interactions between existing components, or wrong values for the model parameters. The researcher hypothesizes a series of new components and tests if simulation models extended with these new components better reproduce the experimental observations. A next step is to experimentally test the validity of the components used to update the simulation model. Typically, the experiments will falsify the new hypothesis or point at further factors that need to be considered in the model, prompting a new round of the *systems biology cycle* (6). In essence this approach does not differ from the traditional empirical approach. The computational model replaces the hypothesis, which biologists typically express in terms of box-and-arrow diagrams (often also called "model"). The computational model helps to check the internal logic of the hypothesis, it predicts unexpected outcomes of the component interactions, and it identifies specific experiments to test the hypothesis.

To build a computational model, most experimental biologists depend on computational biologists to build the models for them, even if they want to test a relatively simple, "back-of-the-envelope" kind of idea. To make constructing and simulating models of biochemical networks more accessible, a range of simulation tools has been made available (see ref. (7) for review). These systems give useful insight in the dynamics of single cells, but they are rarely suitable for studies of developing plant organs because of their multicellularity.

Unraveling the development of a plant organ requires insight at multiple biological scales at the same time: (1) the biochemical networks regulating single cells, (2) the resulting behavior of the cells (i.e., cell expansion, cell division, cell differentiation), (3) the signals received from adjacent cells, (4) the resulting plant tissues and cellular differentiation patterns, and (5) the resulting biomechanics of the whole tissue and its effects on the single cells. For developmental studies a useful technique is cell-based modeling (7–9). The input to cell-based models is the biochemical networks of individual cells, and the resulting cell behaviors; the outputs of such a model are the resulting tissue patterns and shapes.

VirtualLeaf makes it possible for plant biologists to design simple, cell-based models of plant development. To model tissue growth and biomechanics, it uses a Monte-Carlo approach. These tissue growth models are combined with dynamic models of intracellular regulatory networks and intercellular transport of phytohormones, which are simulated using an ordinary differential equation approach (see ref. (7) for details). Although the details of these simulation methods typically remain hidden to the user, they can be modified if necessary without touching the model definition.

To set up a model in *VirtualLeaf*, the first step is to design a *cell-based* hypothesis: what do the cells do (i.e., the cell behaviors), what networks regulate these behaviors (intracellular dynamics), what inputs do the cells obtain from their neighbors and how do these respond to them. The next steps are to implement these rules in *VirtualLeaf*, to simulate the model, and to analyze it. Subheading 2 shows how to prepare your computer for running *VirtualLeaf*. Subheading 3 provides instructions on the basic usage of *VirtualLeaf*, and demonstrates how to build models of a growing plant tissue, a reaction-diffusion mechanism. The section ends by describing a detailed model of auxin transport.

2. Materials

Building a model in *VirtualLeaf* requires basic knowledge of the programming language C++. Basic knowledge of differential equations is useful too (see Note 1). Suitable tutorials can be found on the internet. All required tools have open source versions and can be downloaded for free. Download and install a C++ compiler, the graphical library Qt and the source code of *VirtualLeaf*. *VirtualLeaf* can be run on Windows, Linux, and MacOSX.

2.1. Required Software: C++ Compiler and Libraries

Download and install the Qt Software Development Kit (Qt-SDK) from http://qt-project.org/downloads. The download page presents you first with a license choice, GPL or commercial, and then with list of QT downloads based on operating system and machine architecture. Choose the "Complete Development Environment" appropriate for the operating system you are using. The file you download is a self-extracting archive; when executed it will display a graphical user interface that will guide you through the installation process.

On MacOSX also install the XCode Development environment from the MacOSX DVDs (see Note 16).

2.2. Source Code

Download the source code of *VirtualLeaf* from: http://virtualleaf.googlecode.com/files/VirtualLeaf-v1.0-src.zip.

Unpack the archive to the folder where you wish to install the Virtual Leaf.

Windows:

```
> c:\Documents and Settings\mge\simulations
```

Linux/MacOSX:

```
~/simulations
```

Experienced users can get the latest development version of *VirtualLeaf* using the mercurial version control system. To do so, install mercurial and type the following in a terminal shell:

```
> hg clone https://virtualleaf.googlecode.com/hg/
virtualleaf
```

2.3. Compile the VirtualLeaf Framework

Windows:
Open a Qt Command prompt by choosing "Qt Command Prompt" from the "start" menu, then go to the folder where you have unpacked the source code of *VirtualLeaf*, e.g., (replace "mge" for your own user name)

```
> cd c:\Documents and Settings\mge\simulations
```

Change to the *VirtualLeaf* source directory.

```
> cd VirtualLeaf\src
```

Start the compilation procedure.

```
> set MAKE=mingw32-make
> mingw 32-make
```

Linux and MacOSX:
Open a terminal (on MacOSX: type "Terminal" in Spotlight and press enter; Terminal is in /Applications/Utilities/).

Go to the directory where you unpacked the *VirtualLeaf*.

```
> cd ~/simulations
```

Change to the *VirtualLeaf* source directory.

```
> cd VirtualLeaf/src
```

Start the compilation procedure.

```
> make
```

2.4. Test VirtualLeaf

Once the compilation is complete, you will find the *VirtualLeaf* binary in `virtualleaf/bin` and the models in `virtualleaf/bin/models`. Test *VirtualLeaf* on Windows and Mac by double clicking on the binary; on Linux type `../bin/VirtualLeaf` at the terminal command prompt.

3. Methods

VirtualLeaf implements a suite of biological objects and processes required for modeling plant development, including cells, rigid cell walls, biochemical networks, and transport of chemical signals. You can define the model by describing the properties of these biological objects and the biochemical networks in a small section of C++-code that implements a model definition plugin to *VirtualLeaf*.

Figure 1 outlines the structure of this file. The first line, `#include "vleafmodel.h"` reads a series of definitions required for building a *VirtualLeaf* model. Each code block defines the behaviors of one type of the model element: cells and cell walls.

```
#include "vleafmodel.h"

QString MyModel::ModelID(void) {
        // specify the name of your model here
        return QString( "My Own Model in VirtualLeaf" );
}

// return the number of chemicals your model needs
int MyModel::NChem(void) { return 0; }


// Rules in this method are executed after the cellular mechanics has
// equilibrized
void MyModel::CellHouseKeeping(CellBase *c) {
                // add cell behavioral rules here
}

// To be executed after cell division
void MyModel::OnDivide(ParentInfo *parent_info, CellBase *daughter1, CellBase *daughter2) {
                // rules to be executed after cell division go here
                // (e.g. cell differentiation)
}

// Differential equations describing chemical reactions inside the cells
void MyModel::CellDynamics(CellBase *c, double *dchem) {
        // add biochemical networks for intracellular reactions here
}
```

Fig. 1. Model definition outline. New models are constructed by defining the functions in this model definition file.

- *MyModel::ModelID* allows you to specify a name for your model.
- *MyModel::NChem* specifies the number of chemical species (e.g., proteins, phytohormones, etc.) the model requires.
- *MyModel::CellHousekeeping* is used to specify mechanical changes or state changes of the cells.
- *MyModel::OnDivide* is used to define the rules to be executed after cell division, e.g., the redistribution of the chemicals in the parent cell.
- *MyModel::CellDynamics* and *MyModel::WallDynamics* define the differential equations describing the biochemical networks within the cells and within the walls.
- *MyModel::CelltoCellTransport* defines the rules for active and diffusive transport of chemicals over cell walls.

3.1. Basic Usage of the VirtualLeaf

After starting *VirtualLeaf* with an existing model plugin or your own, the main window appears, and a pop-up window displays version and license information. Dismiss the information window to reveal the default model. The model's name is visible at the top of the window and is included in the list of models visible under the "Models" menu.

1. The simulation can be started and paused by toggling the "Simulation paused" item in the "Run" menu, or more easily by typing "s" on the keyboard.
2. To run a different, predefined model, go to the "Models" menu, and select one of the model descriptions.
3. To change the value of one of the predefined simulation parameters, click on the "Edit parameters" item under the "Options" menu, or type the keyboard shortcut "Ctrl-E" (or "Command-E" on Mac). A window will appear with parameter names followed by their values. Once you have made your changes, save them by clicking the "Write" button in the lower right corner of the screen.
4. You can run the model on different, initial cellular geometries, called "leaves." To select a different "leaf," choose "Read Leaf" from the file menu. Leaves are stored as XML files in the directory "data/leaves." To store a cellular geometry, select "Save Leaf."
5. To store an image of the simulation results, choose "Snapshot" from the File menu. Specify the image format by using the appropriate file extension. Most common image formats are available, including PDF, PNG, and JPEG. The full list of file formats is listed in the field "File type:" of the Snapshot pop-up dialog.
6. To save snapshots at regular intervals, for example to make a simulation movie, in the parameter editing window change the value of "datadir" to an appropriate directory on your file system (under "Data Export"), and toggle the "Start saving movieframes" item from the "View" menu. To change the export interval, modify parameter "storage_stride" from the parameter window.
7. You can also export data in CSV-format (compatible with Excel) to report on the positions, size, and chemical content of cells and chemicals, and to export some basic data on the leaf morphology.
   - To export data on a single morphology, select the "Export cell data" from the "File" menu.
   - To save data at regular intervals, you can make use of the parameters "export_interval" and "export_fn_prefix."
   - A value of zero for "export_interval" means that no data is exported.

3.2. Prepare an Empty Model Plugin

In the next sections, we show how to develop your own model plugin in *VirtualLeaf* (see Note 2). First, you will need to prepare an empty model plugin.

1. In a terminal window, go to the *VirtualLeaf* tutorial directory.

   Windows:

   ```
   > cd virtualleaf\src\TutorialCode
   ```

   Linux/MacOSX:

   ```
   > cd virtualleaf/src/TutorialCode
   ```

2. Copy the empty model template "Tutorial0" to a name of your choice, e.g., "MyModel."

   Windows:
   Use the Finder or Explorer tools, or type the following at the terminal prompt:

   ```
   > copy Tutorial0 MyModel
   ```

   Linux/MacOSX:

   ```
   > cp -pr Tutorial0 MyModel
   ```

3. Remove temporary files generated by the previous compilation.

   ```
   > cd MyModel
   > make clean
   ```

4. Rename all the remaining files using the name you chose.

   Windows:
   Use the Explorer tool.

   MacOSX/Linux:

   ```
   > rename tutorial0 mymodel *
   ```

5. In all remaining files, replace occurrences of the strings "Tutorial0" and "tutorial0" with ones reflecting the name you chose.

6. Rename your new model. Edit the file *mymodel.cpp*, and on line 37 replace the string:

   ```
   return QString( "0: Empty model template (does nothing)" );
   ```

   with one reflecting the name you choose:

   ```
   return QString( "My own model in VirtualLeaf" );
   ```

7. Build your new (empty) project, by typing:

   Windows:

   ```
   > set MAKE=mingw32-make
   > mingw32-make
   ```

   Linux:

   ```
   > make
   ```

   MacOSX:

```
> qmake -spec macx-g++ mymodel.pro; make
```

(or qmake mymodel.pro, then compile from XCode).

8. If still running, quit and restart *VirtualLeaf*. The model should now appear in "Models" menu.

3.3. Tissue Growth

We will first build a simple model of cell division driven tissue growth. Use the model template you have constructed in Subheading 3.2.

1. In the file *mymodel.cpp* append the following code to the `MyModel::CellHouseKeeping` method:

   ```
   c->EnlargeTargetArea(par->cell_expansion_rate);
   ```

   This statement instructs the cells to expand by the value cell_expansion_rate after each relaxation cycle.

2. Recompile the model plugin.

   Windows:

   ```
   > set MAKE=mingw32-make
   > mingw32-make
   ```

   Linux/MacOSX:

   ```
   > make
   ```

3. Restart *VirtualLeaf*, select your updated model from the "Models" menu and start it by pressing "s." You should see a single, expanding cell.

4. To instruct cells to divide after they doubled in size, append the following code to the `MyModel::CellHouseKeeping` method:

   ```
   if (c->Area() > 2 * c->BaseArea() ) {
     c->Divide();
   }
   ```

   The line starting with "`if`" is a conditional statement. When a cell's area grows larger than twice its original area. i.e., `BaseArea`, the cell will divide (see Note 3).

5. Recompile and restart *VirtualLeaf*, select your updated model and press "s" to start it. The cells should expand and divide (Fig. 2a).

6. By default, `Divide()` instructs cells to divide over the shortest principal axis. To add an optional division axis replace `c->Divide()` with:

   ```
   c->DivideOverAxis(Vector(0,1,0));
   ```

   Recompile and start the model as instructed in steps 2 and 3.

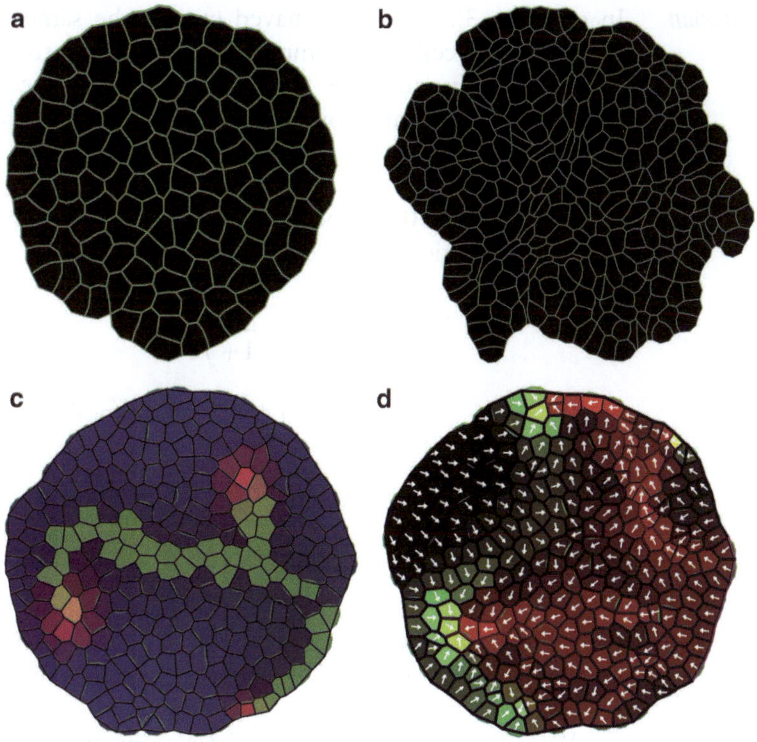

Fig. 2. Simulations. (**a**) Simulation result of the tissue growth model implemented in Subheading 3.3, steps 1–5. In this model, cells expand at a constant rate and divide over their short axis after doubling in size. (**b**) Simulation result of the tissue growth model implement in Subheading 3.3, steps 1–8. In this model, the division axis has a random orientation, resulting in an irregular overall tissue shape. (**c**) Simulation result of the reaction-diffusion model of leaf venation by Meinhardt (10), as implemented in Subheading 3.4. The moving activator-inhibitor front (*red*) traces out the vein shown in *green*. It requires a substrate (*blue*), that the veins consume. (**d**) Simulation result of auxin travelling-wave model on a static domain (13). The auxin concentration is shown in *green*, the concentration of PIN in the cells and at the walls is shown in *red*. The white arrows indicate the polarization directions of PINs. See ref. (13) for details.

7. Let's make the cells divide over an axis of random orientation. Add the following two header files directly after the existing header files to define π and add functionality for random functions.

```
#include "Pi.h"
#include "random.h"
```

8. Replace the `c->DivideOverAxis(axis)` statement with:

```
double orientation = Pi*RANDOM();
Vector axis(sin(orientation),cos(orientation),0.);
c->DivideOverAxis(axis);
```

Recompile and start the model as instructed in steps 2 and 3. The result should look similar to Fig. 2b.

3.4. Reaction-Diffusion and Cell Differentiation

In example 3.3 all cells behaved exactly the same. In an actual plant tissue, cell–cell communication and pattern formation mechanisms instruct cells to behave differently according to the local signal concentrations. This experiment illustrates how to implement a classic reaction-diffusion hypothesis for pattern formation (10).

Meinhardt (10) proposed that leaf venation patterns can be formed by reactions between diffusing chemicals:

$$\frac{d\Upsilon}{dt} = dA - e\Upsilon + \frac{\Upsilon^2}{1 + f\Upsilon^2};$$

$$\frac{dA}{dt} = \frac{cA^2S}{H} - \mu A + D_A\nabla^2 A + \rho_0\Upsilon;$$

$$\frac{dH}{dt} = cA^2S - \nu H + D_H\nabla^2 H + \rho_1\Upsilon;$$

$$\frac{dS}{dt} = c_0 - \gamma S - \varepsilon\Upsilon S + D_S\nabla^2 S,$$

with "Y" a cell differentiation factor, "A" a self-reinforcing activator, "H" an inhibitor, "S" a substrate, and all the other symbols constants.

We will run these biochemical reactions in each of the cells by implementing a set of differential equations that assume mass-action kinetics (see Note 1).

1. First, construct a sufficiently large model tissue to test the reaction-diffusion equations: take the model constructed in Subheading 3.2 and make the cells divide over their shortest axis (step 5 in Subheading 3.3). That is, in the method `MyModel::CellHouseKeeping`, use the division statement (see Note 4):

```
c->Divide();
```

2. Specify the number of chemicals the model considers, by inserting the following code in the method `MyModel::NChem()`:

```
int MyModel::NChem(void) { return 4; }
```

3. Recompile, start *VirtualLeaf* and load your model.
4. Run the model until you have obtained a model tissue with several hundred cells.
5. Switch off cell growth. Click "Cell growth" under the "Options" menu such that the option is unchecked.
6. Give the cells suitable initial values. Open the "Edit Parameters" dialog in the "Options" menu. Under the heading "Auxin transport and PIN1 dynamics," change the first four values of "initval" to 0.001 (within the text box, use the arrow keys to navigate to the front of the list). When done click "Write" on the parameter

dialog and then choose “Reset Chemicals” from the “Edit” menu. If you put your mouse pointer over a cell the values of the chemicals will be shown. Choose “Randomize PIN1 Transporters” from the “Edit” menu to add some noise (see Note 5).

7. Save the tissue template to the *virtualleaf/data/leaves* directory. Choose “Save Leaf” from the “File” menu, navigate to the directory *virtualleaf/data/leaves* and choose a suitable name for your template, e.g., *myleaf.xml.*
8. Use this tissue template as the default for your model. Edit your model’s header file, e.g., *mymodel.h*, and add the following line at the end of the file just before the closing curly-brace (see Note 6).

```
virtual QString DefaultLeafML(void) {
 return QString("myleaf.xml");
}
```

   Replace *myleaf.xml* with the name you chose for your tissue template.

9. Let each cell run the reaction-diffusion equations proposed by ref. (10). Insert them into `MyModel::CellDynamics` method so it becomes:

```
void MyModel::CellDynamics(CellBase *c,
    double *dchem) {
    double Y = c->Chemical(0),
           A = c->Chemical(1),
           H = c->Chemical(2),
           S = c->Chemical(3);
    dchem[0] = (par->d * A - par->e * Y + Y*Y /
               (1 + par->f * Y*Y ) );
    dchem[1] = (par->c * A*A*S/H - par->mu *
                A + par->rho0*Y );
    dchem[2] = (par->c * A*A*S - par->nu*H +
                par->rho1*Y );
    dchem[3] = (par->c0 - par->gamma*S -
                par->eps * Y * S );
}
```

10. Color the cells according to the values of the chemicals. Insert the following code into the `MyModel::SetCellColor` method (see Note 7):

```
double red=c->Chemical(1)/(1.+c->Chemical(1));
double green=c->Chemical(0)/(1.+c->Chemical(0));
double blue=c->Chemical(3)/(1.+c->Chemical(3));
color->setRgbF(red,green,blue);
```

11. Recompile your model, restart *VirtualLeaf* and select your model from the menu.
12. In the "Edit parameters" dialog, set suitable values for the parameters. For example `d=0.002`, `e=0.1`, `f=10`, `c=0.004`, `mu=0.12`, `nu=0.04`, `rho0=0.03`, `rho1=0.0003`, `c0=0.02`, `gamma=0.02` and `eps=0.4`. To save the parameters, click "Write" on the parameter dialog. Choose the "Save leaf" item from the "File" menu and rewrite the template to *virtualleaf/data/leaves/mymodel.xml*. Choose "yes" to overwrite.
13. Start your model. Some cells will turn green, others black, but not much will happen. The reason is that we have not yet implemented chemical diffusion.
14. To implement Fick's law of chemical diffusion, insert the following code into the `MyModel::CelltoCellTransport` method:

```
// Passive fluxes (Fick's law)
for (int c=0;c<NChem();c++) {
   if (w->C1()->BoundaryPolP() ||
       w->C2()->BoundaryPolP())return;
   double phi = w->Length() * ( par->D[c] ) *
                (w->C2()->Chemical(c)-
                 w->C1()->Chemical(c));
   dchem_c1[c] += phi;
   dchem_c2[c] -= phi;
}
```

Here "w" indicates a cell wall separating the two cells, `w->C1()` and `w->C2()`.
The cell wall's length is given by `w->Length()`.
Recompile your model and restart *VirtualLeaf*.
15. Choose suitable diffusion parameters. In the parameter dialog, change the first four values for "D" to: `0, 0.002, 0.018, 0.02`. (see Note 8). Save the template to *virtualleaf/data/leaves/mymodel.xml*.
16. Set `v=1` in one of the cells to initiate the venation pattern. To do so, hover the mouse pointer over the cell whose contents you want to change to display its index number and contents. Open the leaf template file e.g., *virtualleaf data/leaves/mymodel.xml* in a text editor and search for the line starting with `<cell index="#">` where # is the number of the cell you want to change. Near the end of this cell's definition is a `<chem>` tag containing four `<val>` tags. Change the first `<val>` tag's value to `1.0`, i.e.,

```
chem n="4">
```

```
<val v="1.0">
...
...
...
</chem>
```

A more flexible alternative to editing the leaf template file requires a change to the *VirtualLeaf* source code (see Note 9).

Save the file and reopen it in *VirtualLeaf*.

To run the model, press "s" (see Note 10). You should now see a result similar the one shown in Fig. 2c.

17. It is interesting to study the behavior of this model in a growing domain. First define an empty template of a couple of cells—it is easiest to grow it from one cell. To start with an initial single cell again, undo step 8 by inserting two forward slashes "//" (see Note 4) before the definition of the `QString DefaultLeafML`' in file *mymodel.h*. Recompile your model and restart *VirtualLeaf*. Define appropriate parameters and initial conditions by repeating steps 6 and 12 or you will receive an error "step size too small in odeint" because of a division by zero. A quick way to define these values is by reading only the parameters from a previous template. Choose "Read leaf" from the "File" menu and in the file dialog uncheck "geometry," then proceed as usual.
18. Choose "Cell growth" from the "Options menu" to switch on cell growth. Start the simulation until a template of around 8 cells has grown (see Note 11). Repeat step 16 in order to define an initial venature cell and save your growing leaf template.
19. Run the model with the new template. You should now see the pattern develop as the leaf grows out. It may be useful to increase the simulated time per growth cycle for the reaction-diffusion equations. To do so, increase the parameter `rd_dt`.
20. *VirtualLeaf* is particularly suited for modeling mechanisms in which growth and pattern formation feed back on one another. We will implement the effects of chemical concentrations on growth in the `MyModel::CellHouskeeping` method. For example, to prevent vascular cells from expanding, wrap the statement that controls cell expansion:

    ```
    c->EnlargeTargetArea(par->cell_expansion_rate);
    ```

    within a conditional statement like this:

    ```
    if (c->Chemical(0)<0.5) {
     c->EnlargeTargetArea(par->cell_expansion_rate);
    }
    ```

21. You now have seen all functionality in *VirtualLeaf* necessary for implementing reaction-diffusion hypotheses of plant patterning and morphogenesis. You should now be able to experiment with

modifications of existing hypotheses or to implement new reaction-diffusion models. As a suggestion, implement the assumption that the substrate "S," i.e., "Chemical #3," inhibits cell expansion. Another suggestion is to implement a reaction-diffusion hypothesis for trichome patterning (11, 12). The next section shows how to add polar auxin transport to your models.

3.5 Polar Auxin Transport

The previous section implemented a reaction-diffusion hypothesis for leaf venation patterning. Although reaction-diffusion mechanisms are thought to be involved in a range of plant patterning mechanisms, e.g., trichome patterning (11, 12), many recent hypotheses of plant organ patterning assume a role for directed transport of auxin. This section will demonstrate how to implement directed transport mechanisms, starting from the *traveling-wave* hypothesis for leaf venation patterning (13).

The traveling-wave hypothesis is a variant of the *auxin upstream pumping* hypothesis (14, 15). It assumes a membrane bound matrix protein, PIN1, which exports the phytohormone *auxin* toward adjacent cells. PIN1 recycles between the membrane and an intracellular storage, called the *endosome*, and binds preferentially to cell membranes adjacent to cells with a high concentration of auxin.

1. Start with an empty model template. Define the number of chemicals we are using in this model. We will need equations for auxin and for PIN1. Therefore, in *mymodel.cpp*, redefine `MyModel::NChem` as:

```
int MyModel::NChem(void) { return 2; }
```

2. Next we will implement the auxin upstream pumping hypothesis: PIN1 transports auxin actively to adjacent cells; a diffusion term is responsible for downstream auxin transport,

$$\frac{dA_i}{dt} = T_{active}\left(\frac{P_{ji}A_j}{k_a + A_j} - \frac{P_{ij}A_i}{k_a + A_i}\right) + T_{diffusive}\sum_j L_{ij}\left(A_j - A_i\right)$$

where the sum is over all neighbor cells, A_i is the auxin concentration in cell *i*. P_{ij} and P_{ji} are the amounts of PIN1 in cell *i* pumping auxin into cell *j* and *vice versa*, and T_{active} and $T_{diffusive}$ are active and passive transport coefficients, and L_{ij} is the length of the wall between cell *i* and cell *j*.

We will store the concentrations of auxin as "Chemical #0" and the concentration of PIN as "Chemical #1." Transporter proteins and other components that localize within the membranes or within the cell wall matrix, are stored in the Wall class. Insert the following code into the `MyModel::CellToCellTransport` method (see Note 12):

```
void MyModel::CelltoCellTransport(Wall *w,
   double *dchem_c1, double *dchem_c2) {
   for (int c=0;c<NChem();c++) {
```

```
      // diffusive transport
      double phi = w->Length()*(par->D[c])*
       (w->C2()->Chemical(c) -
       w->C1()->Chemical(c));
      dchem_c1[c] += phi;
      dchem_c2[c] -= phi;
     }
      // active transport
      // efflux from cell 1 to cell 2
      double trans12 =
       (par->transport * w->Transporters1(1)*
       w->C1()->Chemical(0)/
       (par->ka + w->C1()->Chemical(0)));
      // efflux from cell 2 to cell 1
      double trans21 =
       (par->transport * w->Transporters2(1)*
       w->C2()->Chemical(0)/
       (par->ka + w->C2()->Chemical(0)) );
      dchem_c1[0] += trans21 - trans12;
      dchem_c2[0] += trans12 - trans21;
}
```

3. Use suitable cell coloring rules, e.g., those defined in Subheading 3.4, step 10. Replace the definition for the "blue" channel by:

   ```
   double blue=0;
   ```

   (see Note 13).

4. To test the implementation, recompile your model and restart *VirtualLeaf*. Read the leaf *tutorial3_init.xml* from the *virtualleafdata/leaves* directory and run the model by pressing "s." This initial condition contains predefined auxin and oriented PINs.

5. Next, implement the PIN1 recycling equations. We define the flux ϕ_{ij} as the translocation of PIN1s from the endosome of cell *i* to its cell membrane adjacent to cell j (for details, see ref. (13)):

$$\phi_{ij} = k_1 \sum_j \frac{P_i A_j f(A_j)}{k_m + P_i} - k_2 \sum_j P_{ij} \quad \text{with } f(A_j) = \frac{A_j R}{k_R + A_j}.$$

Define a new function *PINflux* to calculate ϕ_{ij}. To do so, add the following line of code to the file *mymodel.h* right before the closing curly-brace:

```
virtual double PINflux(CellBase *this_cell,
   CellBase *adjacent_cell, Wall *w);
```

To implement the function, add the following code to the end of *mymodel.cpp*:

```
double MyModel::PINflux(CellBase *this_cell,
   CellBase *adjacent_cell, Wall *w) {
   double adj_auxin =
     adjacent_cell->Chemical(0);
   double receptor_level = adj_auxin *
     par->r/(par->kr + adj_auxin);
   double pin_atwall;
   if (w->C1() == this_cell)
     pin_atwall = w->Transporters1(1);
   else
     pin_atwall=w->Transporters2(1);
   double pin_flux = par->k1 *
     this_cell->Chemical(1) *
     receptor_level /
     (par->km + this_cell->Chemical(1))-
     par->k2 * pin_atwall;
   return pin_flux;
}
```

6. Next implement the following differential equations.

$$\frac{dP_i}{dt} = -\sum_j \phi_{ij} + \alpha A_i - \delta P_i, \frac{dP_{ij}}{dt} = \phi_{ij}.$$

The first equation sums all the net PIN1-fluxes from the membrane to the endosome, and takes it as the change per time unit of the level of PIN1 in the cell. The second equation states that the change in PIN1-level in a cell wall is the flux of PIN1 moving to it.

```
void MyModel::WallDynamics(Wall *w,
   double *dw1, double *dw2){
   dw1[0] = 0.; dw2[0] = 0.;
   dw1[1] = PINflux(w->C1(),w->C2(),w);
   dw2[1] = PINflux(w->C2(),w->C1(),w);
}
```

Similarly, in the method `MyModel::CellDynamics` we specify what comes back from the walls.

```
#include "flux_function.h"
void MyModel::CellDynamics(CellBase *c,
   double *dchem){
   dchem[1] =
      -SumFluxFromWalls(c,
        MyAuxinModel::PINflux)
      + par->pin_prod * c->Chemical(0)
      - par->pin_breakdown * c->Chemical(1);
}
```

7. Merks et al. (13) assume that auxin enters the leaf primordium at its margin with a constant flux, and that all PIN1 is produced in response to auxin stimulation. To implement this assumption, we will need to add the auxin sources to a suitable initial condition.

8. Define the initial condition. Start *VirtualLeaf*, open the new traveling wave model and open the file *tutorial4_init.xml*. Remove all auxin and PIN1 values from the leaf template by changing the first two values of "initval" to zero; click "Write" on the parameters dialog and choose "Reset Chemicals and Transporters" from the "Edit" menu.

9. Next make several of the peripheral walls auxin sources. Make sure "Show Transporters" is checked in the "View" menu. Then, while holding down the *Control* key, click the *outside* of some peripheral walls of the template until they are purple. This indicates that they have turned into auxin sources (see Note 14). Save the template. Alternatively, download a suitable template from http://www.code.google.com/p/virtual-leaf/wiki/Protocols. The template used for the simulations in ref. (13) can be downloaded from this same site.

10. To make the auxin sources work, add the following code to end of the `MyModel::CelltoCellTransport` method in the file *mymodel.cpp*:

```
// Influx at leaf "AuxinSource"
// (as specified in initial condition)
if (w->AuxinSource()) {
   double aux_flux = par->leaf_tip_source *
      w->Length();
   dchem_c1[0] += aux_flux;
   dchem_c2[0] += aux_flux;
}
```

11. Open the parameter dialog to set the parameters used by ref. (13): `D[0]=1e-6` ($T_{diffusive}$; see Note 15); `leaf_tip_source=1e-5` (ϕ_{tip}; see Note 15); `transport=0.08` (T_{active}); `pin_prod =1e-5` (α); `pin_breakdown=1e-8` (δ); `r=100` (R); `kr=100` (k_R); `k1=2e-4` (k_1); `k2=5e-7` (k_2); `km=100` (k_m).

12. Start the simulation. If the calculations seem to progress slowly, increase the value of `rd_dt`. This increases the integrated time between two subsequent plotting steps or cell growth steps. A suitable value for this model would be `rd_dt=1000`.

13. You can now start experimenting with the parameters, e.g., the changes suggested in the Supplements of ref. (13). Interesting effects occur by speeding up the constitutive translocation of PIN1 (`k1=0.2; k2=0.005`), speeding up the production and breakdown of PIN1 (`pin_prod=0.001;`

`pin_breakdown=0.001`) and by increasing the diffusion of auxin (`D=5e-06`). The result should look similar to Fig. 2d. Choose a large leaf template.

4. Notes

1. A discussion on differential equation descriptions of biological networks is out of the scope of this chapter. Useful texts on the subject can be found in refs. (16) and (17).
2. You can skip this step by using the model template "Tutorial0" or by downloading the template "MyModel" from http://www.code.google.com/p/virtualleaf/wiki/Protocols.
3. Optionally, use parameter `"par->rel_cell_div_threshold"` instead of "2":

```
if (c->Area() > par->rel_cell_div_threshold
   *c->BaseArea()){
   c->Divide();
}
```

4. In C++, you can switch off code you want to keep for later re-use, by making it a comment that the compiler will ignore. To do so, place the code between "/*" and "*/," or insert two slashes ("//") at the beginning of the line. In the following two line code snippet, the compiler will interpret the first statement as a comment and perform only the second statement.

```
/* c->DivideOverAxis(axis) */c->Divide();
```

5. In future versions the name of this menu item will be changed to "Randomize Chemicals and Transporters."
6. Only one of these function definitions is allowed per model. So if a definition called `DefaultLeafML` is already present, do not add a second one.
7. `SetRgbF(r,g,b)` is a Qt library function. It takes red, green, and blue color values between 0 (minimal) and 1 (maximal). All other Qt color library functions can be used as well. See http://www.qt.nokia.com for documentation.
8. "*D*" is under "Auxin Transport and PIN1 dynamics." Remember to scroll to the start of the text box using the left arrow keys.
9. As an alternative to editing the leaf XML template, you can redefine function `Cell::OnClick` in file *`VirtualLeaf.cpp`* to change cell contents interactively, i.e., to reset the value of

"Chemical #0" after you have clicked a cell, redefine `Cell::OnClick` as follows:

```
void Cell::OnClick(QMouseEvent *e){
   if (NChem()>0) SetChemical(0,1.);}
```

To effect this change, you will need to recompile the *VirtualLeaf* framework (see Subheading 2.3). Future versions of *VirtualLeaf* will move this functionality to the model definition files.

10. If *VirtualLeaf* becomes unresponsive, decrease the value of parameter `rd_dt`. If the model progresses too slowly, increase the value of `rd_dt`. `rd_dt` is the integration time per display step.

11. If *VirtualLeaf* aborts with the message "stepsize underflow in rkqs, with `h=0,000000` and `htry=0,10000`" you are probably dividing by zero. In this example, `H=0` "Chemical #2." The "h" mentioned in the error message is an unrelated integration parameter. Use an appropriate initial condition with "Chemical #2" set to small positive value, as explained in step 17.

12. In this code, the symbols in the equations are represented as follows. L_{ij}:`w->Length()`; P_{ij} and P_{ji}:`w->Transporters1(1)` and `w->Transporters2(1)`; diffusion coefficient (D): `par->D[c]`; transport coefficient (T): `par->transport`; k_a: `par->ka`.

13. If you used the exact code defined in Subheading 3.4, step 10, you would attempt to read Chemical #3. Here Chemical #3 is undefined, and as a result *VirtualLeaf* may crash. In fact, reading from or writing to undefined memory locations is a common cause of software crashes.

14. In the release of *VirtualLeaf* published with ref. (7), this task of clicking on peripheral walls is quite tedious. It helps to increase the width of the cell walls, by increasing the value of parameter outlinewidth. Also it helps to hide the cells by checking "Hide cells" item in the "View" menu. If you accidentally click the interior walls, it turns red to indicate that you have injected it with a high concentration of PIN. If this happens, simply choose "Reset Chemicals and Transporters" from the "Edit" menu. In a more recent version of *VirtualLeaf* this issue has been solved.

15. To calculate D and leaf_tip_source from the values of $T_{diffusive} = 1.5 \times 10^5 \,\mathrm{m}^{-2}\mathrm{s}^{-1}$ and $\phi_{tip} = 1.5 \times 10^6 \,\mathrm{m}^{-2}\mathrm{s}^{-1}$ mentioned in ref. (13), note that we average length of a cell wall in the template used is around 15 a.u. (arbitrary units). Assuming a cell

wall has an area of around $L_{ij} = 100\mathrm{m}^2 = 10^{-10}\mathrm{m}^2$, we rescale $D = T_{diffusive} \times \frac{L_{ij}}{15} = 10^{-6}$ and $leaf\_tip\_source = \phi_{tip} \times \frac{L_{ij}}{15} = 10^{-5}$.

16. Updates to this chapter, including system specific software and installation issues, are maintained at https://code.google.com/p/virtualleaf/wiki/mmbupdates.

Acknowledgments

This work was financed by the Netherlands Consortium for Systems Biology (NCSB), which is part of the Netherlands Genomics Initiative/Netherlands Organisation for Scientific Research, and by Marie Curie European Reintegration Grant PERG03-GA-2008-230974 to RM.

References

1. Dupuy L, Mackenzie J, Rudge T, Haseloff J (2008) A system for modelling cell–cell interactions during plant morphogenesis. Ann Bot-London 101:1255–1265
2. Grieneisen VA, Scheres B (2009) Back to the future: evolution of computational models in plant morphogenesis. Curr Opin Plant Biol 12:606–614
3. Chickarmane V, Roeder AH, Tarr PT et al (2010) Computational morphodynamics: a modeling framework to understand plant growth. Annu Rev Plant Biol 61:65–87
4. Santos F, Teale W, Fleck C et al (2010) Modelling polar auxin transport in developmental patterning. Plant Biol 12(Suppl 1):3–14
5. Keurentjes JJ, Angenent GC, Dicke M et al (2011) Redefining plant systems biology: from cell to ecosystem. Trends Plant Sci 16:183–190
6. Kitano H (2002) Systems biology: a brief overview. Science 295:1662–1664
7. Merks RMH, Guravage M, Inzé D, Beemster GTS (2011) VirtualLeaf: An open-source framework for cell-based modeling of plant tissue growth and development. Plant Physiol 155:656–666
8. Merks RMH, Glazier JA (2005) A cell-centered approach to developmental biology. Physica A 352:113–130
9. Anderson ARA, Chaplain MAJ, Rejniak KA (eds.) (2007) Single-cell-based models in biology and medicine. Birkhaüser, Basel
10. Meinhardt H (1976) Morphogenesis of lines and nets. Differentiation 6:117–123
11. Benítez M, Espinosa-Soto C, Padilla-Longoria P, Díaz J, Alvarez-Buylla ER (2007) Equivalent genetic regulatory networks in different contexts recover contrasting spatial cell patterns that resemble those in Arabidopsis root and leaf epidermis: a dynamic model. Int J Dev Biol 51:139–155
12. Bouyer D, Geier F, Kragler F, Schnittger A, Pesch M, Wester K, Balkunde R, Timmer J, Fleck C, Hülskamp M (2008) Two-dimensional patterning by a trapping/depletion mechanism: the role of TTG1 and GL3 in Arabidopsis trichome formation. PLoS Biol 6:1166–1177
13. Merks RMH, Van de Peer Y, Inzé D, Beemster GTS (2007) Canalization without flux sensors: a traveling-wave hypothesis. Trends Plant Sci 12:384–390
14. Jönsson H, Heisler MG, Shapiro BE, Meyerowitz EM, Mjolsness E (2006) An auxin-driven polarized transport model for phyllotaxis. P Natl Acad Sci USA 103: 1633–1638
15. Smith RS, Guyomarc'h S, Mandel T, Reinhardt D, Kuhlemeier C, Prusinkiewicz P (2006) A plausible model of phyllotaxis. P Natl Acad Sci USA 103:1301–1306
16. Ellner SP, Guckenheimer J (2006) Dynamic models in biology. Princeton University Press, Princeton
17. Fall CP, Wagner JM, Marland ES, Tyson JJ (eds) (2002) Computational cell biology. Series interdisciplinary applied mathematics, vol 20. Springer, New York

Index

Ive De Smet (ed.), *Plant Organogenesis: Methods and Protocols*, Methods in Molecular Biology, vol. 959, DOI 10.1007/978-1-62703-221-6, © Springer Science+Business Media New York 2013

MIX
Papier aus verantwortungsvollen Quellen
Paper from responsible sources
FSC® C105338

If you have any concerns about our products,
you can contact us on
ProductSafety@springernature.com

In case Publisher is established outside the EU,
the EU authorized representative is:
Springer Nature Customer Service Center GmbH
Europaplatz 3, 69115 Heidelberg, Germany

Printed by Libri Plureos GmbH
in Hamburg, Germany